E. Olmo

C. A. Redi

Chromosomes Today

Volume 13

Ettore Olmo
Carlo Alberto Redi
Editors

Springer Basel AG

Edited by

Prof. Dr. Ettore Olmo
Istituto di Biologia e Genetica
Via Brecce Bianche
60131 Ancona
Italy

Prof. Carlo Alberto Redi
University of Pavia
Dipartimento di Biologia animale / Centro di Studio per l'Istochimica – CNR
Piazza Botta 9
27100 Pavia
Italy

A CIP catalogue record for this book is available from the Library of Congress, Washington D.C., USA

Deutsche Bibliothek Cataloging-in-Publication Data
Chromosomes today : volume 13 / Ettore Olmo ; Carlo Alberto Redi, ed.. - Basel ; Boston ; Berlin : Birkhäuser, 2000
ISBN 3-7643-5799-1

ISBN 978-3-0348-9587-3 ISBN 978-3-0348-8484-6 (eBook)
DOI 10.1007/978-3-0348-8484-6

Originally published by Birkhäuser Verlag, P.O. Box 133, CH-4001 Basel, Switzerland 2000
Softcover reprint of the hardcover 1st edition 2000
Printed on acid-free paper produced from chlorine-free pulp. TFC
Cover design: gröflin. Graphic Design, Basel. www.groeflin.ch

ISBN 3-7643-5799-1

9 8 7 6 5 4 3 2 1

Contents

List of contributors

Almeida-Toledo, Lurdes F. de
Departamento de Biologia, Instituto de Biociências, Universidade de São Paulo, Caixa Postal 11461, 05422-970 São Paulo, Brasil, Tel: +55 11 818 7584, Fax: +55 11 8187553, E-Mail: lftoledo@usp.br

Armstead, I. P.
Institute of Grassland and Environmental Research, Plas Gogerddan, Aberystwyth, Ceredigion, SY23 3EB, Wales, UK

A. Toledo-Filho, Silvio de
Departamento de Biologia, Instituto de Biociências, Universidade de São Paulo, Caixa Postal 11461, 05422-970 São Paulo, Brasil, Tel: +55 11 818 7584, Fax: +55 11 8187553, E-Mail: lftoledo@usp.br

Ballabio, Andrea
Telethon Institute of Genetics and Medicine, San Raffaele Biomedical Science Park, Via Olgettina 58, I-20132 Milano, Italy, Tel: +39 02 21560 202, Fax: +39 02 21560 220

Bernardi, Giorgio
Laboratorio di Evoluzione Molecolare, Stazione Zoologica "Anton Dohrn", Villa Comunale, 80121 Napoli, Italy, Tel: +39-81-5833300, Fax: +39 81 2457294, E-Mail: Bernardi@Alpha.szn.it

Birch, Jillian M.
CRC Paediatric and Familial Research Group, Royal Manchester Children's Hospital, Pendlebury, Manchester M27 1HA, UK

Bowden, L.
Laboratory of Developmental Genetics and Imprinting, Developmental Genetics Programme, The Brabraham Institute, Cambridge CB2 4AT, UK, Tel. +44 1223 496000, Fax: +44 1223 496030

Boyle, John M.
CRC Section of Molecular Genetics, Paterson Institute for Cancer Research, Wilmslow Road, Manchester M20 4BX, UK, Tel: +44 161 446 3062, Fax: +44 161 446 3062

Capanna, Ernesto
Dipartimento di Biologia Animale e dell'Uomo, Università di Roma "La Sapienza", Via A. Borelli 50, I-00161 Roma, Italy, Tel: +39 6 4991 8008, Fax: +39 6 4457516, E-Mail: capanna@ axrma.uniroma1.it

Capriglione, Teresa
Dipartimento di Biologia Evolutiva e Comparata, Via Mezzocannone 8, I-80134 Napoli, Italy. Tel: +39 081 2528946, Fax: +39 081 2528913, E-Mail: caprigli@dgbm.unina.it

Chiavarino, Mauricio A.
Departamento de Biología, Facultad de Ciencias Exactas y Naturales, Universidad de Buenos Aires, 1428 Buenos Aires, Argentina

Constancia, W.
Laboratory of Developmental Genetics and Imprinting, Developmental Genetics Programme, The Brabraham Institute, Cambridge CB2 4AT, UK, Tel. +44 1223, 496000, Fax: +44 1223 496030

Davies, K.
Laboratory of Developmental Genetics and Imprinting, Developmental Genetics Programme, The Brabraham Institute, Cambridge CB2 4AT, UK, Tel. +44 1223, 496000, Fax: +44 1223 496030

Dean, W.
Laboratory of Developmental Genetics and Imprinting, Developmental Genetics Programme, The Brabraham Institute, Cambridge CB2 4AT, UK, Tel. +44 1223, 496000, Fax: +44 1223 496030

De Jong, J. Hans
Wageningen Agricultural University, Graduate School of Experimental Plant Sciences, Laboratory of Genetics, Dreijenlaan 2, 6703 HA Wageningen, The Netherlands, Tel: +31.317.482150, Fax: +31 317 483146; E-Mail: Hans.de Jong@celgen.el.wau.nl

Dutrillaux, Bernard H.
Institute Curie-CNRS, UMR 147, 26 rue d' Ulm, F-75248 Paris Cedex 05; Département de Radiobiologie et Radiopathologie, CEA/DSV, BP No 6, Fontenay-aux-Roses, France, Tel: +33 0142346672, Fax: +33 014234 6674, E-Mail: bdutrileaux@curie.fr

Evans, D. Gareth R.
Department of Medical Genetics, St Mary's Hospital, Hathersage Road, Manchester M13 0JH, UK

Federico, Concetta
Dipartimento di Biologia Animale, University of Catania, via Androne 81, 95124 Catania, Italy

Feil, R.
Laboratory of Developmental Genetics and Imprinting, Developmental Genetics Programme, The Brabraham Institute, Cambridge CB2 4AT, UK, Tel. +44 1223, 496000, Fax: +44 1223 496030

Ferguson-Smith, Malcolm A.
Molecular Cytogenetics Group, University of Cambridge, Centre for Veterinary Science, Madingley Road, Cambridge CB3 0ES, UK, Tel: +44 1223-333692, Fax: +44 1223-339067, E-Mail: maf12@mole.bio.cam.ac.uk

Foresti, Fausto
Departamento de Morfologia, Instituto de Biociências, UNESP, Campus de Botucatu, 18618-000, Botucatu, SP, Brasil

Franco, Brunella
Telethon Institute of Genetics and Medicine, San Raffaele Biomedical Science Park, Via Olgettina 58, I-20132 Milano, Italy, Tel: +39 02-21560-202, Fax: +39 02-21560-220, E-Mail: Franco@tigem.it

Fransz, P. F.
Laboratory of Molecular Biology, Wageningen Agricultural University, Graduate School of Experimental Plant Sciences, Dreijenlaan 2, NL-6703 HAWageningen, The Netherlands

Fukui, Kiichi
Department of Biotechnology, Graduate School of Engineering, Osaka University, Suita 565-0871, Osaka, Japan

Greaves, Martin
CRC Section of Molecular Genetics, Paterson Institute for Cancer Research, Wilmslow Road, Manchester M20 4BX, UK

Hameister, Horst
Abteilung Medizinische Genetik, Universität Ulm, Albert-Einstein-Allee 11, D-89069 Ulm, Germany, Tel: +49 731 5023436, Fax: +49 731 5023438, E-Mail: horst.hameister@medizin.uni-ulm.de

Harper, J. A.
Institute of Grassland and Environmental Research, Plas Gogerddan, Aberystwyth, Ceredigion, SY23 3EB, Wales, UK

Heslop-Harrison, John S.
Department of Cell Biology, John Innes Centre, Colney Lane, Norwich NR4 7UJ, England, Tel: +44 1603 452571, Fax: +44 1603 456844, E-Mail: pat.heslop-harrison@bbsrc.ac.uk

Hong, Chia-sin
Department of Biochemistry and Cellular and Molecular Biology, University of Tennessee, M407 Walters Life Sciences Bldg., Knoxville, Tennessee 37996-0840, USA, Tel: +1 423 974 3612, Fax: +1 423 974 6306

Ikemura, Toshimichi
Department of Evolutionary Genetics, National Institute of Genetics, The Graduate University for Advanced Studies, Mishima, Shizuoka-ken 411-8540 Japan, Tel: +81 559 81 6788, Fax: +81 559 81 6794, E-Mail: tikemura@lab.nig.ac.jp

Jacobsen, E.
Laboratory of Plant Breeding, Wageningen Agricultural University, Graduate School of Experimental Plant Sciences, Dreijenlaan 2, NL-6703 HAWageningen, The Netherlands

James, Louise A.
CRC Section of Molecular Genetics, Paterson Institute for Cancer Research, Wilmslow Road, Manchester M20 4BX, UK

Jiménez, Guillermo
Departamento de Genética, Facultad de Biología, Universidad Complutense, E-28040 Madrid, Spain, Tel: +34 1 3945044, Fax: +34 1 3944844

Jones, R. N.
Institute of Biological Sciences, University of Wales, Aberystwyth, Ceredigion, SY23 3EB, Wales, UK

Jones, T. A. H.
Institute of Grassland and Environmental Research, Plas Gogerddan, Aberystwyth, Ceredigion, SY23 3EB, Wales, UK

Kanaya, Shigehiko
Faculty of Engineering, Yamagata University, Yonezawa, Yamagata-ken 992-8510, Japan

Kelsey, Anna M.
Department of Histopathology, Royal Manchester Children's Hospital, Pendlebury, Manchester M27 4HA, UK

Kelsey, G.
Laboratory of Developmental Genetics and Imprinting, Developmental Genetics Programme, The Brabraham Institute, Cambridge CB2 4AT, UK, Tel. +44 1223, 496000, Fax: +44 1223 496030

King, Ian P.
Institute of Grassland and Environmental Research, Plas Gogerddan, Aberystwyth, Ceredigion, SY23 3EB, Wales, UK; Tel: +44 1970 828 255, Fax: +44 1970 828 357; E-Mail: ian.king@bbsrc.ac.uk

King, J.
Institute of Grassland and Environmental Research, Plas Gogerddan, Aberystwyth, Ceredigion, SY23 3EB, Wales, UK

Kleckner, Nancy
Department of Molecular and Cellular Biology, Harvard University, Cambridge, MA 02138, USA, Tel: +1 617 495 4278, Fax: +1 617 495 0758, E-Mail: kleckner@fas.harvard.edu

Loidl, Josef
Institute of Botany, University of Vienna, Rennweg 14, A-1030 Vienna, Austria, Tel: +43 1 4277-54022, Fax: +43 1 4277-9541, E-Mail: josef.loidl@univie.ac.at

Macgregor, Herbert
Department of Biology, University of Leicester, University Road, Leicester LE1 7RH, UK, Tel: +44 1162523344, Fax: +44116 252 3330, E-Mail: hcm4@le.ac.uk

Maestra, Belén
Departamento de Genética, Facultad de Biología, Universidad Complutense, E-28040 Madrid, Spain

Manzanero, Silvia
Departamento de Genética, Facultad de Biología, Universidad Complutense, 28040 Madrid, Spain, Tel: +34 1 3945044, Fax: +34 1 3944844

McGown, Gail
CRC Section of Molecular Genetics, Paterson Institute for Cancer Research, Wilmslow Road, Manchester M20 4BX, UK

McKee, Bruce D.
Department of Biochemistry and Cellular and Molecular Biology, University of Tennessee, M407 Walters Life Sciences Bldg., Knoxville,

Tennessee 37996-0840, USA, Tel: +1 423 974 3612, Fax: +1 423 974 6306, E-Mail: bdmckee@utk.edu

Morgan, W. G.
Institute of Grassland and Environmental Research, Plas Gogerddan, Aberystwyth, Ceredigion, SY23 3EB, Wales, UK

Murrell, A.
Laboratory of Developmental Genetics and Imprinting, Developmental Genetics Programme, The Babraham Institute, Cambridge CB2 4AT, UK

Naranjo, Carlos A.
Instituto Fitotécnico Santa Catalina (FCAF, UNLP)-Centro de Investigaciones Genéticas (UNLP-CONICET-CIC), C.C.4 1836 Llavallol, Buenos Aires, Argentina

Naranjo, Tomás
Departamento de Genética, Facultad Biología, Universidad Complutense, Ciudad Universitaria, E-28040 Madrid, Spain, Tel: +34 1 3945040, Fax: +34 1 394 4844, E-Mail: toranjo@euemax.sim.ucm.es

Nash, J. V.
Institute of Grassland and Environmental Research, Plas Gogerddan, Aberystwyth, Ceredigion, SY23 3EB, Wales, UK

Nicklas, R. Bruce
Departments of Zoology and Cell Biology, Duke University, Box 91000, Durham, NC 27708-1000, USA, Tel: +1 919 6138196, Fax: +1 919 613 8177, E-Mail: bnicklas@acpub.duke.edu

O'Brien, Patricia C. M
Molecular Cytogenetics Group, University of Cambridge, Centre for Veterinary Science, Madingley Road, Cambridge CB3 0ES, UK, Tel: +44 1223-333692, Fax: +44 1223-339067

Ohmido, Nobuko
Laboratory of Rice Genetic Engineering, Hokuriku National Agricultural Experiment Station, Joetsu, 943-0193, Niigata, Japan

Ohno, Mizuki
Department of Evolutionary Genetics, National Institute of Genetics, The Graduate University for Advanced Studies, Mishima, Shizuoka-ken 411-8540, Japan; Institute of Genetic Information, Kyushu University, Fukuoka 812-0054, Japan

Pimpinelli, Sergio
Istituto Pasteur, Fondazione Cenci Bolognetti, Dipartimento di Genetica e Biologia Molecolare, Università di Roma "La Sapienza", Piazzale Aldo Moro 5, I-00185 Roma, Italy, Tel: +39 6 49912876, Fax: +39 6 44 56866, E-Mail: pimpinelli@axcasp.caspur.it

Poggio, Lidia
Departamento de Biología, Facultad de Ciencias Exactas y Naturales, Universidad de Buenos Aires, 1428 Buenos Aires, Argentina

Puertas, María J.
Departamento de Genética, Facultad de Biología, Universidad Complutense, E-28040 Madrid, Spain, Tel: +34 1 3945044, Fax: +34 1 3944844, E-Mail: majetas@eucmax.sim.ucm.es

Reik, Wolf
Developmental Genetics and Imprinting, Developmental Genetics Programme, The Babraham Institute, Cambridge CB2 4AT, UK, Tel: +44 1223 496000, Fax: +44 1223 496030, E-Mail: wolf.reik@bbsrc.ac.uk

Rens, Willem
Molecular Cytogenetics Group, University of Cambridge, Centre for Veterinary Science, Madingley Road, Cambridge CB3 0ES, UK, Tel: +44 1223-333692, Fax: +44 1223-339067

Rettenberger, Günther
Abteilung Medizinische Genetik, Universität Ulm, Albert-Einstein-Allee 11, D-89069 Ulm, Germany; Labor Drs. Mehnert, Wegener Strasse 15, D-89231 Neu-Ulm, Germany

Roberts, L. A.
Institute of Grassland and Environmental Research, Plas Gogerddan, Aberystwyth, Ceredigion, SY23 3EB, Wales, UK

Rosato, Marcela
Departamento de Biología, Facultad de Ciencias Exactas y Naturales, Universidad de Buenos Aires, 1428 Buenos Aires, Argentina

Saccone, Salvatore
Dipartimento die Biologia Animale, University of Catania, via Androne 81, I-95124 Catania, Italy

Taylor, Steve
Institute of Grassland and Environmental Research, Plas Gogerddan, Aberystwyth, Ceredigion, SY23 3EB, Wales, UK

Tenzen, Toyoaki
Department of Evolutionary Genetics, National Institute of Genetics, The Graduate University for Advanced Studies, Mishima, Shizuoka-ken 411-8540 Japan, Tel: +81 559 81 6788, Fax: +81 559 81 6794

Thomas, H. M.
Institute of Grassland and Environmental Research, Plas Gogerddan, Aberystwyth, Ceredigion, SY23 3EB, Wales, UK

Thorncroft, Mary
CRC Section of Molecular Genetics, Paterson Institute for Cancer Research, Wilmslow Road, Manchester M20 4BX, UK

Varley, Jennifer M.
CRC Section of Molecular Genetics, Paterson Institute for Cancer Research, Wilmslow Road, Manchester M20 4BX, UK, Tel: +44 161 446 3062, Fax: +44 161 446 3062, E-Mail: jvarley@picr.man.ac.uk

Vogel, Walther
Abteilung Medizinische Genetik, Universität Ulm, Albert-Einstein-Allee 11, D-89069 Ulm, Germany

Wako, Toshiyuki
Laboratory of Rice Genetic Engineering, Hokuriku National Agricultural Experiment Station, Joetsu, 943-0193, Niigata, Japan

Walter, J.
Max-Planck Institute of Molecular Genetics, Berlin, Germany

Watanabe, Yoshihisa
Department of Evolutionary Genetics, National Institute of Genetics, The Graduate University for Advanced Studies, Mishima, Shizuoka-ken 411-8540 Japan, Tel: +81 559 81 6788, Fax: +81 559 81 6794

Wennekes-van Eden, J.
Laboratory of Plant Breeding, Wageningen Agricultural University, Graduate School of Experimental Plant Sciences, Dreijenlaan 2, 6703 HAWageningen, The Netherlands

Yamagata, Tesushi

Department of Evolutionary Genetics, National Institute of Genetics, The Graduate University for Advanced Studies, Mishima, Shizuoka-ken 411-8540 Japan, Tel: +81 559 81 6788, Fax: +81 559 81 6794

Yang, Fengtang

Molecular Cytogenetics Group, University of Cambridge, Centre for Veterinary Science, Madingley Road, Cambridge CB3 0ES, UK, Tel: +44 1223-333692, Fax: +44 1223-339067

Yoo, Siuk

Department of Biochemistry and Cellular and Molecular Biology, University of Tennessee, M407 Walters Life Sciences Bldg., Knoxville, Tennessee 37996-0840, USA, Tel: +1 423 974 3612, Fax: +1 423 974 6306

Zabel, P.

Laboratory of Molecular Biology, Wageningen Agricultural University, Graduate School of Experimental Plant Sciences, Dreijenlaan 2, NL-6703 HAWageningen, The Netherlands

Zhong, X.-B.

Yale University, School of Medicine I-310 SHM, New Haven, CT 06520-8005, USA

Yamaguchi, Tsutomu
Department of [illegible] [illegible] National [illegible] The
[illegible]
[illegible] Japan; [illegible]

[illegible]
[illegible]

[illegible]
[illegible]

[illegible]
[illegible]

Group photo of the members of the 13th International Chromosome Conference at the University of Ancona.

Dedication

In the months that have preceded and followed the 13th Conference, three outstanding Italian cytogeneticists have disappeared.

Mario Benazzi, a pioneer of Evolutionary Cytogenetics, died in December 1997 after a long and fruitful life mainly devoted to the study of Plathelmintes cytogenetics.

Alessandro Morescalchi, the outstanding and respected student of Amphibian cytogenetics, with whom E. Olmo shared work and friendship for many years, died untimely in March 1997.

Marco Milani-Comparetti, an outstanding medical cytogeneticist and expert in bioethics, a member of the local Organising Committee, died in May 1999.

We wish to dedicate to them this 13th volume of Chromosomes Today.

Preface

The 13th International Chromosome Conference that was held in Ancona and Numana from 8th to 12th September 1998 was the last of this century and offered a unique opportunity to draw a balance of more than one hundred years of chromosome research.

The participation of over 300 researchers and the high scientific quality of the contributions represent a significant index of the important results obtained so far and of the ever growing interest raised by cytogenetic studies. This also helped us to understand the importance of the insight that led C.D. Darlington to found, in Oxford in 1964, what developed into the International Chromosome Conference so that scientists working on different aspects of chromosome research would have the opportunity of "discussing the fundamental problems of chromosomes with one another".

The present volume of the series Chromosomes Today contains all but one of the invited lectures given at the 13th International Chromosome Conference. The programme, the poster presentations, the poster abstracts and the full list of the organisers and of the members of the Advisory Committee have been published in Cytogenetics and Cell Genetics vol. 81 (2): 91–168, September 1998.

In the previous volume of the series it was observed that molecular biology has become definitively incorporated into cytogenetics. The articles collected in this volume are a witness to the fruitfulness of this integration.

A new, important aspect that emerged from the 13th International Chromosome Conference are the advances made in the new, powerful methods for studying chromosomes derived from the amazing synergy of molecular biology, optical science and computing technology. However, the important methodological developments and the new applicative opportunities that have occurred in our field have not diminished the attention towards some essential issues such as the composition and organisation of the interphase nucleus, the interaction of homologous chromosomes in meiotic events, the regulation of chromosome distribution during division, or the role of genome structural and molecular variations during evolution. The relevant recent advances in all these issues are represented and discussed in this volume of Chromosomes Today.

The organisation of the 13th International Chromosome Conference has been a demanding undertaking repaid by great satisfaction. For their hard work and their commitment we are grateful to the Standing Committee, the International Scientific Advisory Committee, the local Organising Committee, many volunteer helpers, the University of Ancona and several other bodies which provided both financial and organisational support.

The 14th International Chromosome Conference will be held at Wuerzburg University, Germany, under the guidance of Michael Schmid. We are looking forward to the further advances that will have been made in chromosome studies at the beginning of the third millennium and hope that the International Chromosome Conference will continue to be one of the most important opportunities for scientists from the different fields of cytogenetics to meet and to discuss and exchange data.

The Editors

Introduction

Ernesto Capanna

Chromosomes Today, Vol. 13
ed. by E. Olmo and C.A. Redi

Chromosomes yesterday: a century of chromosome studies

Ernesto Capanna

Department of Animal and Human Biology, Rome University "La Sapienza", Via A. Borelli 50, I-00161 Roma, Italy

The birth of cytogenetics

When the organising committee of the thirteenth chromosome conference asked me to prepare the opening lecture, I realised that it would be the last one this century. In fact the next one will be held at the beginning of the third millennium A.D., in 2001. While thinking along these lines it occurred to me just how important this century has been in the development of cytogenetics!

This century started with the rediscovery of Mendel's laws. The story [1] goes that on the eighth of May 1900 William Bateson (1861–1926) was travelling from Cambridge to London by train. During the journey he read an interesting old article which had been published 34 years earlier by a certain "Father Gregory", a priest who cultivated peas. The work had been mentioned in a paper published by the Dutch geneticist Hugo de Vries (1848–1935) [2]. Carl Correns (1864–1933) and Erich von Tschermak (1871–1962) were two other biologists who, like Bateson and de Vries, took interest in this article. The time was right to understand Mendel's genetics. In fact there had been a lot of discoveries after the Augustinian monk's works on plant hybrids had been read by the sceptical Professor von Naegeli. The intuition of Walter Stanboroug Sutton (1876–1916) – a young graduate student of Columbia University, who at the time was only 24 – became of prime importance. He had worked in the summer of 1902 at the *Stazione Zoologica* in Naples with two of the most important cellular biologists of that time, Edmund Beecher Wilson (1856–1939) and his fraternal friend Theodor Boveri (1862–1915). These two were principally responsible for the rebirth of interest in genetics at the end of the last century and the beginning of the present one. Many circumstances, therefore, favoured Sutton: the beautiful sunny Neapolitan weather, the advice of two important biologists, but above all his careful study of the gametogenesis of *Brachystola magna*, an orthopteran [3]. This insect, like most orthopterans, has few but large chromosomes, which are easy to individualise. In this way Sutton was able to pin down the key aspects of what was to constitute the

Mendelian-chromosome theory, i.e. the chromosome as the physical basis of Mendelian heredity.

Some historians of science assert that the credit for this idea should go to Thomas Harrison Montgomery (1873–1912) who, in 1901 [4], studied the gametogenesis of hemipterans. In reality it is true that Montgomery described certain aspects which revealed a relationship between the behaviour of chromosomes and that of Mendelian alleles, for example the presence of homologous pairs which made up the chromosome set, and the segregation of the homologous pairs in meiotic anaphase. However, he was not explicit like Sutton in recognising that chromosomes were carriers of Mendelian alleles. The priority of the discovery is of no importance to our discussion. The important thing to underline is that the *birth* of the century saw the *birth* of cytogenetics. Sutton announced this birth when he wrote "*We have seen reason [...] to believe that there is a definite relationship between chromosomes and allelomorphs [...]. It is conceivable that the chromosomes may be divisible into smaller entities (somewhat as Weismann assumes), which represent the allelomorphs*" [5].

Nineteenth century roots of cytogenetics

The explosion of interest in chromosomes and heredity, in fact, was based on a background of scientific data which had been accumulated during the last few years of the nineteenth century. The origins of cytogenetics stem back to that incredibly fertile 25 year period that revolutionised our way of considering biological phenomena. This golden era started Matthias Jakob Schleiden (1804–1881) and Theodor Schwann (1810–1882) with a cell theory, formulated in 1839. Charles Darwin's (1809–1882) "*The Origin of Species*" was published in 1859 to be followed by Johann Gregor Mendel's (1822–1884) publication on plant hybrids in 1866. Each publication was innovative enough in itself to constitute a genuine scientific revolution.

Thomas D'Aquino, whilst referring to Aristotle's philosophies, wrote that he felt like a dwarf on a giant's shoulder. I do not know whether D'Aquino was right in thinking this concerning Aristotle, but I do think that all modern day biology rests on the shoulders of the giant that Mendel, Darwin, Schleiden, and company, built at the end of the last century.

The first cytogenetic roots took hold in the 1860s and were based on non-Mendelian heredity theories. It was a transition period between "*mystical hypotheses*" on heredity and "*more or less scientific*" theories. This sarcastic opinion on heredity hypotheses at the end of the last century is not the view of any modern-day epistemologist but comes from a lecture of J. Arthur Thompson delivered to the Royal Society of Edinburgh in1889 [6] entitled *The History and Theory of Heredity*. The hypotheses sustained by Hippocrates and later by the most important founders of physiology in the eighteenth century could be considered more theological and philosophical than scientific. However, these theories were supported by geniuses such as Georges Buffon (1707–1788), Albrecht von Haller (1708–1778) and Charles Bonnet (1720–1793). Even Darwin's theory of Pangenesis [7] went no further than being "*more or less scientific*". The theory, in fact, was defined officially as the "*provisional theory of Pangenesis*". Francis Galton (1822–1911) strongly criticised his cousin Charles' theory of Pangenesis and wrote "*The doctrine of Pangenesis, pure and simple, is incorrect*". Galton was, without a shadow of doubt, the first to consider heredity from a statistical and therefore scientific standpoint [8]. Despite the fact that he used the strange *gemmulae* hypothesis of Pangenesis, he managed to produce some surprisingly concrete results. In a letter to Darwin dated December 19th 1875, Galton maintained that the factors determining heredity do not mix during hybridisation, but remain separate [9, 10]. He went on to expand this concept by stating that if a certain characteristic is regulated by two, three or more *gemmulae*, for example in black and white colours, the frequencies of grey colour phenotype would show a Pascal's triangle type distribution in succeeding generations.

Carl Wilhelm von Naegeli (1817–1891) proposed his own heredity theories [11] at nearly the same time as Galton. He imagined the base structures and inherent components of the protoplasm to be *micellae*, a derivative of Georges Buffon and Baron von Haller's molecules, albeit not exactly the same. These *micellae* are found in idioplasma that act to control the structure, the form and the function of the cell and, in this way determine all the important properties of organisms. In the reproductive cells of the reproductive organs of a plant, the *idioplasma* contains a complete set of micellar threads, that are transmitted to the progeny [12]. Naegeli's hypothesis helps us to see where August Weismann's (1834–1914) "*Continuity of the Germ Plasma*" comes from. A couple of names and ideas came from Naegeli's ideas, but as we will see later, the value of Weismann's innovative concepts were in another league.

August Weismann and the germinal plasma

Many fundamental discoveries were made in the short space of time between Naegeli's and Weismann's theories. This period, in fact, marked the birth of experimental embryology. Thanks to Wilhelm Roux's (1850–1924) experiments and those of the brothers Richard (1850–1937), and Oscar Hertwig (1849–1922), Hans Driesch (1867–1941), Edmund Wilson and Edwin Grant Conklin (1863–1952) the idea of embryonic prelocalisation, embryonic determination, and equipotential development, came into being around this time. Thus, already in 1881 Weismann proposed that a separation between germinal (totipotent) and somatic (differentiated) cells must take place during the development of organisms. In his studies on the Hydromedusae life cycle he described the migration of germ cells to the gonads of hydroids [13].

Based almost entirely on speculative background Weismann formed his theory of *Kontinuität des Keimplasmas* – i.e. "Continuity of Germ Plasma" – a line of immortal germ cells transmitted through the progeny [14]. In reality, the early identification of the blastomeres that determine the germinal line came a decade later in 1910 in the form of Theodor Boveri's work in *Ascaris megalocephala* [15], and in 1911 in Karl Amma's work in various species of Copepoda of the genus *Cyclops* [16]. The cells which are capable of transmitting hereditary characteristics to the progeny are easily identifiable in the embryo, so there is no need to imagine migratory elements, produced by the soma, like gemmulae, micellae or idioplasma.

The essential element needed to complete the theory was to localise these germinal plasma within the cellular structure. Weismann's attention turned towards the nucleus and for good reason. The great cytologist Rudolf Virchow's (1821–1902) aphorism from 1855, "*Omnis cellula a cellula*", any cell from a cell, acted as a stimulus. In fact many cytologists started studying the phenomenon of cell division so that in 1879, Walther Flemming (1843–1905) in animals [17] and Eduard Strasburger (1844–1912) in plants [18] described mitosis. Many scientists then studied the process resulting in Wilhelm von Waldeyer's (1836–1921) coding of the stages, structures and nomenclature of "*Karyokinesis*" [19].

Going back only a few years, Oscar Hertwig [20] described karyogamy in the sea urchin, i.e. the fusion of male and female nuclei in the formation of the progeny's hereditary structure. August Weismann therefore had all the necessary elements to form a coherent and well-constructed theory which gave credible explanations to all the cell physiology problems, including heredity. We will limit the discussion of our problem to the identification of the physical base of heredity and to do so we will utilise a plan (Fig. 1) that Weismann himself suggested in one of his works in 1887 [21]. The visible parts of chromo-

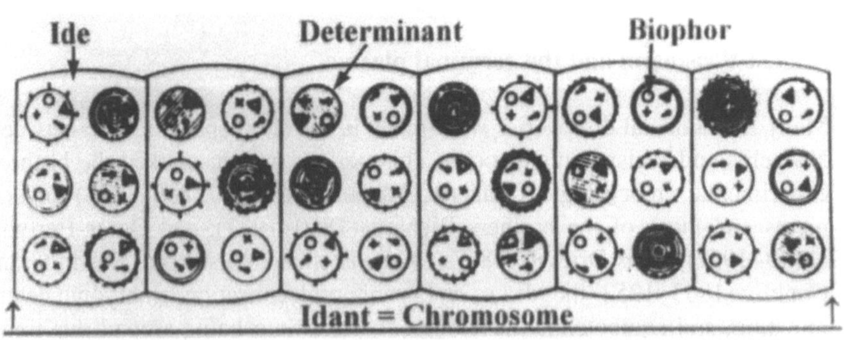

Figure 1. August Weismann's model of chromosome organisation and localisation of the hereditary units.

somes which today are called chromomeres were named "*idi*" by Weismann, coming from Naegeli's "*idioplasma*". Consequently he gave the name of "*idantes*" to chromosomes because they are the carriers of *ides*. He suggested that the *idi* contained different *determinants*, a name which – not by chance – evoked the concepts of experimental embryology. The determinants were formed from *biophors*. The word in German is *Lebensträger* and means carriers of life. They are the physiological and hereditary units which control the differentiation and functioning of cells. "*Different biophors constitute the different features of the cell*" [quoted in 12]. We can say therefore that the biophors are much more similar to genes in modern genetics than to Mendelian alleles.

However there remained a problem with Weismann's hypothesis. If, as Oscar Hertwig had demonstrated, karyogamy takes place in every egg's fertilisation, why does the number of determinants not double after every reproductive cycle? In 1882 Weismann wrote "*There must be a form of nuclear division in which the ancestral germplasm contained in the nucleus is distributed in the daughter nuclei in such a way that each of them receives only half of the number contained in the original nucleus*" [quoted in 22]. A year later Édouard van Beneden (1846–1910) described meiosis in the oogenesis of *Ascaris megalocephala* [23] (Fig. 2) so closing the cycle. In one of Weismann's notebooks, dated 1884 [24], we find a plan of meiosis where the chromosomes of *Ascaris* can be clearly recognised due to their shape and number.

There have been many observations on the cytology of this nematode parasite of the horse. It would not be an exaggeration to say that just about every nineteenth century cytologist has studied the gametogenesis and embryology of this roundworm. We could, in fact, say that *Ascaris megalocephala* was the first model animal for cytogenetics: a *Drosophila* before the *Drosophila* (Fig. 3). In 1890, Theodor Boveri, in his third work entitled "*Zellen studien*" [25] wrote "*The original number of chromosomes is reduced by half. This is no longer just a hypothesis but fact*".

Debate on meiotic reduction continued for some time, mixing with embryological and evolutionary theories. Hertwig and Strasburger sustained the idea of a simple "quantitative" reduction. Roux, Weismann and Boveri, however, favoured the idea of a "qualitative" division, which separates the different hereditary "determinants" in equal parts. It was Boveri, indeed, who realised the importance of what he termed " chromosome conjugation", which we now name synapsis [26]. The significance of the tetrads and the exact way the elements which make them up are separated was demonstrated clearly by Hermann Henking (1858–1942) and Johannes Rückert (1854–1923) [27, 28]. They confirmed, in fact, this second hypothesis. Johannes Rückert has also another claim to fame, in that it was he who described the lampbrush chromosomes in the nuclei of selachian oocytes [29].

Certainly, the images which we have from Rückerts (Fig. 4) are quite poor in quality, especially if we compare them to the masterly works of Harold G.

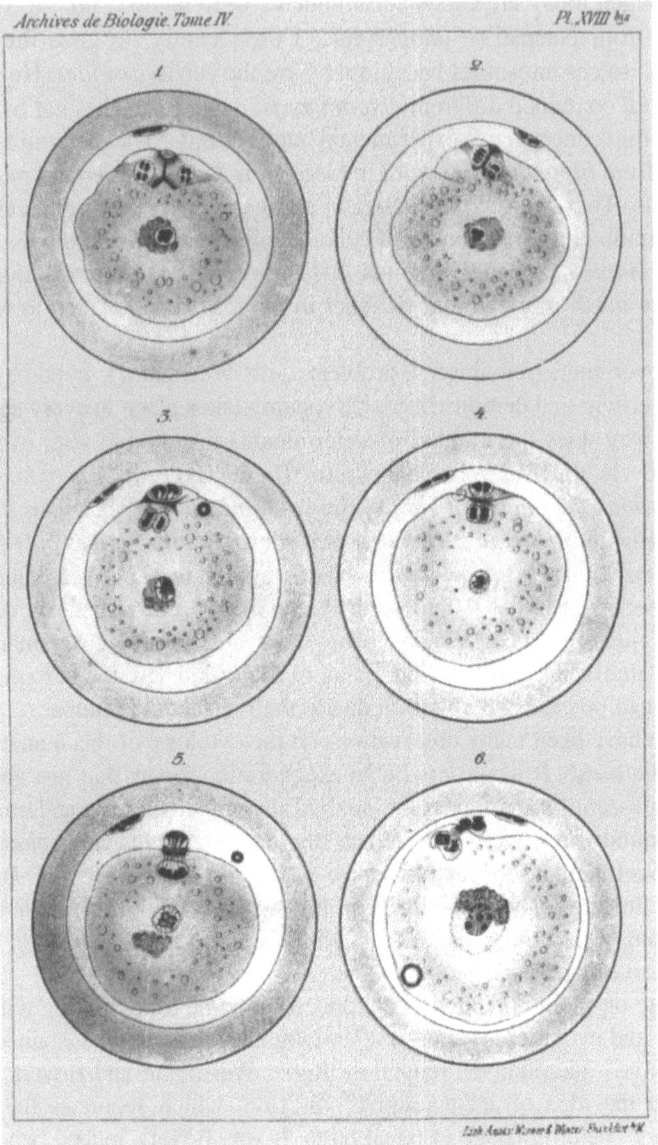

Figure 2. A plate from van Beneden's paper on the maturation and fertilisation of *Ascaris* egg where the meiotic process was first described [21].

Callan (1917–1993) and Joseph G. Gall, which they obtained thanks to a technique which still today remains unsurpassed. The lampbrush chromosomes have been, for many years, a fine instrument for research on functional cytogenetics and for detailed cell taxonomy.

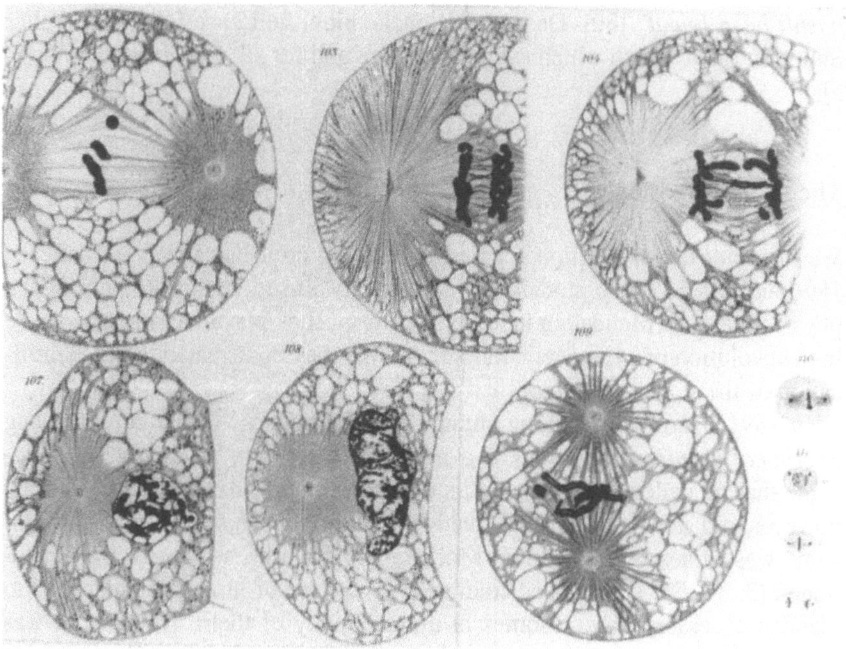

Figure 3. A plate from Theodor Boveri's "Zellen-Studien Hefte 4. *Ueber die Natur der Centrosomen*" (1900), showing very fine pictures of chromosomes in *Ascaris megalocephala* egg clevage.

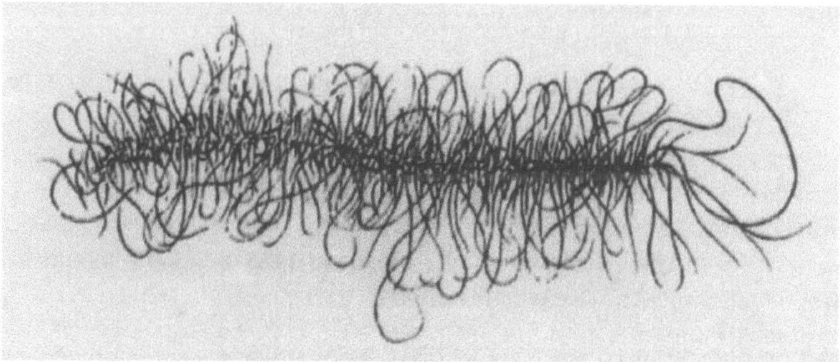

Figure 4. The first picture of a lampbrush chromosome in Rückert's paper on the development of Selachian ovaries [27]

At the end of the century, Edmund Wilson published the first edition of his famous book entitled "*The Cell in Development and Inheritance*". There he wrote "*Considering the origin of the tetrads and the latest interpretation of their reduction, we now find ourselves in a labyrinth of observations and hypotheses, each one in conflict with the other, from which a solution still hasn't been found*" [30]. Despite this declaration, he himself directed Walter Sutton in the research which eventually, once and for all, found the answer [3, 5].

The sex chromosomes

We must not fail to mention a series of research projects carried out between 1890 and 1905. These studies are not directly linked to the problem of the physical basis of Mendelian heredity. However, they played an important role in the evolution of cytogenetic thought, in that they contributed to the identification of the sex chromosomes.

In 1891, Hermann Henking studied spermatogenesis in the common bug *Pyrrhocoris apterus* and noticed a peculiar "extra" or "accessory" chromosome that he called X to indicate its unknown nature [31] (Fig. 5). Henking, however, did not entirely understand how important this accessory chromosome was in sex determination. Montgomery's works, which we have mentioned [3, 32–34,], contained studies of 42 species of insects; he managed to identify accessory chromosomes in a great many of them. However, it was

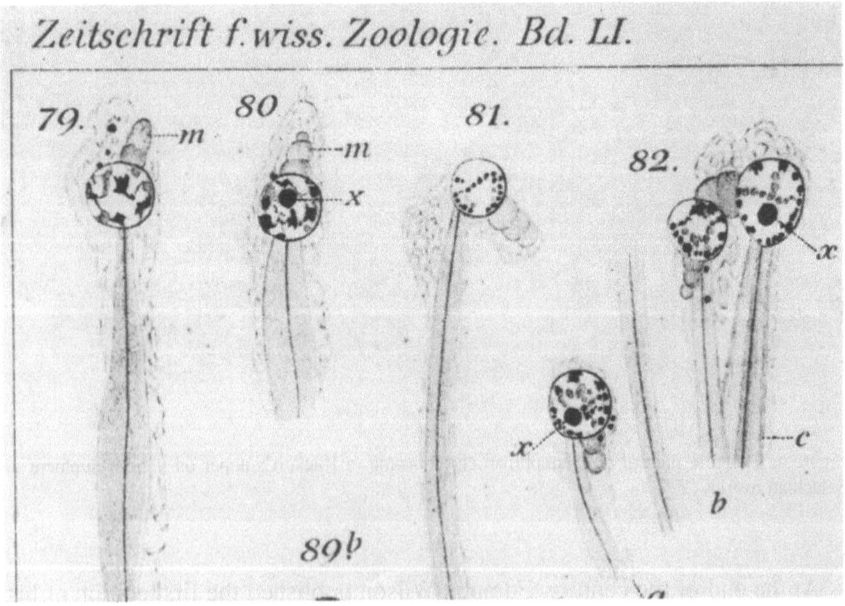

Figure 5. Part of a plate from the paper of Hermann Henking on the spermatogenesis of *Pyrrhocoris apterus* [29] where the X chromosome is observed in spermatozoon nuclei.

Erwin McClung (1870–1946) who, in 1901, clearly explained the role of the "accessory chromosomes" in sex determination [35, 36]. His theory explained the relationship between the two kinds of spermatozoa (produced by males) and the uniformity of the ova (produced by females) and the correct sex ratio which this phenomenon inevitably produces. Between 1904 and 1905 the identification of the sex chromosomes was concluded, above all, due to the fundamental works of Nettie M. Steven (1861–1912) [37] and Edmund Wilson [38].

Theodor Boveri and the chromosome individuality

All that was left to do at that point was to clarify a couple of odds and ends; minor and simple details for us in this day and age, but almost insurmountable "conceptual constraints" for those who studied chromosomes in times gone by. One of the many problems, was the univocal identity of the chromosome and its persistence during interphase. The fact that chromosomes disappear and reappear in every cellular cycle seemed to suggest that they were not responsible for the entire cell physiology and heredity. Some people supposed that the whole nuclear complex was made up of a continual *spirema*, which could be seen during early prophase, in both mitosis and meiosis, where it broke up more or less randomly into chromosomes. I would like to mention in this concern a paper of Kristine Bonnevie [39]. She gave an interpretation of the persistence of chromosomes during interphase based on the theory of the continuous spirema. Theodor Boveri made a significant contribution resolving the enigma [40]. Once again, his experiments took place at the Zoological Station, in 1902. That year, we remember, saw the presence of Edmund Wilson and Walter Sutton in Naples. The foundation by Anton Dohrn (1840–1909) was to create a biologist's paradise where grand old masters and young enthusiasts could meet to discuss common problems. In Naples, Boveri fertilised sea urchin eggs with a concentrated suspension of sperm. Under these conditions two sperms often fertilised one egg. He found there were three sets of chromosomes and three mitotic apparatuses in these eggs, the chromosomes being randomly distributed to each blastomere so that each cell received different kinds and a different number of chromosomes. Out of more than 1500 embryos produced in this way not one developed normally. This evidence clearly demonstrated that each individual element of the chromosomal set governed different parts of the development process. I remember, in this concern, the fine experiments carried out by Marcella Boveri [41], who actively collaborated with her husband. Marcella O'Grady, an American biologist, on the recommendation of Edmund Wilson, came to Würzburg in 1896 to work with Theodor Boveri, and she married him in 1897 [42]. Marcella Boveri took over her husband's scientific work after his premature death; it was she who edited a review of their joint research on the development of merogonic hybrid embryos of sea urchin in a ponderous posthumous paper in 1918 [43].

The fly room at Columbia University

In 1908, a crucial year for cytogenetics, the first bottle of fruit flies entered Thomas Hunt Morgan's (1866–1945) laboratory. Morgan had started his scientific career as a zoologist and his thesis had been on the phylogeny of the Picnogonida, written under the guidance of Professor Brooks at the Johns Hopkins University. In 1891, however, he met Hans Driesch at the *Stazione Zoologica* in Naples. Both were 25 and both were very enthusiastic. So Morgan abandoned zoology and moved on to experimental embryology. We have already said that in the late nineteenth and early twentieth centuries evolution, embryology and heredity were considered a joint problem, so it is no wonder that Morgan's new interests also included evolution and heredity. As an evolutionist, to start with, he was anti-Darwinian and in favour of the mutationism of Hugo de Vries with a bent for Lamarkism. The Dutch scientist imagined that a new species could come into being in only one generation due to mutation [44]. So to test and expand the mutation theory of de Vries to the animal world, Morgan chose *Drosophila.* This choice proved to be even more effective than he could possibly have ever imagined. In 1910, he published an article in *Science* entitled "*Sex limited inheritance in* Drosophila" [45]. He mentioned the "*white eyes*" mutation and the fact that it only occurred in males. In 1910, Morgan was called to Columbia University by Edmund Wilson. Morgan knew all about Wilson's work on the role of chromosomes in sex determination, but Morgan was very cautious in interpreting his own discovery. In his earlier works, he named this kind of inheritance "*sex-limited*" rather than "*sex-linked*". At that time, digamety of *Drosophila* was well known, but yet he did not dare to suggest that the allele for "white eyes" was localised on the accessory X chromosome. In 1910, Morgan believed that the Mendelian factor for *Drosophila* eye colour was transmitted together with the X chromosomes, but not as a part of it. A year later [46], he obtained white eyed females from backcrossing and also new *Drosophila* mutations, some of which were "sex-limited" and others not. He realised that horizons were broadening for the study of heredity.

The fruit fly was ready to fly

In the following years, a host of *Drosophila* mutations were recorded in scientific works that reached fever pitch. The *Fly Room* at Columbia University started filling up with many enthusiastic graduate and post-graduate students. To name but a few: Calvin Blackman Bridges (1889–1938), Alfred Harry Sturtevant (1891–1970), Hermann Joseph Muller (1890–1967), and from other parts of the scientific world, Curt Stern (1902–1981) from Germany, Otto Mohr (1886–1967) from Norway and finally Theodosius Dobzhansky (1900–1975) from Russia.

I would like to point out the remarkable speed at which the *Drosophila* Group at Columbia University worked. In 1915, only four years after the first study, Morgan, Sturtevant, Bridges and Muller published the book "*The Mechanism of Mendelian Heredity*" [47]. What else was now necessary to give the Mendelian-chromosome theory absolute credibility? The last requirement was to see all the alleles in a line like lead soldiers, along a chromosome. Evidence was achieved by Theophilus Shichel Painter (1889–1969) and Calvin Bridges who transferred Morgan's statistical maps to the polytene chromosomes of salivary glands [48].

Salivary chromosomes were nothing new, having been first described with a dipteran (*Chironomus plumosus*) in 1881 by Édouard-Gérard Balbiani (1825–1899) (Fig. 6) who described inside the nuclei of the salivary glands "*un corps pâle, qui a la forme d'un cordon cylindrique diversement enroulé sur lui-même [...] et continu dans toute sa longeur*" [49]. Evidence for the chromosomal nature of polytene elements had been established first in 1933, by Emil Heitz (1892–1965) and Hans Gustav Emil Bauer (1904–1988) [50].

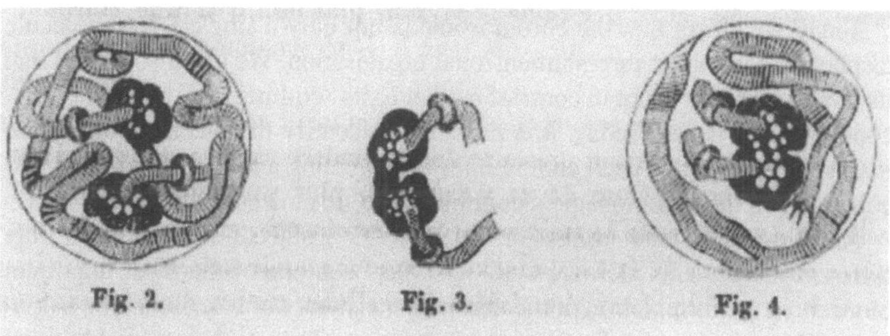

Figure 6. Balbiani's pictures of nuclei of salivary glands in *Chironomus* [47]

However, at the end of the thirties, thanks to Painter's techniques [51] the salivary chromosomes became an indispensable cytogenetic tool, which is still fundamental today in molecular cytogenetics. However, they also led researchers to believe in the illusory hope of producing the first pictures of genes in the subtle bands under the microscope. This was, indeed, the conceptual error of discontinuity, seen microscopically in the polytene chromosomes from salivary glands of Diptera, as well as in the lampbrush chromosomes from Amphibian oocytes.

However, there was a certain disagreement

The success of Morgan's group didn't convince the entire genetics community. I will leave the lesser important objectors aside, but three great names in the world of genetics are certainly worth mentioning.

Already from 1900 onwards, William Ernest Castle (1867–1962) had introduced the fruit fly at Harvard to experiment on the effects of consanguinity. Now, using the same insect, Morgan obtained incredible results. However, this was not the reason why Castle held certain reservations on compiling a map using recombination frequencies. His doubts were based on valid theories. Morgan and his scholars supposed that the chromosome structure was a linear array of genes, the genes being considered as separate entities. Castle, on the other hand, considered the chromosome to be one single molecule possessing a complex three-dimensional structure [52]. Time eventually proved him right on this. Muller, first in 1920, by means of a statistical approach [53] and then, in 1927, from evidence of mutagenesis induced by X-rays [54] definitively demonstrated the linear arrangement of the genes and their discontinuity along the chromosome.

Today we affirm that the chromosome is not only a single DNA molecule displaying a complex three-dimensional organisation. We can also affirm that its molecular unity is not in contrast neither to its "co-linearity" nor to the "discontinuity" of genes. Today, it is possible to identify the precise punctuation marks which separate each coding unit, thus creating a kind of functional discontinuity in the chromosome's continuity and its molecular unity.

Richard Benedict Goldschmidt's (1873–1958) objections were of entirely different nature. This German cytologist has been one of the most important pioneers in the formulation of the chromosome theory of heredity at the beginning of the century. In the "twenties", however, his interests turned to gene physiology, and he was one of the founders of the so called "gene theory" [55]. He envisaged the hereditary factors as being dynamic and functional, a theory that did not lend itself very well to the idea of fixed hereditary structures anchored to structural material. For Goldschmidt, chromosomes were temporary containers of genes, the genes moving constantly in and out between metaphase and interphase.

William Bateson had been sceptical about the possibility of allele mapping on chromosomes right from the beginning. William Coleman [56] the historian, wrote that this attitude was due to an extremely conservative personality with an experimental and not theoretical background, so typical of some English geneticists at the end of the nineteenth century. Their work was linked mainly to breeding and selection. In reality, Bateson's scepticism was purely based on logic. He insisted that Morgan's theories were dependent on too many ancillary assumptions. In 1921 however, William Bateson passed through New York to visit Morgan at Columbia University. Here, Calvin Bridges gave so many explanations and with such enthusiasm that he completely convinced Bateson, so that already by 1922 [57], he had become a fervent supporter of the ideas of Morgan and his collaborators.

The roots of human cytogenetics

Theophilus Painter, who was one of the leading cytogeneticists of *Drosophila*, can lay another claim to fame: he was the first to face the thorny problem of the human karyotype, producing one which, though incorrect, was a good approximation.

The human diploid chromosome set, obtained by tissue sectioning, had previously been reported at rather low numbers and by consensus of several authors was estimated at 24. This number had also been confirmed by several eminent scientists such as Flemming [58] and Montgomery [59], and the issue remained thus for a long time. The techniques used to determine this diploid number were absolutely not suitable. Nevertheless, von Winiwarter using sections of human testicle [60, 61] established the human diploid number to be $2n = 47–48$ with XO sex chromosomes. These findings were contested; Michael Guyer [62], in order to accredit both diploid numbers observed, i.e. 24 and 48, attributed the difference to the different persons studied. I quote:" *It must be borne in mind, however, that Montgomery and I worked on tissues of negroes and von Winiwarter on those of a white man. I am, at present, engaged in a study of material of two different white men and although not ready to make a detailed statement I can say with assurance that the number of chromosomes is considerably in excess of those found in my negro material.*" (!)

This explanation was considered plausible by T.H. Morgan [63] who wrote: "*It is with great interest I note in the last paragraph of Guyer's paper a hint (or is it intended as an announcement?) that the white man has more chromosomes than the negro, a point of view I mentioned in the book under discussion* (i.e. Morgan's "Heredity and Sex") *as a possible way of harmonizing Guyer's results with those of v. Winiwarter. If the suggestion is established, some revision may be necessary concerning the Mendelian expectation of the inheritance of skin colour in the black-white cross.*"

Painter's study [64] was quite accurate: the material was obtained from "*three inmates of an insane asylum, two black peoples and a young white man. In all cases the cause of the removal of testes was excessive self-abuse*". In Painter's opinion, the male sex chromosome complement was XY and therefore, the human diploid number was 48 both in males and in females. It is interesting to compare drawings produced by the two authors because the small unpaired X chromosome of von Winiwarter is in fact the Y chromosome. Several years later the human diploid number was established as $2n = 46$ using cells grown in culture by Joe Hin Tjio and Albert Levan [65]. Charles Ford [66] noted that in a large number of cases the sex chromosomes underwent precocious disjunction, so that during diakinesis of the primary spermatocyte, 22 bivalents and two univalents appear, which simulate 24 bivalents. This could have been the origin of the incorrect estimate of both von Winiwarter and Painter. If we count the bivalents which Painter reported in his drawings (Fig. 7), often we note only 22 autosomal bivalents, and the XY bivalent is quite clearly defined indicating a diploid number of 46. In spite of this, Painter assumed as correct the number he observed in sectioned spermatogonial metaphase plates, i.e. 48, and he supposed that in the sections of primary spermatocyte diakinesis, another bivalent was covered by other elements of the diakinesis (Fig. 7)

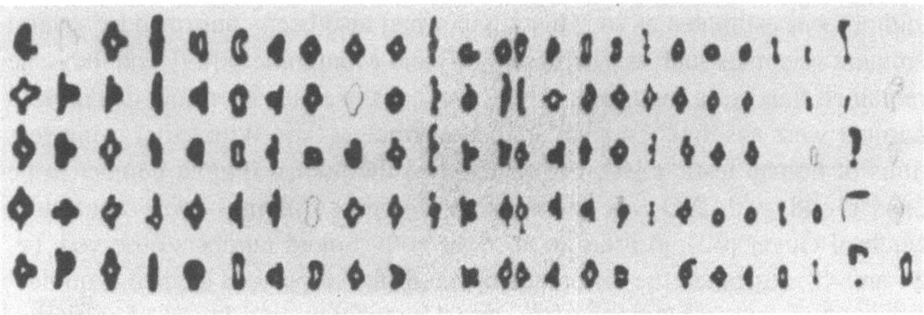

Figure 7. Figures from the paper of Theophilus Painter on human chromosomes [62]. Bivalents from different primary spermatocyte diakinesis are presented: elements outlined with a dotted contour are the hypothetical bivalents which he assumed to be covered by other elements of the section

This period of von Winiwarter and Painter might be considered the heroic period of the patriarchs of karyotyping, and these researchers may be considered, as T.C. Hsu [67] called them, the "*Knights of the Dark Ages*". From this dark age, however, we have emerged thanks to the discovery of hypotonic shocking of cells in culture [68]. The development of human and medical cytogenetics is a chronicle of contemporary researches, pertaining to modern cytogenetics, which was the subject of a beautiful lecture by Fergusson Smith [69] at the 11th Chromosome Conference.

Chromosomes and evolution

Theodosius Dobzhansky worked with Morgan and the fruit fly group after he had accumulated experience in population genetics in his native Russia, at a time when Sankt Petersburg was called Petrograd. He was profoundly impressed by the mathematical models of genic equilibrium formulated – in the late twenties and early thirties – by Ronald Aylmer Fisher (1890–1962), John Burdon Haldane (1892–1964), Sewall Wright (1898–1989) and Sergej

Sergeevic Cetverikov (1880–1959), who was his tutor in Petrograd. Dobzhansky was different from the others in the *Drosophila* group in that he was particularly interested in natural population characterisation and cytological comparison of *Drosophila* species. To cut a long story short, the famous "*Genetics and Origin of Species*" was published in 1937 [70] and was to form the genetic foundations of the Evolutionary "*Modern Synthesis*" Theory also proposed by Ernst Mayr and George Gaylord Simpson (1902–1984).

The group at the Kol'tsov Institut, in Petrograd, of Sergej Cetverikov, Nikolaj Petrovich Dubinin and Nikolaj Vladimirovich Timofeeff-Ressovsky (1900–1981) also worked on the *Drosophila* karyotype comparisons up to the time when Stalinist ideologists destroyed the group. *Drosophila* cytogenetics in Italy originated from the friendship between Timofeeff-Ressovsky and two young geneticists of that period: Carlo Jucci (1897–1962) and Adriano Buzzatti Traverso (1913–1983).

The number of karyotype comparisons continued to multiply both in plants and animals. Karyotype analysis became more meticulous and geneticists began looking for the mechanisms responsible for chromosome rearrangement and their consequences on evolutionary processes. This is the field I have been working on for almost forty years. I feel particularly attached to Michael J.D. White (1910–1983) and Robert Matthey (1900–1982) from whom I learnt the trade. They were different in personality, with different approachs to the role of chromosomes in evolution. Matthey's "*Les Chromosomes des Vertébrés*" [71] dealt with the problem at the macroevolutionary level if you allow this definition of the cytotaxonomic approach. White, on the other hand, followed a microevolutionary path. In 1948 he discussed the problem of the role of chromosomes in speciation in his work "*Animal Cytology and Evolution*"[72], followed by his last work "*Modes of Speciation*" [73], in 1978. Many Italian cytogeneticists remember Michael White in his frequent visits to Italy. We remember his perfect Italian, spoken with a Tuscan accent, which he picked up as a boy while attending primary school between Sienna and Florence.

Chromosome structure

Morgan himself admitted in his Nobel lecture that before the early Thirties the problem concerning chromosome structure had not been approached in the right way: "*Now that we have located genes on chromosomes are we justified in regarding them as material units, as chemical bodies, of higher order than molecules? Frankly, these are questions for which the working geneticist has not much concern himself.*" Then after a brief report on current hypotheses at that time he concluded: "*Therefore it makes no difference in the actual work in genetics which point of view is taken*" [74].

Morgan was certainly wrong about this. We well know today how important the precise understanding of the chromosome molecular structure has been in genetics. Fortunately the problem was tackled by cytologists, biochemists and those geneticists working on the nature and function of the gene. In 1928, Emil Heitz managed to observe two types of chromatin in plants [75] namely euchromatin and heterochromatin according to their position on chromosomes. Emil Heitz then carried out similar experiments with polytene chromosomes of dipterans [76, 77]. Obviously this evidence was very exciting, because it was possible to correlate the cytochemical organisation of the chromosome to the mechanisms of expression of genes [78, 79].

This intellectual challenge was accepted by Torbjörn Caspersson (1910–1997) who, from 1934 onwards [80, 81], decided to study the chromosome's nucleic-acid composition, the nucleic acid cycles and the relationship between this cycle and the cycles of protein syntheses. In a nutshell, he took into account all the necessary experimental ideas which would allow him to draw the conclusion that genes are formed from molecules of nucleic acid. However, the theoretical and experimental evidence supporting the theory that genes were nucleo-proteins was so strong that even Caspersson was loath to contradict the idea. He wrote: "*If one assumes that the genes consist of the known substances, there are only the proteins to be considered, because these are the only known substances which are specific for the individual. On that assumption a protein structure in chromosomes takes on a very great interest*" [82]. Such an emphasis on proteins, this kind of chemical Mendelism, this dynamic genetics, codified by the works of Goldschmidt [55] and Beadle [83] caused a conceptual error which bewildered for many years researches on the nature of the gene: the confusion between the gene and its product. Caspersson's observations had important impact on a cytogeneticist of great personality, Cyril Dean Darlington (1903–1981), the founder of our Chromosome Conferences.

Before 1939, Darlington was interested in the mechanics of meiosis [84]. He studied the forces which act in the coupling and separation of chromosomes. In this sense he claimed to be Newtonian, and, in fact, he quoted some of the great physicist's sentences to help explain meiosis. In 1939 the Eighth International Genetics Conference was held in Edinburgh. Caspersson's work circulated at this assembly, although Caspersson was not able to attend. Despite this, Darlington was very enthusiastic about Caspersson's observations, and from then on, he made more and more precise references to the chemical constitution of the chromosome and the cycle of nucleic acids. This resulted in the publication of "*Chromosome Chemistry and Gene Action*" in *Nature* in January 1942 [85].

But this knowledge acquired in fifty years of study was about to change profoundly. In 1953, Francis Crick and James Watson published two articles in *Nature* [86, 87] on the double helix structure of DNA. However, to suggest this was a sudden dramatic discovery would be quite misleading. It would be the same as saying that Newton formulated his universal law of gravity from the simple fact an apple fell on his head! There had been 10 years of study to move on from the idea of genes being proteins to the discovery of the double helix. Many mighty minds were involved in this progress. Biophysicists and biochemists such as William Thomas Astbury (1898–1961), Linus C. Pauling, Jean Brachet (1909–1988), Maurice Wilkins, and geneticists of microrganisms such as Oswald Theodor Avery (1877–1955) and Alfred Hershey and Martha Chase. How the idea of the double helix evolved can be found in a fascinating classic book by Robert Olby [88] entitled "*The Path to the Double Helix*".

The value of the "Scientific Revolution" inherent in the discovery of the Double Helix and the Genetic Code is quite obvious.

After 1953 genetics and, above all cytogenetics, has never been the same. The way we consider biological phenomenon of heredity, development and differentiation is completely different today. Cytogenetics started the molecular dissection of chromosomes. In 1953 we moved from yesterday's chromosome studies to today's molecular cytogenetics.

I apologise for any omission. In fact too many events and too many famous persons have not been quoted in this review which, I must say, has unfortunately been pitifully too short when you think it has covered more than 100 years of studies of several hundreds of scientists. However, to conclude, I would like to quote a phrase which sums up perfectly the history of cytogenetics in the twentieth century. Giuseppe Montalenti (1904–1990), that unforgettable *maestro*, started the concluding paragraph of one of his last works by saying: "*Il futuro è cominciato ieri*", "the future began yesterday" [89].

Acknowledgements

I am particularly indebted to my colleagues Sheila Counce and Helmut Zacharias for their comments, suggestions and valuable bibliographic help concerning the "golden era" of embryology and genetics. I thank Gilberto Corbellini, my epistemologist friend, for the pleasing discussions we had concerning the nineteenth century roots of modern biology.

References

1 Corbellini G (1997) *Le grammatiche del vivente. Storia della biologia molecolare*. Laterza, Bari
2 de Vries H (1900) Sur la loi de disjunction des hybrides. *C R Acad Sci Paris* 130: 845–847
3 Sutton W (1902) On the morphology of chromosome groups in *Brachystola magna*. *Biol Bull* 4: 24–39
4 Montgomery TH (1901) A study of the chromosomes of the germ-cells of metazoans. *Trans Amer Phil Soc* 20: 154–236
5 Sutton W (1902) The chromosomes in heredity. *Biol Bull* 4: 231–248
6 Thompson JA (1889) The history and theory of Heredity. Proc Royal Soc Edinburgh [sess. 1888–1889] 16: 91–116
7 Darwin CR (1871) Pangenesis. *Nature* 3: 369–378
8 Galton F (1872) A theory of Heredity. *Contemp Rev* 27: 80–95
9 Olby RC (1966) *The Origin of Mendelism*. Constable, London
10 Mayr E (1982) *The Growth of Biological Thought. Diversity, Evolution, and Inheritance*. The Belknap Press of Harward Univ. Press, Cambridge (MA), and London

11 von Naegeli C (1884). *Mechanisch-physiologische Theorie der Abstammungslehre*. München
12 Hall TS (1969) *Ideas of Life and Matter. Studies in the history of general physiology. Volume two: from the Enlightenment to the end of the nineteenth century*. University Chicago Press, Chicago and London
13 Weismann A (1883) *Die Enstehung der Sexualzellen bei den Hydromedusen. Zugleich ein Beitrag zur Kenntnis des Baues und der Lebenserscheinungen dieser Gruppe*. Fischer, Jena
14 Weismann A (1885) *Die Kontinuität des Keimplasmas als Grundlage einer Theorie der Vererbung*. Fischer, Jena
15 Boveri T (1910) Die Potenzen der *Ascaris* Blastomeren bei abgeänderter Furchung. Zugleich ein Beitrag zur Frage qualitativ ungleicher Chromosomen-Theilung. *In*: *Festschrift für R. Hertwig*, vol 3, 131–214
16 Amma K (1911) Über die Differenzierung der Keimbahnzellen bei den Copepoden. *Arch Zellf* 6: 497–576
17 Flemming W (1879) Beiträge zur Kenntniss die Zelle und ihrer Lebenserscheinungen. Theil II. *Arch Mikr Anat* 18: 151–258
18 Strasburger E (1879) Ueber ein zu Demonstrationen geeignetes Zelltheilungs-Object. *Sitzungsbericht med-naturwiss Ges Jena* 18: 93–112
19 Waldeyer W (1888) Über Karyokinese und ihre Beziehungen zu den Befruchtungsvorgängen. *Arch Mikr Anat* 32: 1–112
20 Hertwig O (1875) Beiträge zur Kenntniss der Bildung, Befruchtung und Theilung des thierischen Eies. *Morphol Jahrb* 1: 347–434
21 Mondella F (1971) La biologia alla fine dell'Ottocento. *In*: L Geymonat (ed.): *Storia del pensiero filosofico e scientifico*, vol. 5. Garzanti, Milano, 648–710
22 Churchill FB (1987) The Life Sciences in Germany. From heredity theory to Vererbung. The transmission problem, 1850–1915. *Isis* 78: 337–364
23 van Beneden E (1883) Recherches sur la maturation de l'oeuf et la Fécondation. *Ascaris megalocephala*, *Arch Biol* 4: 265–638
24 Churchill FB (1969) August Weismann and the break from tradition. *J Histol Biol* 3: 139–177,
25 Boveri T (1890) Zellen-Studien. Heft 3. Ueber das Verhalten der chromatischen Kernsubstanz bei der Bildung der Richtungskörper und bei der Befruchtung. *Jena Z Med Naturwiss NF* 17) 24: 314–401
26 Goldschmidt RB (1956) *The golden age of Zoology: Portraits from memory. Recollections of a Zoologist*. Univ. Washington Press, Seattle and London
27 Rückert J (1892). Ueber die Verdoppelung der Chromosomen in Keimbläschen der Selachiereies. *Anat Anz* 8: 44–52
28 Rückert J (1894) Die Chromatinreduktion bei der Reifung der Sexualzellen. *Ergeb Anat Entwickelungsgesch* 3: 517–583
29 Rückert J (1892) Zur Entwickelungsgeschichte des Ovarialeis bei Selachiern. *Anat Anz* 7: 107–158
30 Wilson EB (1896) *The Cell in Development and Inheritance*. Columbia Univ Press, New York
31 Henking H (1891) Untersuchungen über die ersten Entwicklungsvorgänge in der Eiern der Insekten. 2. Über Spermatogenese und deren Beziehung zur Eientwicklung bei *Pyrrhocoris apterus*, L. *Z wiss Zool* 51: 685–736
32 Montgomery TH (1898) The spermatogenesis in *Pentatoma* up to the formation of spermatid. *Zool Jahrb Anat* 12: 1–88
33 Montgomery TH (1904) Some observation and consideration upon the maturation phenomena of the germ-cells. *Biol Bull* 6: 136–157
34 Montgomery TH (1905) The spermatogenesis of *Syrbula*, and *Lycosa* with general consideration upon chromosome reduction and the heterochromosomes. *Proc Acad Natur Sci Phila* 57: 162–205
35 McClung CE (1901) Notes on the Accessory Chromosomes. *Anat Anz* 20: 220–226
36 McClung CE (1902) Accessory Chromosome – sex determinant? *Biol Bull* 3: 43–84
37 Stevens N. (1905–1906) Studies on spermatogenesis with especial references to the accessory chromosome. *Public Carnegie Inst Washington* 36 (part I, 1905; part II, 1906)
38 Wilson EB (1905) The chromosomes in relation to the determination of sex in insects. *Science* 22: 500–502
39 Bonnevie K (1908) Chromosomenstudien: Individualität der Chromosomen. *Arch Zellforsch* 1: 450–514
40 Boveri T (1902) Ueber merpolige Mitose als Mittel zur Analyse des Zellkerns. *Verh Phys-Med Ges Würzburg NF* 35: 67–90

41 Boveri M (1903) Ueber Mitosen bei einseitiger Chromosomenbindung. *Jena Z Med Naturwiss* 37: 401–446
42 Balzer F (1962) Theodor Boveri: Leben und Werk eines grossen Biologen. Wissenschaftliche Verlagsgesellschaft M.B.H. Stuttgart
43 Boveri Th (1918) Fehlerquellen bei Merogonieversuchen und die Entwicklungsfähigkeit merogonischer und partiell-merogonischer Seeigelbastarde. *Arch Entwick-Mech* 44: 417–471
44 de Vries H (1901 and 1903) *Die Mutationtheorie: Versuche und Beobachtungen über die Entstehung von Arten in Pflanzenreich.* 2 vol, Von Veit and Co., Leipzig
45 Morgan TH (1910) Sex limited Inheritance in *Drosophila. Science* 32: 120–122
46 Morgan TH (1911) An attempt to analyze the constitution of chromosomes on the basis of sex-limited inheritance in *Drosophila. J Exp Zool* 11: 365–413
47 Morgan TH, Sturtevant AH, Muller HJ, Bridges CB (1915) *The Mechanism of Mendelian Heredity.* H Holt, New York
48 Bridges CB (1935) Salivary chromosome maps. *J Hered* 26: 60–64
49 Balbiani EG (1881) Sur la structure du noyau des cellules salivaires chez les larves de *Chironomus. Zool Anz* 4: 637–641
50 Heitz E, Bauer H (1933) Beweise für die Chromosomennatur der Kernschleifen in den Knäuelkernen von *Bibio hortulanus* L. *Z Zellforsch mikrosk Anat* 17: 67–82
51 Painter TS (1933) A new method for the study of chromosome rearrangements and the plotting of chromosome maps. *Science* 78: 585–586
52 Castle WE (1919) Is the arrangement of genes in the chromosome linear? *Proc Natl Acad Sci USA* 5: 25–32
53 Müller HJ (1920) Are the factors of Heredity arranged in a line? *Amer Naturalist* 54: 97–121
54 Müller HJ (1927) Artificial transmutation of the Gene. *Science* 66: 84–87
55 Goldschmidt R (1938) *Physiological Genetics.* McGraw-Hill, New York
56 Coleman W (1970) Bateson and Chromosomes: Conservative thought in Science. *Centaurus* 15: 228–314
57 Bateson W (1922) Evolutionary faith and modern doubts. *Science* 4: 55–61
58 Flemming (1898) Ueber die Chomosomenzahl beim Menschen. *Anat Anz* 14: 171–174
59 Montgomery TH (1912) Human spermatogenesis. Spermatocytes and spermatogenesis. A study in Inheritance. *Proc Acad Natur Sci Phila* 15: 3–22
60 de Winiwarter H (1912) Etudes sur la spermatogenèse humaine. *Arch Biologie* 27: 147–149
61 de Winiwarter H (1921) La formule chromosomiale dans l'espèce humaine. *C R Soc Biol* 85: 266–267
62 Guyer MF (1914) Accessory chromosomes in man. *Science* 39: 721–722
63 Morgan TH (1914) Chromosomes of the white man and the negro. *Science* 39: 827–828
64 Painter TS (1922) Studies in mammalian spermatogenesis. II. The spermatogenesis of man. *J Exp Zool* 37: 291–338
65 Tjio JH, Levan A (1956) The chromosome number of man. *Hereditas* 42: 1–6
66 Ford CE, Hamerton JL (1956) The chromosomes of man. *Nature* 178: 1020–1023
67 Hsu TC (1979) *Human and Mammalian Cytogenetics, an Historical Perspective.* Springer, New York
68 Hsu TC (1952) Mammalian Chromosomes *in vitro.* 1. The Karyotype of Man. *J Hered* 43: 167–172
69 Ferguson-Smith MA (1993) From chromosome number to chromosome map: the contribution of human cytogenetics to genome mapping. *Chromosome Today* 11: 3–19
70 Dobzhansky Th (1937) *Genetics and the Origin of Species.* Columbia University Press, New York
71 Matthey R (1945) *Les Chromosomes des Vertébrés.* Rouge, Lausanne
72 White MJD (1948) *Animal Cytology and Evolution.* Cambridge Univ. Press, London
73 White MJD (1975) *Modes of Speciation.* WH Freeman, San Francisco
74 Morgan TH (1934) The relation of genetics to physiology and medicine. *In*: *Nobel Lectures. Including Presentation, Speeches, and Laureates' Biographies. Physiology and Medicines 1922–1941.* Elsevier, Amsterdam (1965), 313–328
75 Heitz E (1928) Das Heterochromatin der Moose. *Jahrb Wiss Bot* 69: 762–818
76 Heitz E (1933) Über totale und partielle somatische Heteropyknose, sowie strukturelle Geschlechtschromosomen bei *Drosophila funebris. Z Zellforsch Mikr Anat* 19: 720–742
77 Heitz E (1934) Die somatische Heteropyknose bei *Drosophila melanogaster* und ihre genetische Bedeutung (Cytologische Untersuchungen und Dipteren. III). *Z Zellforsch Mikr Anat* 20: 235–287

78 Heitz E (1934) Ueber α- und β-heterochromatin sowie konstanz und Bau der Chromomeren bei *Drosophila. Biol Zbl* 54: 588–609
79 Heitz E (1935), Chromosomenstruktur und Gene. *Z induktive Abstammungs- und Verebungslehre* (now: *Mol General Genetic*) 70: 402–447
80 Caspersson T (1934) Druchfiltrierung von Thymonucleinsäure. *Biochem Z* 270: 161–163
81 Caspersson T (1937) Aufbau der Structuren des Zellkernes. *Acta Med Skand* 73, Suppl. 8: 1–151
82 Caspersson T, Hammarsten E, Hammarstem H (1935) Interactions of proteins and nucleic acid. *Trans Faraday Soc* 31: 367–389
83 Beadle GW (1945) Genetics and metabolism in *Neurospora. Physiol Rev* 25: 643–663
84 Darlington CD (1937) *Recent Advances in Cytology.* 2nd ed. J & A Churchill, London
85 Darlington CD (1942) Chromosome chemistry and gene action. *Nature* 147: 66–69
86 Watson JD, Crick FHC (1953) Molecular structure of nucleic acid. *Nature* 171 [4356]: 737–738
87 Watson JD, Crick FHC (1953) Genetical implications of the structure of the desoxyribonucleic acid. *Nature* 171 [4361]: 964–967
88 Olby R (1974) *The Path of the Double Helix.* Macmillan, London
89 Montalenti G (1985) *L'Evoluzione del concetto di Gene. In: La vita e la sua storia. Stato e prospettive degli studi di genetica.* Scientia, Milano

Part one
Chromosome organisation

Giorgio Bernardi

Sergio Pimpinelli

J.S. Heslop-Harrison

Toshimichi Ikemura

Chromosomes Today, Vol. 13
ed. by E. Olmo and C.A. Redi

The prometaphase bands of human chromosomes: compositional features and gene distribution

Giorgio Bernardi[1], Concetta Federico[2] and Salvatore Saccone[2, 3]

[1] *Laboratorio di Evoluzione Molecolare, Stazione Zoologica Anton Dohrn, Villa Comunale, I-80121 Napoli, Italy*
[2] *Dept. Biologia Animale, University of Catania, via Androne, 81, I-95124 Catania, Italy*
[3] *Present address: DIPROVAL, Sez. Allevamenti Zootecnici, University of Bologna, via F.lli Rosselli 107, I-42100 Reggio Emilia, Italy*

Introduction

The human genome is a mosaic of long, compositionally homogeneous DNA segments, the isochores, that can be partitioned into five families. Two GC-poor families (L1 and L2) representing 63% of the genome, and three GC-rich families (H1, H2 and H3) representing 24%, 7.5% and 4–5% of the genome, respectively [1]. Gene concentration increases with increasing GC levels, and reaches a 20-fold higher level in H3 compared to L isochores [2]. *In situ* hybridization of DNA from different isochore families provides, therefore, information on the chromosomal distribution of genes. Using this approach, three subsets of R(everse) or G(iemsa)-negative bands, $H3^+$, $H3^*$ and H3 , containing large, moderate, and no detectable amounts, respectively, of the gene-richest H3 isochores were identified at a resolution of 400 bands [3]. $H3^+$ bands largely overlap with the most heat-denaturation-resistant bands [4], the chromomycin-A3-positive, 4,6-diamindino-2-phenylindole (DAPI)-negative bands [5], the bands with the highest CpG island concentrations [6], and the earliest replicating bands [7]. Here we have defined the $H3^+$ bands at a 850-band resolution, and have thus identified the human genome regions, having an average size of 4 Mb, that are endowed with the highest gene density.

Results and discussion

DNA from the H3 isochore family was hybridized to human prometaphase chromosomes using the biotin/avidin system under conditions in which repeated sequences were competed out (Materials and Methods were already described in ref. [8]). The distribution of hybridization signals on chromosomes (see Fig. 1 for an example) showed that H3 isochores are only located on a small number of R_{850} bands (Figs 2–5) and on none of the G_{850} bands

Figure 1. (See color plate I between pages 56 and 57.) Human chromosomes 7, 11 and 19 hybridized with the biotin-labeled DNA from the H3 isochore family, at different levels of resolution. The hybridized regions were visualized by fluorescein (yellow signals) and chromosomes were red-stained with propidium iodide. Each panel presents chromosomes with a band resolution ranging from about 300 to about 850.

Figures 2–5. (See color plates II–IV between pages 56 and 57.) Ideograms of human chromosomes at a 850 band resolution [16] showing the $H3^+$ bands as red bands.

Figure 6. (See color plate IV between pages 56 and 57.) Ideogram of human chromosome 11 at 400 (left) and 850 (right) band resolution showing the chromosomal regions containing H3 isochores. Red, yellow and white bands on the left chromosome indicate the $H3^+$, $H3^*$ and $H3^-$ bands. Red bands on the right chromosome indicate the regions hybridizing to the H3 isochores.

(R_{400}, G_{400}, R_{850}, G_{850} indicate the R and G bands at resolutions of 400 and 850 bands, respectively). In fact, H3 hybridization signals covered almost all, many, and only a few of the R_{850} bands derived from R_{400} $H3^+$ bands, $H3^*$ bands, and $H3^-$ bands, respectively (see Fig. 6 for an example).

Indeed, 23 out of the 28 R_{400} $H3^+$ bands only yielded R_{850} bands containing H3 isochores, whereas only some of the R_{850} bands originating from the other five R_{400} $H3^+$ bands showed H3 hybridization signals. For example (Fig. 6), the R_{400} $H3^+$ band 11q13 is resolved into three R_{850} $H3^+$ bands (q13.1, q13.3, and q13.5), and two G_{850} bands (q13.2, and q13.4), whereas the $H3^+$ band 11p15 was one of the five exceptions, with only one (11p15.5) of the three derived R_{850} bands showing hybridization signals. In no case did G_{850} bands derived from the R_{400} $H3^+$ bands show hybridization signals.

In contrast, only some of the R_{850} bands derived from 23 out of 31 R_{400} $H3^*$ bands showed hybridization signals. For example, the $H3^*$ band 11q23 (Fig. 6) yielded only one R_{850} $H3^+$ band (11q23.3; in fact, only the distal part of it was $H3^+$), whereas the other R_{850} band (11q23.1) was $H3^-$. The remaining eight $H3^*$ bands show the features observed in the vast majority of $H3^+$ bands in that all the derived R_{850} bands were $H3^+$.

As far as the R_{400} $H3^-$ bands are concerned, the higher resolution allowed the identification of 20 bands containing H3 isochores that had not been detected at the lower resolution [3]. The majority of these bands were located close to other $H3^+$ or $H3^*$ bands (see bands 5q33. 1, 6p21.1, and 12q24.13) and were very thin (see bands 1p13.3, and 7p13). Only some of the R_{850} bands derived from these 20 R_{400} $H3^-$ bands exhibited hybridization signals (see band 11p11.2 in Fig. 6). Moreover, in a number of cases, the signal was thinner than the corresponding R_{850} bands, indicating that only part of the R_{850} band contained H3 isochores.

Incidentally, previous work [3] had shown that H2 and H3 isochores co-localize on metaphase chromosomes, with only four exceptions (the telomeric bands 3q29, 6q27, 13q34, and 20p13) which were $H2^+$ and $H3^-$. Now these bands were shown to be $H3^+$, indicating also in these cases a co-localization of H3 and H2 isochores.

Finally, G bands did not reveal the presence of H3 isochores, the only exceptions being two G_{400} bands, 1p36.2 and 19q13.4, which yielded two R_{850} $H3^+$ bands, 1p36.22, and 19q13.42, respectively.

The present results lead to several conclusions. (i) Since the colocalization of H2 and H3 isochores (which represent 12% of the human genome) in R_{850} $H3^+$ bands appears now to be the rule, the fraction of these isochores in those bands (which represent 17% of the total genome) correspond to the majority, 70%, of the DNA contained in them. (ii) In some cases, however, the coverage of R_{850} $H3^+$ bands by hybridization signals is overestimated. For example, the present experiments suggest that almost 50% of band Xq28 is $H3^+$, whereas compositional mapping has shown that only 5% is formed by H3 isochores [9]. (iii) In a number of R_{850} $H3^+$ bands, H3 hybridization coverage was limited to a fraction of the band. This indicates that the present results provide information concerning a resolution higher than 850 bands; thus, they may correspond, in many cases, to the practical highest resolution that can be attained, namely 1250 bands. (iv) 83% of the bands, namely the R_{850} $H3^-$ and the G_{850} bands, present low or very low gene concentrations; since genome size is remarkably constant in mammals and since such regions are conserved in syntenic regions of chromosomes from mammalian orders that diverged about 100 millions years ago [10–14], this suggests some functional role for the gene-poor majority of the genome and the corresponding intergenic regions. (v) Finally, the present results are relevant for the choice of the regions of the human genome that deserve sequencing priority.

Interestingly, these regions correspond to gaps in the physical map of the human genome [3, 15]. The difficulty experienced in cloning these regions into YACs and/or in avoiding high levels of chimerism and deletion is most probably related to their high recombination level, a property which apparently is conserved when these regions are cloned in yeast.

References

1 Bernardi G (1995) The human genome: organization and evolutionary history. *Annu Rev Genet* 29: 445–476
2 Zoubak S, Clay O, Bernardi G (1996) The gene distribution of the human genome. *Gene* 174: 95–102
3 Saccone S, Cacciò S, Kusuda J, Andreozzi L, Bernardi G (1996) Identification of the gene-richest bands in human chromosomes. *Gene* 174: 85–94
4 Dutrillaux B (1973) Nouveau système de marquage chromosomique *Chromosoma* 41: 395–402
5 Ambros PF, Sumner AT (1987) Correlation of pachytene chromomeres and metaphase bands of human chromosomes, and distinctive properties of telomeric regions *Cytogenet Cell Genet* 44: 223–228
6 Craig M, Bickmore WA (1994) The distribution of CpG islands in mammalian chromosomes. *Nat Genet* 7: 376–382
7 Federico C, Saccone S, Bernardi G (1998) The gene-richest bands of human chromosomes replicate at the onset of the S-phase. *Cytogenet Cell Genet* 80: 83–88
8 Saccone S, Federico C, Solovei I, Croquette MF, Della Valle G, Bernardi G (1999) Identification of the gene-richest bands in human prometaphase chromosomes. *Chromosome Res* 7: 379–386

9 De Sario A, Geigl EM, Palmieri G, D'Urso M, Bernardi G (1996) A compositional map of human chromosome band Xq28. *Proc Natl Acad Sci USA* 93: 1298–1302
10 Saccone C, Cacciò S, Perani P, Andreozzi L, Rapisarda A, Motta S, Bernardi G (1997) Compositional mapping of mouse chromosomes and identification of the gene-rich regions. *Chromosome Res* 5: 293–300
11 Scherthan H, Cremer T, Arnason U, Weier HU, Lima-de-Faria A, Fronicke L (1994) Comparative chromosome painting disclouses homologous segments in distantly related mammals. *Nat Genet* 6: 342–347
12 Raudsepp T, Frönicke L, Scherthan H, Gustavsson I, Chowdhary BP (1996) Zoo-FISH delineates conserved chromosomal segments in horse and man. *Chromosome Res* 4: 218–225
13 Morescalchi MA, Schempp W, Consigliere S, Bigoni F, Wienberg J, Stanyon R (1997) Mapping chromosomal homology between humans and the black-handed spider monkey by fluorescence *in situ* hybridization. *Chromosome Res* 5: 527–536
14 Chowdhary BP, Raudsepp T, Frönicke L, Scherthan H (1998) Emerging patterns of comparative genome organization in some mammalian species as revealed by Zoo-FISH. *Genome Res* 8: 577–589
15 Chumakov IM, Rigault P, Le Gall I, Bellanné-Chatelot C, Billault A, Guillov S, Soularve P, Guasconi G, Poullier E, Gros I et al (1995) A YAC contig map of the human genome. *Nature* 377 (suppl): 175–297
16 Francke U (1994) Digitized and differentially shaded human chromosome ideograms for genomic applications. *Cytogenet Cell Genet* 6: 206–219

Chromosomes Today, Vol. 13
ed. by E. Olmo and C.A. Redi

The functional and structural organization of *Drosophila* heterochromatin

Sergio Pimpinelli

Istituto Pasteur, Fondazione Cenci Bolognetti, Dipartimento di Genetica e Biologia Molecolare, Università di Roma "La Sapienza", Piazzale Aldo Moro 5, 00185 Roma, Italy

Introduction

Heterochromatin is a strongly conserved component of the eukaryotic chromosome wich was defined for the first time by Heitz in 1928 [1]. Understanding its biological meaning remains a major problem for chromosome biologists.

The permanently heterochromatic chromosome regions are often large and constitute very substantial proportions of the genomes of higher organisms, yet they appear inconsequential from a functional, or genetic, point of view. Thus, for example, in *Drosophila melanogaster* about 33% of the genome is composed of heterochromatin. However, only a few (about 30 loci) of the many hundreds of genetic functions revealed by mutational analysis are located in heterochromatic regions (for review see [2]). Further evidence of the paucity of genetic function in heterochromatin is the observation that, while euchromatic aneuploidy almost invariably results in morphological abnormalities, or lethality, heterochromatic duplications and deficiencies, even if quite large, are mostly without substantial effects [3]. Moreover, studies of the phenomenon of position effect variegation (PEV) in *Drosophila* revealed that heterochromatin can inactivate genes when it is abnormally juxtaposed to euchromatin in chromosome rearrangements (for review see [4]), leading to the view that heterochromatin itself is genetically inactive. The properties of PEV, the paucity of heterochromatic gene-encoding regions, and the finding that heterochromatin is comprised predominantly of non-coding, repetitive DNA sequences (for review see [2]) have led many scientists to view heterochromatin as a homogeneous, tightly compacted and genetically inert chromosomal domain whose molecular organization renders it incompatible with gene expression. The issues that then emerge are, why do the genomes of higher organisms contain such large amounts of genetic material of seemingly little biological significance and what are the evolutionary forces that result in the maintenance and stability of the heterochromatin over evolutionary time? Two genera of explanations, by no means mutually exclusive, are usually propounded. One is that heterochromatin has no function and it is maintained wholly by stochastic processes, or because it is "selfish" and is maintained by virtue of processes internal to itself with no benefit accruing to the organism. The other answer is that heterochromatin contains biological functions. In that case, the apparently inconsequent nature of the heterochromatin might be due either to a peculiar organization of the genetic material producing the few functions that exist and/or to a difficulty in detecting heterochromatic functions by ordinary genetic analysis.

Drosophila melanogaster offers unique opportunities to obtain insights into these problems. Its sophisticated genetics and cytogenetics permit isolation of chromosome rearrangements involving specific heterochromatic segments that can be employed to question the biological role of the genetic material they define. Moreover, the sophisticated genetic tools of *Drosophila* permit the identification and the analysis of genetic systems in which the heterochromatic functional elements are "cryptic" in that they can be detected only in the presence of mutations in other euchromatic loci (for review see [2, 5]).

The heterochromatin of *Drosophila*

The literature relating to heterochromatin in *D. melanogaster* is vast and has been recently reviewed by different authors [2, 4, 6]. Briefly, constitutive heterochromatin comprises about 33% of the genome of *Drosophila melanogaster.* The entire Y chromosome, about 50% of the X chromosome and 25% of the autosomes are heterochromatic [7]. More recently, sensitive chromosome banding and FISH (Fluorescent *In Situ* Hybridization) techniques to

Figure 1. A cytological map of *Drosophila melanogaster* heterochromatin showing the localization of the major heterochromatic functional elements, transposon clusters and different satellite DNAs. The entirely heterochromatic Y chromosome, the X, second, third and fourth chromosome heterochromatin are schematically represented. The diagrams represent the heterochromatic regions of larval neuroblast prometaphase chromosomes stained with Hoechst or DAPI fluorochromes. Euchromatin is depicted as a broken line. Filled segments indicate bright fluorescence, cross-hatched segments indicate intermediately bright fluorescence, hatched segments indicate dull fluorescence and open segments indicate no fluorescence. c indicates the centromere position; the location of the fourth chromosome centromere has not been precisely determined. The horizontal thin lines above the chromosomes indicate the locations of satellite DNAs [62–63]. The 1.688 g/cm^3 satellite DNA contains only one repeated DNA sequence of 359 bp. Each of the other satellites includes different short repeats. Five simple sequences have been cloned from the 1.672 g/cm^3 satellite DNA: the AATAT repeats that are the main component of this satellite and the less abundant AATAG, AATAC, AATAAAC and AATAGAC repeats [62]. The 1.686 g/cm^3 satellite consists predominantly of AATAACATAG sequences and contains the AAGAC repeats as a minor component; however, the AAGAC repeats are rather abundant representing about 2.4% of the *D. melanogaster* genome [62]. The two main components of the 1.705 satellite DNA are the AAGAG and the AAGAAGAG repeats [62]. The horizontal thick lines above each diagram indicate the location of the TE hybridization signals [9]. The first row of lines below the Y chromosome indicates the location of the Y chromosome fertility factors *(kl-5, kl-3, kl-2, kl-1, ks-1 and ks-2), the bobbed (bb)* and *crystal (cry)* loci. The second row of lines below the Y designates the *ABO* elements. The lines below diagrams of the X and the second chromosome heterochromatin indicate the X-linked *bb* locus and the *compensatory response (cr)* locus, the heterochromatic encoding genes (*v.l.*, *f.s.*), the *Enhancer of Segregation Distortion (E(SD))* and *Responder (Rsp)* loci and the *ABO* elements. The third chromosome carries 12 vital loci which are not yet precisely mapped.

elaborate a cytogenetic map of *Drosophila* heterochromatin have been developed [2]. The mitotic heterochromatin can now be resolved into sixty-one distinct bands, designated h1–h61 (Fig. 1).

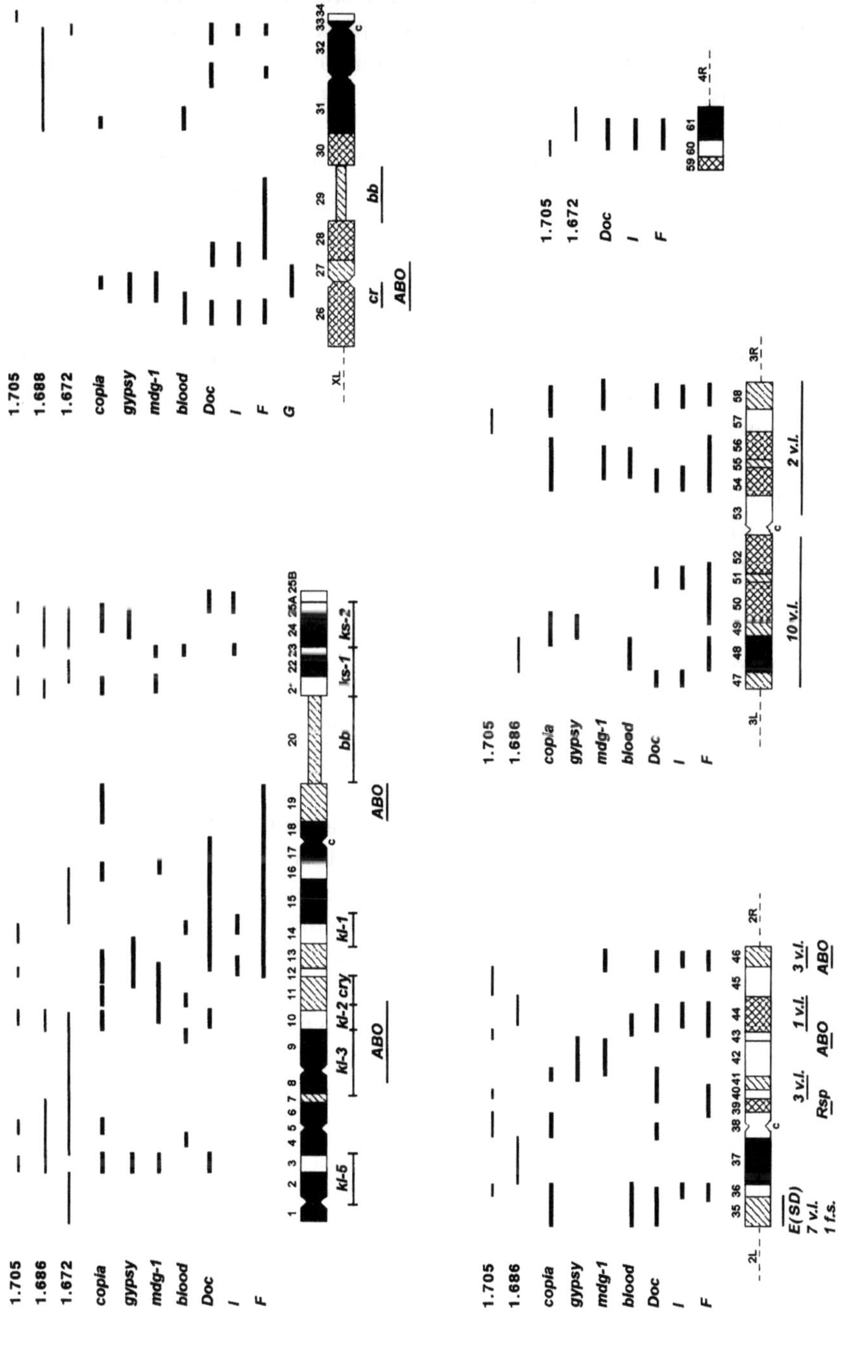

Substantial progress has already been made on determining the genetic organization of *Drosophila* heterochromatin. In particular, it has been shown that heterochromatin harbors genes encoding ribosomal RNAs, canonical coding genes, and the Y-linked fertility factors (for review see [2]). Other studies have shown that many heterochromatic regions once considered devoid of genetic information do contain functional repeated elements. These elements, that also have multiple chromosomal locations, are not detectable by ordinary genetic analysis but only through their interaction with euchromatic mutations (for review see [8]).

The relationships between the cytological bands, the locations of thirty genetically defined heterochromatic loci, the major satellite DNA clusters and twelve different middle repetitive DNA families have been determined [2, 9] (see Fig. 1).

Another very interesting aspect of the heterochromatin is the position effect variegation phenomenon which appears to be due to repressive mechanisms very similar to those used to maintain developmentally critical genes in a repressed state in some tissues during development. This similarity has furthered the interest in studies of PEV as a good experimental model to understand how the repressed state of chromatin is established and maintained during development. Such studies, on one hand, are giving information on the regulation of gene expression during development and, on the other hand, they are adding substantial information on the structural organization of heterochromatin.

The heterochromatic lethal loci

An extensive genetic analysis of chromosome two pericentric heterochromatin has led to the identification of seven vital loci to the left and six to the right of the centromere, and the suggestion that these loci correspond to unique DNA sequences [10, 11]. Most of the ethyl methanesulphonate (EMS)-induced lethals die at very late larval or pupal stages and for some of them the lethality is due to extensive cell death in the larval imaginal discs. A maternal effect gene, *concertina (mat(2)cta)*, has also been identified [12]. Two recent sets of data have given further insights on the cytological and molecular organization of the heterochromatic loci. First, an accurate cytological mapping, using many heterochromatic rearrangements, [13] has shown that the heterochromatic loci are not randomly distributed. All mutations of the left arm are located in the most distal heterochromatic region at the eu-heterochromatic border (h35). In the right arm, two loci are located in the heterochromatic region of the eu-heterochromatic border (h46) and the other four loci are located very close to the centromere. Second, the *light* gene, one of the heterochromatic genes of the left arm, was analysed molecularly [14, 15]. This analysis showed that *light* is a single-copy gene. An unusual feature of this gene is that the introns and flanking regions are a heterogeneous array of middle repetitive sequences. This gene and the other heterochromatic genes when relocated in distal euchromatic regions, by chromosome rearrangements, exhibit position effect variegation (for review see [4]). It has been shown that some modifiers that suppress position-effect variegation of euchromatic genes behave as enhancers of variegation of *light* and other heterochromatic genes [16]. This suggests that the wild-type product of modifiers of PEV is required for normal expression of heterochromatic genes, but the opposite effects on variegation of the euchromatic *versus* the heterochromatic genes also suggests that heterochromatic genes are regulated differently than euchromatic genes.

Since *light* is located in the most distal heterochromatic region of 2lh, at the border with euchromatin, the results obtained with this gene may not apply to the molecular organization of the genes located close to the centromere, and thus proximal to many different satellite DNAs. However, it has been shown recently that these genes are located in heterochromatic blocks that mainly contain middle repetitive DNA [9, 13]. In addition, a molecular analysis of one of these genes, the *rolled* gene, has shown that this gene is located in islands of middle repetitive DNA and has a similar molecular organization to *light*. [17].

The heterochromatin of the third chromosome was also analysed [18, 19]. So far, two genes have been located in the heterochromatin of the right arm and ten in that of the left arm. In this case as well, interallelic complementation between some EMS-induced lethals and recovery of temperature sensitive mutants have suggested that the heterochromatic genes correspond to unique-sequence DNA.

The molecular analysis of many of these heterochromatic genes, and an extensive search for and mapping of homologs in different species, will be important to understand whether the heterochromatic location of "ordinary" genes has a particular evolutionary significance.

The Y-linked fertility factors

The genetic and cytological analysis of the entirely heterochromatic Y chromosome has revealed that this chromosome contains six genetic loci necessary for male fertility, collectively called fertility factors. The phenotypic effects of mutations of the fertility factors consist of the induction of a spectrum of abnormalities resulting in a general disorganization of the sperm tails (for review see [2]). An extensive cytogenetic analysis of the fertility factors using cytologically characterized Y chromosome rearrangements revealed striking structural features for at least three of them. *kl-5, kl-3 and ks-1*, as defined by non-complementing chromosomal rearrangement breakpoints, possess an extraordinarily large size of about 4 Mb that is larger than euchromatic loci by two orders of magnitude [20]. These loci produce giant lampbrush-like loops in primary spermatocytes consisting of DNA axes associated with transcribed RNA on which large amounts of proteins are accumulated [21–23].

Two alternative, but not necessarily mutually exclusive, views are currently propounded regarding the functional role of the fertility factors. According to one model the fertility factors that form loops in primary spermatocytes serve to store the large quantity of sperm proteins necessary for the correct development of sperm flagella. The alternative view considers that the fertility factors are canonical coding genes with very large introns made up of repetitive sequences. Data to support the first view is derived from the accumulation of a tektin-like protein on the *kl-3* loop that, later in sperm development, will be incorporated into the sperm flagella [23]. On the other hand, the second view was suggested by the findings that in males deficient in *kl-5* or *kl-3,* two high-molecular weight polypeptides and the outer dynein arms of the peripheral doublets of the axoneme are simultaneously absent [24]. This latter view seems also to be supported by recent, strong evidence that *kl-5* contains a gene coding for a dynein heavy chain polypeptide and that is expressed only in primary spermatocytes [25].

Interaction systems among euchromatic genes and heterochromatic elements

In *Drosophila melanogaster* three genetic systems involving specific interactions between euchromatic and cryptic heterochromatic genetic elements are known: the Segregation Distortion, the *crystal-Stellate* and the *abo-ABO* systems [for a detailed review see 8]. The first system seems to be composed of individually dispensable elements. The second may represent an intriguing case of a balanced selfish system in which the absence, or the partial deletion, of each of its elements causes an alteration of the normal meiotic phenotype with negative impact on fitness. The *abo-ABO* system contains elements which appear to be required for normal embryogenesis.

The segregation distortion

Segregation Distortion in *D. melanogaster* was discovered by Hiraizumi in 1956 and is virtually ubiquitous in natural populations. Males heterozygous for an *SD (Segregation Distorter)* second chromosome and an SD^+ homolog, transmit an excess of *SD* chromosomes due to disfunction of SD^+ bearing sperm. Genetically, Segregation Distortion depends on the interaction of three main elements. Two of these, *Sd (Segregation distorter)* and *E(Sd) (Enhancer of Sd)* act to induce dysfunction of sperm carrying a sensitive allele of the third element *Rsp (Responder).* While *Sd* maps to a euchromatic region of the polytene chromosome map (37D2-6), both *E(Sd)* and *Rsp* have heterochromatic locations (see Fig. 1). *E(Sd)* maps in 2L heterochromatin and *Rsp* is located in 2R heterochromatin. In addition, a molecular analysis of the *Rsp* locus has shown that it corresponds to a 120 bp AT-rich repeat (for the entire topic see [26])

Intriguingly, the genetic analysis has also suggested that all these loci are dispensable and can produce an alteration of the normal phenotype only by their interaction, perhaps identifying a special case of a selfish genetic system. However, a complete molecular analysis of the *Sd, E(Sd)* and *Rsp* elements will be needed to elucidate their nature and the mechanism of their interaction, and might also give insights into the maintenance of this system in nature.

The crystal-stellate interaction

Males of *Drosophila melanogaster* lacking the Y-linked *crystal (cry)* [5] locus show crystal aggregates in the nuclei and cytoplasm of primary spermatocytes [27]. In addition, these males also exhibit abnormal condensation and segregation of meiotic chromosomes and meiotic drive [28]. Molecular studies have shown that *crystal* corresponds to tandemly repeated sequences that are 800 bp long after CfoI digestion [29]. Intriguingly, the X-linked interacting locus *Stellate* is also made up of tandem repeats that are partially homologous to the *crystal* repeats [30]. The *Ste* sequences, which contain two introns, are abundantly and correctly transcribed only in testes of cry^- males and cDNA sequence analysis has shown that these sequences encode a 19,500 dalton protein with striking homology to the β-subunit of Casein-kinase 2 (CK2) [30]. Interestingly, this enzyme is able to aggregate into filamentous structures and, among its multiple functions, is involved in topoisomerase II activity which is essential for chromosome condensation and segregation (for review see [31]).

Recently, it has been shown the crystals observed in cry^- spermatocytes are mainly composed of the *Ste* protein. In addition, it was also shown that this protein is capable of binding the CK2 α-subunit *in vitro* [32]. These data have suggested that the *Ste* protein produced in cry^- testes would compete with the β-subunit disrupting the activity of this enzyme. As a consequence, topoisomerase II function could also be affected, thus altering the proper chromosome condensation and segregation.

Although *cry* formally appears as an indispensable locus by genetic analysis, the absence of *Ste* expression in wild type [32], and the absence of its homologs in closely-related species [29] suggest that the *cry-Ste* interaction may correspond to a particular selfish genetic system. That is, the *cry-Ste* system seems to be a novel type of selfish genetic system whose components are normally silent, but if perturbed can cause alterations in normal meiotic processes.

The abo-ABO interaction

abo (abnormal oocyte), is a recessive maternal effect mutation that is located in the euchromatic left arm of the second chromosome [33]. Homozygous *abo* mothers lay eggs with a markedly reduced probability of hatching. As is the case for some other maternal effect mutations, the maternal defect can be partially rescued by the presence in the embryo of the normal abo^+ allele. The peculiarity of this maternal effect, however, is that the survival of the defective eggs also depends on the amount of heterochromatin, either in the embryo or in the mutant mother [33]. Extensive cytogenetic analysis has shown that the heterochromatic amelioration of the maternal defect depends on multiple specific heterochromatic regions, called *ABO*, that are located in the heterochromatin of the X, Y, and second chromosomes (Fig. 1) ([34] and unpublished observations). The observation that *ABO* can be partially inactivated by small deletions [34] suggested that these elements are made up of repetitive subunits. From the functional point of view, the *ABO* elements seem to act very early during embryogenesis [35].

The existence of the *ABO* elements has general implications regarding the functional role of heterochromatin. *ABO* is located in heterochromatic regions that are entirely dispensable in wild-type crosses and homozygous deficiences of this element have no effect on abo^{+} individuals. However, recent experiments have shown that simultaneous deletion of four *ABO* regions results in embryonic lethality (unpublished observations) thus suggesting that the apparent dispensability of each separate region, could be due to their multiple locations. Although the molecular organization of the *ABO* sequences is still unknown, significant progress in the molecular analysis of the euchromatic *abo* gene has been made. The isolation and the molecular characterization of *abo* has shown that it encodes a nuclear protein with a specific binding site on polytene chromosomes ([36] and unpublished observations).

In conclusion, our knowledge of these systems shows that heterochromatin possesses indispensable functions, controlled in part by a complicated structural organization. On the other hand, it seems equally evident that part of this material contains genetic elements that, although able to produce phenotypic effects, do not correspond to canonical functions. Nevertheless, the comparison between the three systems described here suggests a very interesting notion. These systems, apart from their functional prerogatives, share some analogous structural characteristics. They are composed of interacting heterochromatic and euchromatic loci. The heterochromatic loci contain middle repetitive DNA that, at least for some, is apparently derived from transposons. This suggests a general model for the origin and evolutionary conservation of heterochromatic middle repetitive sequences. According to this model, different transposable element families, after accumulation in heterochromatin, may progressively aquire novel functional properties, with positive impact on fitness, by establishing interactions with euchromatic loci. The euchromatic-interacting loci could correspond to either canonical genes or similar transposable sequences. The conceptual consequence of this model is the existence of a functional link between the heterochromatin and euchromatin.

Position effect variegation (PEV)

A functional link between the euchromatin and heterochromatin is also supported by the studies of PEV which revealed an interesting relationship between heterochromatin, gene expression and chromosome behavior.

Position effect variegation (PEV) is a well known case of *cis* inactivation of a normal euchromatic gene when relocated in, or very close to heterochromatin. The first evidence of PEV was provided by Muller [37] in *Drosophila melanogaster* and successively described also in other species. One of the best examples is the PEV of the *white* gene in *D. melanogaster* when this gene, normally located near the telomere of the X chromosome, is transferred by chromosome rearrangements to a new position close to heterochromatin. In this location, *white* undergoes a *cis*-heterochromatin inactivation that occurs only in a proportion of the cells during development giving, for example in the eyes, a mosaic phenotype of mutant and wild-type areas (for review of the entire topic see [4]).

A large amount of genetic, cytogenetic and molecular data have provided a general picture of the probable mechanism of gene inactivation by heterochromatin. It has been shown in *Drosophila melanogaster* that the euchromatic regions containing the variegating gene lose their normal morphology in polytene chromosomes, appearing heterochromatinized (for review see [38]), and this is accompained by a decrease in transcription [39]. In *D. melanogaster*, genetic, physical and chemical factors that can modify PEV are known (for review see [40]). The most classic suppressor of PEV is the entirely heterochromatic Y chromosome. Studies performed on three different examples of PEV have shown that the intensity of suppression is related to the amount of Y heterochromatin present in the genome and does not depend on any mappable factor [41]. Our experiments on the same three cases using free duplications of different sizes covering the entire X chromosome heterochromatin and the entire heterochromatic right arm of the second chromosome showed that, also in this case, the intensity of suppression was related to the amount of heterochromatin rather than some mappable factor (unpublished observations). These results are consistent with the hypothesis that the Y chromosome competes for the limited amount of proteins responsible for the heterochromatinization processes, diluting these proteins at the variegating sites [42]. A few chemicals have been found capable of modifying PEV. Among them, butyrate strongly suppresses PEV of *white* [43]. It is well known that butyrate inhibits the histone deacetylase enzyme [44] producing an accumulation of tetracetylated H4 histones [45, 46] that are associated with active chromatin (for review see [47]). Moreover, experiments using carnitine, a compound with a similar structure to butyrate, have shown that this substance also suppresses PEV [48]. Evidence that histones are involved in producing PEV was also independently provided [49, 50].

Many genetic modifiers of PEV have been isolated and their analysis provided a means to identify trans-acting components affecting chromatin structure or function. Extensive genetic analyses have led to the isolation of a large number of mutations that alter the variegation of the white allele, wm^4 (for review see [38, 51]). The molecular analysis of about 15 modifiers has revealed an interesting and diverse set of proteins. One of the most extensively studied is HP1, encoded by the *Su(var)2-5* gene [54], which was first identified because of its localization to the chromocenter of salivary gland nuclei [52, 53]. Recent studies have shown that HP1 can be involved in the epigenetic silencing of repeated transgenes [55]. Most important, it has recently been shown that HP1 is also a telomere "cap" protein that is required for telomere stability [56]. Other modifiers are known to be transcriptional regulators of euchromatic genes. For example, several affect homeotic gene expression [51] and one, the GAGA factor, encoded by the *Trithorax-like* gene [57], affects the expression of homeotic and segmentation genes, as well as the Adh, heat shock and histone genes [58–60]. Our recent cytogenetic mapping by immunostaining of mitotic chromosomes of several proteins provides a basis for the effects on both types of chromatin domains. Preliminary results demonstrate that different classes of chromosomal proteins such as structural proteins or transcriptional factors, are localized in euchromatin and in heterochromatin with specific patterns that, in many cases, correspond to those observed for different transposon families or different classes of satellite DNA (unpublished observations). The localization patterns of these proteins raise several intriguing questions. Is the localization in both euchromatin and heterochromatin a common feature of modifier proteins? Are the DNA sequences and protein motifs responsible for localization of a modifier protein at euchromatic sites distinct from those that mediate its binding in heterochromatin? Is the localization of modifier proteins in both chromatin domains functionally significant? One possibility is that a modifier protein has two separate, non-overlapping functions. A particularly interesting example of this possibility is the Rap1 protein of *S. cerevisiae* (for review see [61]). Rap1 binds repetitive sequences at the telomere. It also acts as either a positive or negative transcriptional regulator, depending upon chromosomal context. Another possible role for dual localization of modifier proteins may be an involvement in regulating the pool of protein available for gene transcription. That is, heterochromatic and euchromatic binding sites could compete for a limiting quantity of protein; the relative binding affinities would then influence the ability of the protein to function in either domain. This type of model implies, again, a type of functional interdependency between heterochromatin and euchromatin.

The origin and evolution of heterochromatin and functional consequences for the genomes

It has been shown that the heterochromatin of *D. melanogaster* contains at least eleven different types of satellite DNAs [62]. Approximately 70% of het-

Figure 2. (See color plate V between pages 56 and 57.) Cytological localization of AGAGAG satellite sequences (a) and the *Doc* transposable element (b) along *D. melanogaster* heterochromatin by fluorescent *in situ* hybridization (FISH) and CCD camera analysis. To facilitate the localization of the FISH signals (red) with respect to the DAPI banding pattern on the merged figures, the DAPI and the FISH signals are also separately shown. Y = Ychromosome; X = X chromosome; the numbers indicate the different chromosome pairs.

erochromatin is made up of these sequences. Analysis of the chromosomal localization of the different satellite DNAs revealed specific patterns of distribution [63, 64] (Fig. 1, and for examples see Fig. 2a).

A recent investigation on the distribution of eleven different TE families in the *D. melanogaster* heterochromatin, has shown that nine of these TEs *(copia, gypsy, mdg-1, blood, Doc, I, F, G* and *Bari-1)* are preferentially clustered into one or more discrete heterochromatic regions in chromosomes of the Oregon-R laboratory stock [9] (Fig. 1, and for examples see Fig. 2b). Moreover, a high resolution fluorescent *in situ* hybridization (FISH) analysis of geographically distant strains revealed that the locations of these heterochromatic TE clusters are highly conserved [9].

The data showing a direct involvement of mobile elements in forming a conspicuous portion of heterochromatin suggests the intriguing possibility that transposable elements could also be involved in generating short sequences of heterochromatin that, by some sort of reiteration mechanism, could give origin to satellite DNAs. Two reasons seem to support the possibility that some, if not all, the basic satellite sequences correspond to evolutionary retrotransposon footprints. First, in *Drosophila,* several different basic satellite sequences are richly represented in the LTRs of many of the retrontransposon families (unpublished observations). Second, there are examples of chromosomal insertions of sequences derived from the LTRs of this class of transposons. A very interesting case was found in species of the rodent *Ctenomys.* In these species the major satellite DNA is made up of several consensus sequences of retroviral origin such as CCAAT and TATA boxes, binding sites for the CCAAT/enhancer binding protein (C/EBP) that are characteristic of retroviral long terminal repeats (LTR) [65]. Thus, it is not unreasonable to suppose that short transposon sequence insertions could be a potential starting point for their amplification leading to the generation of satellite DNAs.

Proteins interacting with transposons and satellite DNA are also localized in the euchromatin and can be both structural and regulatory factors

It is known that transposon-like sequences are usually very rich in protein binding motifs, as for example the enhancers and other regulatory elements (for review see [66]) that are evolutionary related to the *cis*-regulatory elements contained within the control regions of euchromatic canonical genes [66]. Also, many of the short sequences composing the different types of satellite DNA are identical or very similar to many of these regulatory sequences. For example, the sequence of the 1.705 highly repetitive satellite DNA of *D. melanogaster* is identical to GAGA boxes, and it has been shown that the GAGA factor is strongly accumulated in the mitotic heterochromatic regions highly enriched by this type of DNA [67]. In addition, as mentioned in the PEV section, our immunomapping of several chromosomal proteins has shown that these proteins are localized in euchromatin and in heterochromatin with specific patterns that, in many cases, overlap those observed for different transposon families or different classes of satellite DNA (unpublished observations). On this basis, it appears that structural and/or regulatory proteins that bind euchromatic sequences are capable of binding to transposon and satellite DNA sequences. This greatly simplifies models of the origin and formation of heterochromatin because it is not necessary to invoke a coevolution of both repetitive sequences and specific heterochromatic proteins. In this regard, it is important to note that, while the list of proteins that are present in both the heterochromatin and euchromatin is growing rapidly, no one protein that is specific for heterochromatin has been found so far in *Drosophila*. The best candidate for heterochromatic specificity was the HP1 protein that heavily stains the chromocenter of the polytene chromosomes. However, our recent analysis of the immunofluorescence pattern has shown that this protein is located at about 190 euchromatic sites, including all the telomeres ([56] and unpublished observations).

Conceptually, this model contains some evolutionary and functional implications. The limited quantitative availability of chromosomal proteins could regulate the copy number of both heterochromatic satellite and middle repetitive DNA sequences. A disproportionate increase in such sequences may, in fact, accumulate several key regulatory proteins in heterochromatin, thus decreasing their availability for the regulation of normal euchromatic gene expression at various loci with a negative impact on fitness. This mechanism establishes, once more, a functional connection between heterochromatin and euchromatin with heterochromatin regulating euchromatic gene expression by controlling the chromatin structure.

Conclusions

The studies mainly performed in *Drosophila* have given ample evidence that heterochromatin is both functionally and structurally highly heterogeneous. Genetic analyses have shown that heterochromatin contains diverse biological activities, including protein-encoding genes. In addition, considerable progress in our understanding of the distribution and types of sequences that comprise the heterochromatin of *Drosophila* have also been made. Intriguingly, heterochromatin and euchromatin seem to share similar, if not identical, sequences. In addition, recent cytogenetic mapping of several proteins that are transcriptional regulators of euchromatic genes has shown that these proteins have euchromatic and heterochromatic binding sites. These data seem to suggest that the difference between heterochromatin and euchromatin is mainly based on differences in the quantity of similar DNA sequences, that in heterochromatin could be highly repeated (satellite DNAs), rather than on qualitative differences in DNA sequences. This model implies a functional interdependency between heterochromatin and euchromatin.

Acknowledgements

I wish to thank L. Fanti, M. Berloco and G. Giovinazzo for their contribution to the heterochromatin project and their permission to cite our unpublished observations. I wish to thank also K. Golic for critical reading of the manuscript. I apologize to those scientists who were only cited through reviews.

References

1 Heitz E, (1928) Das Heterochromatin der Moose. *Jahrb Wiss Bot* 69: 762–818
2 Gatti M, Pimpinelli S (1992) Functional elements in *Drosophila melanogaster* heterochromatin. *Annu Rev Genet* 26: 239–275
3 Lindsley DL, Sandler L, Baker BS, Carpenter ATC, Denell RE, Hall JC, Jacobs PA, Miklos GLG, Davis BK, Gethmann RC et al (1972) Segmental aneuploidy and the genetic gross structure of the *Drosophila* genome. *Genetics* 71: 157–184
4 Weiler KS, Wakimoto BT (1995) Heterochromatin and gene expression in *Drosophila. Annu Rev Genet* 29: 577–605
5 Pimpinelli S, Bonaccorsi S, Gatti M, Sandler L (1986) The peculiar genetic organization of *Drosophila* heterochromatin, *Trends Genet* 2: 17–20
6 Lohe AR, Hilliker AJ (1995) Return of the H-word (heterochromatin). *Curr Opin Genet Develop* 5: 746–755
7 Pimpinelli S, Santini G, Gatti M (1978) ^{3}H-Actinomycin-D binding to mitotic chromosomes of *Drosophila melanogaster. Chromosoma* 66: 389–395
8 Palumbo G, Berloco M, Fanti L, Bozzetti MP, Massari S, Caizzi R, Caggese C, Spinelli L, Pimpinelli S (1995) Interaction systems between heterochromatin and euchromatin in *Drosophila melanogaster. Genetica* 94: 267–274
9 Pimpinelli S, Berloco M, Fanti L, Dimitri P, Bonaccorsi S, Marchetti E, Caizzi R, Caggese C, Gatti M (1995) Transposable elements are stable structural componenents of *Drosophila melanogaster* heterochromatin. *Proc Natl Acad Sci USA* 92: 3804–3808
10 Hilliker AJ, Holm DG (1975) Genetic analysis of the proximal region of chromosome 2 of *Drosophila melanogaster.* I. Detachment products of compound autosomes. *Genetics* 81: 705–721
11 Hilliker AJ (1976) Genetic analysis of the centrometric heterochromatin of chromosome 2 of *Drosophila melanogaster:* Deficiency mapping of EMS-induced lethal complementation groups. *Genetics* 83: 765–782
12 Schupbach T, Weischaus E (1989) Female sterile mutations of the second chromosome of *Drosophila melanogaster.* I. Maternal effect mutations. *Genetics* 121: 101–117
13 Dimitri P (1991) Cytogenetic analysis of the second chromosome heterochromatin of *Drosophila melanogaster. Genetics* 127: 553–564
14 Devlin RH, Holm D, Morin K, Honda B (1990) Identifying a single-copy DNA sequence associated with the expression of a heterochromatic gene, the *light* locus of *Drosophila melanogaster. Genome* 33: 405–415
15 Devlin RH, Bingham B, Wakimoto BT (1990) The organization and expression of the *light* gene of *Drosophila melanogaster. Genetics* 125: 129–140
16 Hearn MG, Hedrick A, Grigliatti TA, Wakimoto BT (1991) The effect of modifiers of position-

effect variegation on the variegation of heterochromatic genes of *Drosophila melanogaster. Genetics* 128: 785–797
17 Biggs WH, Zavitz KH, Dickson B, van der Straten A, Brunner D, Hafen E, Zipursky SL (1994) The *Drosophila rolled* locus encodes a MAP kinase required in the sevenless signal transduction pathway. *EMBO J* 13: 1628–1635
18 Marchant GE, Holm DG (1988) Genetic analysis of the heterochromation of chromosome 3 in *Drosophila melanogaster.* I. Products of compound autosome detachment. *Genetics* 120: 503–517
19 Marchant GE, Holm DG (1988) Genetic analysis of the heterochromatin of chromosome 3 in *Drosophila melanogaster*. II. Vital loci identified through EMS mutagenesis. *Genetics* 120: 519–532
20 Gatti M, Pimpinelli S (1983) Cytological and genetic analysis of the Y chromosome of *Drosophila melanogaster*. I. Organization of the fertility factors. *Chromosoma* 88: 349–373
21 Bonaccorsi S, Pisano C, Puoti F, Gatti M (1988) Y chromosome loops in *Drosophila melanogaster. Genetics* 120: 1015–1034
22 Bonaccorsi S, Gatti M, Pisano C, Lohe A (1990) Transcription of a satellite DNA on two Y chromosome loops of *Drosophila melanogaster. Chromosoma* 99: 260–266
23 Pisano C, Bonaccorsi S, Gatti M (1992) The *Kl-3* loop of the Y chromosome of *Drosophila melanogaster* binds a tektin-like protein. *Genetics* 133: 569–579
24 Goldstein LSB, Hardy RW, Lindsley DL (1982) Structural genes on the Y chromosome of *Drosophila melanogaster*. *Proc Natl Acad Sci USA* 79: 7405–7409
25 Gepner J, Hays TS (1993) A fertility region on the Y chromosome of *Drosophila melanogaster* encodes a dynein microtubule motor. *Proc Natl Acad Sci USA* 90: 11132–11136
26 Temin RG, Ganetzky B, Powers PA, Lyttle TW, Pimpinelli S, Dimitri P, Wu C-I, Hiraizumi Y (1991) Segregation distortion in *Drosophila melanogaster*: genetic and molecular analyses. *Amer Naturalist* 137: 287–331
27 Meyer GF, Hess O, Beermann W (1961) Phasenspezifische Funktionsstrukturen in Spermatocytenkernen von *Drosophila melanogaster* und ihre Abhangigkeit vom Y Chromosom. *Chromosoma* 12: 676–716
28 Hardy RW, Lindsley DL, Livak KJ, Lewis B, Siversten L, Joslyn GL, Edwards J, Bonaccorsi S (1984) Cytogenetic analysis of a segment of the Y chromosome of *Drosophila melanogaster*. *Genetics* 107: 591–610
29 Livak KJ (1984) Organization and mapping of a sequence on the *Drosophila melanogaster* X and Y chromosomes that is transcribed during spermatogenesis. *Genetics* 107: 611–634
30 Livak KJ (1990) Detailed structure of the *Drosophila melanogaster Stellate* genes and their transcripts. *Genetics* 124: 303–316
31 Cardenas ME, Gasser SM (1993) Regulation of topoisomerase II by phosphorylation: a role for casein kinase II. *J Cell Sci* 104: 219–225
32 Bozzetti MP, Massari S, Finelli P, Meggio F, Pinna L, Boldyreff B, Issinger O-G, Palumbo G, Ciriaco C, Bonaccorsi S et al (1995) The peculiar genetic organization of the *cry-Ste* parasite system of *Drosophila melanogaster. Proc Natl Acad Sci USA* 92: 6067–6071
33 Sandler L (1970) The regulation of sex-chromosome heterochromatic activity by an autosomal gene in *Drosophila melanogaster*. *Genetics* 64: 481–493
34 Pimpinelli S, Sullivan W, Prout M, Sandler L (1985) On biological function mapping to the heterochromatin of *Drosophila melanogaster*. *Genetics* 109: 701–724
35 Tomkiel JE, Pimpinelli S, Sandler L (1992) Rescue from the *abnormal oocyte* maternal effect lethality by *ABO* heterochromatin in *Drosophila melanogaster*. *Genetics* 128: 583–594
36 Tomkiel J, Fanti L, Berloco M, Spinelli L, Wakimoto BT, Pimpinelli S (1995) Developmental genetical analysis and molecular cloning of the *abnormal oocyte* gene of *Drosophila melanogaster. Genetics* 140: 615–627
37 Muller HJ (1930) Types of visible variations induced by X-rays in *Drosophila*. *J Genet.* 22: 299–334
38 Grigliatti T, (1991) Position-effect variegation. an assay for nonhistone chromosomal proteins and chromatin assembly and modifying factors. *Methods Cell Biol* 35: 587–627
39 Rushlow CA, Bender W, Chovnick A (1984) Studies on the mechanism of heterochromatic position effect at the *rosy* locus of *Drosophila melanogaster. Genetics* 108: 603–615
40 Singh PB (1994) Molecular mechanism of cellular determination: their relation to chromatin structure and parental imprinting. *J Cell Sci* 107: 2653–2668
41 Dimitri P, Pisano C (1989) Position effect variegation in *Drosophila melanogaster*. Relationships

between suppression effects and the amount of Y chromosome. *Genetics* 122: 793–800

42 Zuckerkandl E (1974) Recherches sur les properties et l'activite biologique de la chromatine. *Biochemie* 56: 937–954

43 Mottus R, Reeves R, Grigliatti TA (1980) Butyrate suppression of position-effect variegation in *Drosophila melanogaster. Mol Gen Genet* 178: 465–469

44 Candido EPM, Reeves R, Davie JR (1978) Sodium butyrate inhibits histone deacetylation in cultured cells. *Cell* 14: 105–113

45 Riggs MG, Whittaker RG, Neumann JR, Ingram VM (1977) n-Butyrate causes histone modification in HeLa and Friend erythroleukemic cells. *Nature* 268: 462–464

46 Sealy L, Chalkley R (1978) The effect of sodium butyrate on histone modification. *Cell* 14: 115–121

47 Pfeffer U, Vidali G (1991) Histone acetylation: recent approaches to a basic mechanism of genome organization. *Int J Biochem* 3: 277–285

48 Fanti L, Berloco M, Pimpinelli S (1994) Carnitine suppression of position-effect variegation in *D. melanogaster. Mol Gen Genet* 244: 588–595

49 Moore GD, Sinclair DA, Grigliatti TA (1983) Histone gene multiplicity and position-effect variegation in *Drosophila melanogaster. Genetics* 105: 327–344

50 Moore GD, Procunier JD, Cross DP, Grigliatti TA (1979) Histone gene deficiencies and position-effect variegation in *Drosophila. Nature* 282: 312–314

51 Reuter G, Spierer P (1992) Position effect variegation and chromatin proteins. *Bioessays* 14: 605–612

52 James TC, Elgin SCR (1986) Identification of nonhistone chromosomal protein associated with heterochromatin in *Drosophila* and its gene. *Mol Cell Biol* 6: 3862–3872

53 James TC, Eissenberg JC, Craig C, Dietrich V, Hobson A, Elgin SCR (1989) Distribution patterns of HP1, a heterochromatin-associated nonhistone chromosomal protein of *Drosophila. Eur J Cell Biol* 50: 170–180

54 Eissenberg JC, James TC, Foster-Hartnett DM, Hartnett T, Ngan V, Elgin SCR (1990) Mutation in a heterochromatin-specific chromosomal protein is associated with suppression of position-effect variegation in *Drosophila melanogaster. Proc Natl Acad Sci USA* 87: 9923–9927

55 Fanti L, Dorer DR, Berloco M, Henikoff S, Pimpinelli S (1998) Heterochromatin protein 1 binds transgene arrays. *Chromosoma* 107: 286–292

56 Fanti L, Giovinazzo G, Berloco M, Pimpinelli S (1998) The heterochromatin protein 1 prevents telomere fusions in *Drosophila. Mol Cell* 2: 527–538

57 Farkas G, Gausz J, Galloni M, Reuter G, Gyurkovics H, Karch F (1994) The *Trithorax-like* gene encodes the *Drosophila* GAGA factor. *Nature* 371: 806–808

58 Bhat KM, Farkas G, Karch F, Gyurkovics H, Gausz J, Schedl P (1996) The GAGA factor is required in the early *Drosophila* embryo not only for transcriptional regulation but also for nuclear division. *Development* 122: 1113–1124

59 Becker PB (1994) The establishment of active promoters in chromatin. *Bioessays* 16: 541–547

60 Granok H, Leibovitch BA, Shaffer CD, Elgin SC (1995) Chromatin. Ga-ga over GAGA factor. *Curr Biol* 5: 238–241

61 Shore D (1994) RAP1: a protean regulator in yeast. *Trends Genet* 10: 408–412

62 Lohe AR, Brutlag DL (1986) Multiplicity of satellite DNA sequences in *Drosophila melanogaster. Proc Natl Acad Sci USA* 83: 696–700

63 Bonaccorsi S, Lohe A (1991) Fine mapping of satellite DNA sequences along the Y chromosome of *Drosophila melanogaster:* relationships between satellite sequences and fertility factors. *Genetics* 129: 177–189

64 Lohe AR, Hilliker AJ, Roberts PA (1993) Mapping simple repeated DNA sequences in heterochromatin of *Drosophila melanogaster. Genetics* 134: 1149–1174

65 Pesce CG, Rossi MS, Muro AF, Reig OA, Zorzópulos J, Kornblihtt AR (1994) Binding of nuclear factors to satellite DNA of retroviral origin with marked differences in copy number among species of the rodent *Ctenomys. Nucl Acid Res* 22: 656–661

66 McDonald JF (1990) Macroevolution and retroviral elements. *Bioscience* 40: 183–191

67 Raff JW, Kellum R, Alberts B (1994) The *Drosophila* GAGA transcription factor is associated with specific regions of heterochromatin throughout the cell cycle. *EMBO J* 13: 5977–5983

Chromosomes Today, Vol. 13
ed. by E. Olmo and C.A. Redi

RNA, genes, genomes and chromosomes: repetitive DNA sequences in plants

J.S. (Pat) Heslop-Harrison

Department of Cell Biology, John Innes Centre, Norwich, NR4 7UJ, UK

The RNA world and its relics

In reviewing the nature and evolution of key repetitive DNA sequences in plants, one can start by considering the most evolutionarily ancient of nucleic acid sequences, at the start of the evolutionary process of modern organisms. Work by several groups since the late 1980s has shown that many features of "life" existed in a world where RNA was a key molecule, and this evolved to the present DNA-based living world. This "RNA world" included RNA-catalyzed reactions, making and breaking chemical bonds, transferring energy between different molecular structures, and self-replication mechanisms. Auto-catalytic RNA sequences (ribozymes) have been known since the 1980s. Gesteland and Atkins [1] have pointed out the enormous richness of apparent "relics" from the RNA world that have been discovered in present day organisms.

In this review, I aim to discuss some of the major components – structural and genetical – of plant and other eukaryotic genomes, but particularly those with relationships to RNA sequences.

The ribosomal genes

Ribosomes are a critical feature of the cell, in combination with proteins, providing the protein synthesis machinery that translates mRNA into protein. The rRNA transcripts are among the most abundant sequences present in the cell. In eukaryotic nuclei, the ribosomes are made up of four different rRNA molecules, the 5S, 5.8S, 18S and 25S rRNA molecules. The molecules encoded by the genomes of prokaryotes, mitochondria and chloroplasts are slightly smaller than those encoded in eukaryotic nuclei, where the rRNA genes are present in hundreds, and sometimes thousands, of tandemly repeated copies. The 5S genes, with intergenic spacers, are present normally at one group of loci, while the 45S unit, comprising the 18S, 5.8S and 25S rRNA genes and intergenic spacers are present at other loci. Both are present at one or a small number of chromosomal sites in a haploid chromosome set, but the number of sites is not correlated with genome size or other features of an organism. Secondary constrictions at the nucleolar organizer regions (NORs) are the major chromosomal sites of the 45S rDNA and represent a clear morphological feature of chromosomes under the microscope. The repeated copies of rDNA present at these sites may comprise several percent of all the DNA in the genome: 8% in *Arabidopsis thaliana*.

rDNA and rRNA are abundant sequences in the cell, and it is now clear that the rRNA molecules themselves have enzymatic activity. Ribozymes were first found as components of eukaryotic gene introns, where the RNA intron catalyzed its own excision to give the processed mRNA molecule – a self-splicing function. In bacterial transcription systems, Nitta and colleagues [2] have shown that proteins are not a key functional component of the ribosome but the critical peptidyl transferase activity required for peptide bond formation during protein synthesis is a function of the larger rRNA molecule. These experiments showed that RNA has both the capacity to carry the genetic information and the enzyme activity needed to convert that information into proteins [3].

The rRNA gene sequences are highly conserved, and a DNA probe for ribosomal genes from one eukaryotic species will hybridize to homologous genes in essentially all other species, whether animal, fungal or plant. In molecular cytogenetic studies, rDNA fragments are perhaps the most widely used probes for *in situ* hybridization, with their location reported in 50 or more species. The high conservation and repetition at a small number of loci on the chromosomes makes the probes very efficient for *in situ* hybridization (see [4] for details of methods). Both the location and number of sites vary even between closely related species; many species have major loci and others with smaller numbers of copies of the repeat unit, or fragments of the unit, and these are often not known to be transcribed. In banana, the single site of the 45S rDNA per genome provides a quick maker for the ploidy of tissue culture or hybrid material [5]. In the important tribe Triticeae, including many cereal crops and their wild relatives, there is a single pair of sites of the 45S rDNA in rye ($2n = 2x = 14$), while barley ($2n = 2x = 14$) has two major pairs of loci but minor loci on three more chromosome pairs [6, 7]. In cultivated wheat, a hexaploid of relatively recent origin ($2n = 6x = 42$, with A, B and D genomes), the two major loci of 45S rDNA are present on B genome chromosomes (1B and 6B), while sites on the A and D genomes are much reduced (Fig. 1, arrows). The numbers and locations of sites of rDNA loci are useful features to identify morphologically similar chromosomes sites, while the diversity and relatively rapid evolution of the loci makes them useful chromosomal markers within a species and evolutionary markers between species [8, 9] (Fig. 4). There is evidence that sites may "move" in chromosomal location without involvement of translocations or other chromosomal rearrangements [10].

Increasingly, there is a need to link genetical maps made using molecular markers with the physical maps of chromosomes, for several reasons. Where chromosomal introgressions, translocations or even recombinants are used for plant breeding, it can be valuable to know which chromosome is involved and where the physical recombination events are happening along the chromosome arms. *In situ* hybridization of rDNA, which can often be mapped genetically by polymorphisms in their internal sequences, is an efficient method to link some chromosomes on the map, and can allow correlation of at least one physical site with a genetical distance [11, 12]. Data can be extended using additional probes for *in situ* hybridization of gene probes.

Viruses

RNA viruses are present in the DNA world. In plants, many of these replicate by sequestering the DNA-based replicative machinery of the cell. The mechanisms that viruses use to complete their life cycle are now understood in considerable detail for some DNA viruses (e.g. [13]), and it is likely that similar mechanisms are used for RNA viruses.

In mammals, retroviruses (RNA viruses that can be copied into DNA and incorporated into the nuclear genome and amplified with it) represent up to 6% of the nuclear DNA [14]. There is increasing data suggesting that such viruses may exist in plants, but not with the high abundance of those in mammals [15]. Some plants, however, contain a related class of viruses to the retroviruses, where the infective form is a double-stranded circular DNA molecule, the pararetroviruses [16]. New data from *in situ* hybridization shows that the sequences of the pararetrovirus Banana Streak Virus (BSV) are integrated into the nuclear genome of banana [17]. This technique provides a unique method for looking at the integration of this virus since alternative methods examined DNA and it is very difficult to exclude the possibility of episomal forms of the

---

Figure 1. (See color plate VI between pages 56 and 57.) A line of bread wheat ($2n = 6x = 42$) with a pair of recombinant chromosomes from the wild species *Aegilops umbellulata. In situ* hybridization of genomic DNA from *Ae. umbellulata* shows the two recombinant chromosomes in bright green, up to the position of the 45S rDNA locus (red) (arrowheads). The satellite, beyond the constriction at the NOR (red) is dull green and originates from the wheat chromosome. The recombinant chromosome is mostly 1U, which carries a major rDNA site. Major (on chromosomes 1B and 6B) and minor rDNA sites (chromosomes 5D, 1A) are detected by red sites of fluorescence (arrows)

Figure 2. (See color plate VI between pages 56 and 57.) *In situ* hybridization of a banana streak virus probe to chromosomes of the banana variety Obino L'Ewai ($2n = 3x = 33$). (a) shows DAPI staining of the chromosomes, while (b) shows (arrow and arrowhead) major and minor sites of the BSV probe on chromosomes as bright red fluorescence.

Figure 3. (See color plate VI between pages 56 and 57.) A metaphase preparation of *Arabidopsis thaliana* ($2n = 10$) showing the centromeric location of the major tandemly repeated sequence present around the centromeres (pink-orange) of all ten chromosomes (blue) after *in situ* hybridization. The sequence may be related to centromere function.

Figure 4. (See color plate VI between pages 56 and 57.) Two nuclei of *A. thaliana* hybridized with 45S rDNA, showing the four major sites representing some 8% of all the DNA in the genome.

virus occurring within the nucleus and being extracted along with DNA. In a triploid banana cultivar ($2n = 3x = 33$), we can identify both major and minor sites of BSV (Fig. 2). Using hybridization to extended DNA fibres isolated from the variety, we can characterize the organization of the integrated BSV sequences at the two types of loci and show that multiple copies of at least parts of the BSV genome are present in the nucleus.

The RNA world becomes the DNA world

Information encoded in RNA is no longer the major information store for modern life forms, and one must consider why DNA has become dominant. RNA is an unstable molecule and very prone to degradation when compared to DNA. Although many aspects of RNA degradation may be under genetic control in eukaryotes (the half-life or turnover rate of RNA is an important factor in gene expression studies), it is unlikely that an RNA molecule of chromosomal length could replicate as a unit and be transcribed. Thousands of genes on an RNA chromosome could not be organized and their replication and expression controlled in a systematic manner. The inert nature of DNA allows a long, stable and organized information store, and has another related advantage. DNA itself is not known to have significant enzymatic activity, allowing control of genes at the transcriptional level. Control of RNA sequences with ribozymal activity would not be possible, and the complex feedback loops so critical to balanced regulation of genetic activity in DNA-based organisms would be impossible.

RNA and nucleic acid replication

Recent data indicate that RNA molecules can replicate in an enzyme-free environment by spontaneous base-pairing of mononucleotides with existing long RNA molecules, followed by spontaneous polymerization [18]. RNA is also an important component of nucleic acid replication in the DNA-based world: semi-conservative DNA replication is primed by RNA molecules (making the Okazaki fragments). In bacteria, an RNA polymerase, primase, copies about five nucleotides of the DNA before a DNA polymerase continues replication. In eukaryotes, subunits of DNA polymerase-alpha first synthesize an RNA primer, and the same enzyme then extends DNA from the primer, before the major DNA polymerase, DNA polymerase-delta, continues DNA replication from the RNA-DNA primer [19]. Finally, the RNA primer is removed, replaced with DNA and ligated to form a single continuous DNA molecule. The priming function of RNA seems to be central to DNA replication, and is important to the mechanisms of chromosome end-replication and retrotranposon characteristics discussed below.

Reverse transcriptase and retroelements

Because of the advantages of the stability of genes encoded in DNA over RNA, it is probable that enzymatic conversion of RNA into DNA, and the ability to transcribe DNA into RNA, were very important evolutionary steps. This movement between the RNA world and DNA world requires an RNA-dependent DNA polymerase – a reverse transcriptase – as well as a DNA-dependent RNA polymerase to generate the mRNA required for translation. The discovery of reverse transcriptase in the 1970s was a surprise because its presence contradicted the central dogma. However, recent surveys of plant genomes are showing that reverse transcriptase genes, or at least their remnants, are among the most abundant genes in the genome because of their presence in retroelements. Retroelements are sequences that are present in the nuclear genome (and also known in prokaryotes and organelles; [16]). The RNA transcript (which has two open reading frames in most retroelements) serves two functions: it is translated to a number of proteins including reverse transcriptase, and it is a template for reverse transcription into DNA by that enzyme. The DNA copies are then able to reinsert into the host genome with the involvement of further enzymes.

In some plants [20, 21] and animals [14], retroelements comprise about 50% of the total DNA present, although many are degenerate, including stop-codons or frameshifts in the open reading frames. There are two major classes of nuclear retroelements, the retrotransposons that have long terminal repeats (LTR) and the non-LTR retroposons. In the human, and presumably most mammalian genomes, non-LTR retroposons are very abundant: LINEs – long interspersed nuclear elements – represent 17% of the total DNA, with up to 100,000 copies. However, only some 4000 of these are full length, and fewer – 15 to 30 – are active [22]. Another class of non-LTR elements, the S1 dispersed nuclear elements (SINEs), are related to LINEs but with many deletions: they include the ALU sequences and represent a similar portion of the human genome as LINEs. Although retroviruses, as discussed above, are widespread in mammals, LTR retrotransposons of the *gypsy* and *copia* groups have not been detected. Contrasting with this, in plants the LINE [23] and SINE [24] retroelements are present in thousands of copies, although most data suggests that retrotransposons of the *gypsy* and *copia* groups are more abundant in most species. San Miguel et al. [20] and Pearce et al. [21] have shown that some 50 or 40% of the genomes of maize and *Vicia faba*, respectively, are represented by the *copia* group of retrotransposons alone.

Typically, retrotransposons show a dispersed distribution in the genome by *in situ* hybridization (Figs 7, 8; [25]) with, in general, depletion in some chromosomal regions such as the ribosomal DNA sites, centromeres and some heterochromatic (e.g. in rye-origin chromosomes in Fig. 7) sequences. There are, though, many exceptions and there seem to be controls on copy number and insertion sites of retroelements, whether positive or negative through selection [26]. In *Arabidopsis thaliana*, for example, the *copia* retroelements seem to cluster near the centromere, which might occur because of the high density of genes and regulatory sequences along chromosome arms, where disruption could not be tolerated.

Key functional regions of the reverse transcriptase domains of *copia* group retrotransposons are extremely conserved [27], and Flavell et al. [28] and Hirochika and Hirochika [29] have shown how degenerate PCR primers made to conserved domains can be used to amplify fragments of the gene from many species of plants and animals. Although the proteins are highly conserved in function, through reverse transcriptase, the sequence between the functional domains may show a high level of variation, enabling differences to be highlighted between different genomes. It is now clear that some retroelement families are essentially specific to one genome in plants of hybrid origin: The wheat × rye amphidiploid Triticale (Fig. 7) and hexaploid oat, *Avena sativa* ($2n = 6x = 42$, with ACD genomes; Fig. 8) provide good examples. In Triticale, the two wheat genomes and rye, and in oat, the three *Avena* genomes have a common ancestry. In Triticale, a rye-origin retroelement probe hybridizes much more strongly to the rye-origin chromosomes than those of the wheat genomes. Katsiotis et al. [30] have shown that a retroelement probe isolated from a diploid A genome species hybridizes only weakly to C-genome chromosomes. The same hybridization of retroelement probes to both the A and D genome chromosomes confirms their very close relationship [31–33].

The variation in the reverse transcriptase domains of retroelements isolated by PCR can be used to generate phylogenetic trees of relationships between

---

Figure 5. (See color plate VI between pages 56 and 57.) A metaphase and interphase nucleus of rye (a, DAPI staining), probed with the tandem repeat pSc119.2 (b, green; the two rDNA sites are also visible red). Many sites colocate with DAPI positive heterochromatin bands.

Figure 6. (See color plate VI between pages 56 and 57.) Rye DNA digested for various times with micrococcal nuclease and probed with pSc119.2 after size separation. The 170 bp ladder of individual nucleosomes is separated, and the 120 bp sequence probes only the major bands on the gel, showing, with other data [60] that the sequence indicates a fixed packing arrangement around the nucleosomes.

Figure 7. (See color plate VI between pages 56 and 57.) A metaphase of Triticale ($2n = 6x = 42$), a wheat × rye amphidiploid hybrid (a, DAPI staining of chromosomes), hybridized with *copia* group retroelement fragments from rye (C, red hybridization signal; b shows a superimposition of a and c). The retroelements hybridize relatively uniformly along the length of the 14 rye chromosomes, but are excluded from the DAPI positive sites of tandem repeats near the ends of many chromosomes (see also Fig. 5). No wheat-rye translocations are seen.

Figure 8. (See color plate VI between pages 56 and 57.) A metaphase of hexaploid oat, *Avena sativa* ($2n = 6x = 42$; ACD genomes; a, DAPI staining). A retroelement probe hybridized to the 28 chromosomes of the A and D genomes (red, c, with superimposition of a and c in b), showing the close relationship of these chromosomes, and the relatively uniform organization of retroelements along the chromosomes. In this species, there are also reciprocal translocations involving terminal chromosome segments between the A/D and C genome chromosomes, shown by dark ends on red chromosomes or red ends on otherwise dark chromosomes (b).

Figure 9. (See color plate VI between pages 56 and 57.) The chromosome model showing the organization and location of different classes of repetitive sequences in plants.

different elements [21]. Retroelements have an important feature in that they can modify gene expression by their movement, amplification and reinsertion into genes and regulatory sequences, whether in plants [34] or human [35]. In conclusion, retroelements are major and ubiquitous genomic structures involved in shaping the plant genome and determining its size, organization and aspects of its evolution.

Centromeres

The compartmentalization of the genome within cells, whether prokaryotic or eukaryotic, requires partitioning of the genome into daughter cells after replication mitosis, or its equivalent in bacteria. The mechanics and function of the DNA elements involved in chromosome segregation in bacteria [36, 37] and eukaryotic yeast [38–40] are well understood. In mammals, and to a lesser extent plants, microscopic structures involved are well known [41]. The function of the DNA at the centromere is highly conserved: the DNA is, indirectly, attached to microtubules (a very conserved feature of cells), and the centromeres enable segregation of chromatids to opposite spindle poles at mitosis and meiosis. However, despite the conserved function, surprisingly little else seems conserved at the DNA level, and the critical domains of DNA sequence associated with centromeric function are the subject of intense work [42, 43]. In mammals, there is considerable work on the alpha satellite, a sequence representing several percent of the genome present as tandemly arrayed units of a 170 bp monomer that is localized at many centromeres [44]. Mammalian artificial chromosomes including synthetic arrays of alpha-satellite DNA have been constructed and shown to segregate [45], demonstrating that the alpha-satellite DNA is sufficient for nuclear kinetochore assembly and chromosome segregation. However, this sequence is neither required for segregation, nor always involved in segregation when present in mammals: the human Y-chromosome and chromosomes of the African Green Monkey do not have the sequence (see [44, 46]), while only one centromere is active in some human chromosomes with two sites of tandem arrays of the alpha satellite sequence. There has been relatively little molecular and structural analysis of centromeric DNA in higher plants, although many repetitive DNA sequences have been located at or near the centromere by *in situ* hybridization (Fig. 3) (Crucifers: [47, 48]; Legumes: [49]; grasses: [50–53]) but little is published with respect to the function of these sequences. Centromeric regions of many other plant species stain strongly with fluorescent or colorimetric dyes suggesting that they consist of highly repetitive DNA sequence motifs. Although any function of these sequences remains unknown, it has been suggested that, like the human alpha-satellites, centromeric repetitive sequences may be related to centromere function because of their location [54], their sometimes widespread species distribution, and their high repetition. The sequence is present in blocks of tandem repeats longer than 1Mb [55], and retroelements or retroelement fragments are interspersed between repeats [25, 56]. We consider that the inclusion of DNA sequence boxes defined for centromere function in mammals and yeasts within the centromeric repetitive sequence is significant. Within all plants, some sequences present at the centromere are species-specific [57], while others are widespread across groups of species such as the grasses [51, 52]. In the grasses, sequences so far localized to centromeres have many similarities to retroelements, and *gypsy* retroelements have been found at the centromeres of barley [58], suggesting another role for these abundant sequences. In mammals, the role of retrotransposon-derived centromeric sequence domains has also been considered [43].

There are two further features of eukaryotic chromosomes: the packaging of DNA with proteins, and the presence of special structures at the ends, the telomeres.

Chromatin packaging

A further conserved structural feature of the eukaryotic chromosome is the packing of the DNA into the chromosome, avoiding the presence of stretches of naked DNA fibre. The nucleosome core, first described by Klug and colleagues [59], involves the wrapping of two turns of DNA around the octamer of histone proteins. Using the enzyme micrococcal nuclease, the DNA can be cleaved between nucleosomes, and the resulting nucleosomes size separated on a high resolution gel, giving the typical ladder of 180 base pairs. In rye, wheat and their relatives, we have shown that different tandemly repeated DNA units show this nucleosomal organization [60]. Although tandem repeat monomer units may vary extensively, we found that the sequences showed fixed phasing with respect to the nucleosome packing. For example, one unit of a 350 bp repeat would be packaged around two nucleosomes, while three units of a 120 bp repeat would be packaged around two nucleosomes. Figure 5 shows an example of *in situ* hybridization of a 120 bp repeat to wheat, while Figure 6 shows hybridization of the same probe to micrococcal nuclease digests of genomic wheat DNA: the presence of discrete bands rather than a smear shows that the monomer is wrapped uniformly around the nucleosomes of about 170 bp, and the pSc119.2 monomer is cut at the same small number of sites by the nuclease.

Telomeres and chromosome-end replication

The structure and function of eukaryotic telomeres is now well known, and the story of the discovery of the enzymes and structure can be followed through previous volumes of *Chromosomes Today* from 1983 to the resolution of their structure and replication in human [61] and *Drosophila* [62]. Like the centromere, the function of telomeres is highly conserved. Briefly, variants of the short nucleotide motifs (TTAGGG) in mammals and (TTTAGGG) in plants [63] are the most frequent sequences at the ends of chromosomes, and have been found in most species of plants [64] and animals [61], although a few species such as *Drosophila* and *Allium* [64] have different structures. In most species, the telomere is added enzymatically to the physical ends of chromosomes by the enzyme telomerase. Telomerase is another good example of a reverse transcriptase, including within its structure a RNA core [65] which is copied to give the simple sequence repeat at the end of the chromosomes.

The plant chromosome model

Many other classes of repetitive DNA sequences are known in plants. In particular, simple sequence repeats, nucleotide motifs 1–4 nucleotides long, repeated hundred or thousands of times, are very abundant, and different motifs seem to give different characteristic locations in the genome [66, 67] (Fig. 5). Other repetitive sequences include tandemly repeated elements and dispersed repeats of no known function, many of them showing characteristics of fast evolution including specificity to species or small species groups. Bringing all these together, with the structural and functional sequences described in this paper, we can generate an integrated chromosome model of the genome ([68]; Fig. 9). The study of molecular cytogenetics and chromosome structure is central to understanding evolution and brings together the different structural, functional and genetical characteristics of the chromosomes. The results are undoubtedly of value to understanding evolution, mitosis and meiosis, as well as recombination, genome organization and gene expression.

Acknowledgements

I thank EC Framework IV grants for support of aspects of this work. I am also grateful to the many collaborators for their continuing assistance, and particularly those who provided the pictures used: Alexandra Castilho, Julian Osuji, Nuno Neves, Alexander Vershinin, Steven Pearce, Gill Harrison, Andreas Katsiotis and Thomas Schmidt.

References

1 Gesteland RF, Atkins JF (1993) *The RNA World*. Cold Spring Harbor Press, New York
2 Nitta I, Kamada Y, Noda H, Ueda T, Watanabe K (1998) Reconstitution of peptide bond formation with *Escherichia coli* 23S ribosomal RNA domains. *Science* 281: 666–669
3 Allen TA (1999) *Genomes*. Bios, Oxford

4 Schwarzacher T, Heslop-Harrison JS (2000) Practical in situ hybridization. Bios, Oxford 203 + XII pp
5 Osuji JO, Crouch J, Harrison G, Heslop-Harrison JS (1998) Molecular cytogenetics of *Musa* species, cultivars and hybrids: location of 18S-5.8S-25S and 5S rDNA and telomere-like sequences. *Ann Bot* 82: 243–248
6 Leitch IJ, Heslop-Harrison JS (1992) Physical mapping of the 18S-5.8S-26S rRNA genes in barley by *in situ* hybridization. *Genome* 35: 1013–1018
7 Pedersen C, Rasmussen SK, Linde-Laursen I (1996) Genome and chromosome identification in cultivated barley and related species of the Triticeae (*Poaceae*) by *in situ* hybridization with the GAA-satellite sequence. *Genome* 39: 93–104
8 Castilho A, Heslop-Harrison JS (1995) Physical mapping of 5S and 18S-25S rDNA and repetitive DNA sequences in *Aegilops umbellulata*. *Genome* 38: 91–96
9 Taketa S, Harrison G, Heslop-Harrison JS (1999) Comparative physical mapping of the 5S and 18S-25S rDNA in nine wild *Hordeum* species and cytotypes. *Theor Appl Genet* 98: 1–9
10 Dubcovsky J, Dvořák J (1995) Ribosomal RNA multigene loci: Nomads of the triticeae genomes. *Genetics* 140: 1367–1377
11 Castilho A, Miller TE, Heslop-Harrison JS (1997) Analysis of a set of homologous group 1 wheat-*Aegilops umbellulata* recombinant chromosome lines using genetic markers. *Theor Appl Genet* 94: 293–297
12 Leitch IJ, Heslop-Harrison JS (1993) Physical mapping of four sites of 5S rDNA sequences and one site of the alpha-amylase-2 gene in barley (*Hordeum vulgare*). *Genome* 36: 517–523
13 Liu L, Saunders K, Thomas C, Davies JW, Stanley J (1999) Bean yellow dwarf virus repA but not rep binds to maize retinoblastoma protein and the virus tolerates mutations in consensus binding motif. *Virology* 256: 270–279
14 Smit AFA (1996) The origin of interspersed repeats in the human genome. *Curr Opin Genet Develop* 6: 743–748
15 Kumar A (1998) The evolution of plant retroviruses: moving to green pastures. *Trends Plant Sci* 3: 371–374
16 Hull R, Covey SN (1996) Retroelements: propagation and adaptation. *Virus Genes* 11: 105–118
17 Harper G, Osuji JO, Heslop-Harrison JS, Hull R (1999) Integration of banana streak badnavirus into the *Musa* genome: molecular and cytogenetic evidence. *Virology* 255: 207–213
18 Robertson HM, Ellington AD (1998) How to make a nucleotide. *Nature* 395: 223–225
19 Foiani M, Luccini G, Plevani P (1997) The DNA polymerase-alpha-primase complex couples DNA replication, cell-cycle progression and DNA-damage response. *Trends Biochem Sci* 22: 424–427
20 San Miguel P, Tikhonov A, Jin Y-K, Motchoulskaia N, Zakharov D, Melake-Berhan A, Springer PS, Edwards KJ, Lee M, Avramova Z et al (1996) Nested retrotransposons in the intergenic regions of the maize genome. *Science* 274: 737–738
21 Pearce SR, Harrison G, Wilkinson M, Li D, Heslop-Harrison JS, Flavell AJ, Kumar A 1995. The *Ty1-copia* group retrotransposons in *Vicia* species: copy number, sequence heterogeneity and chromosomal localisation. *Mol Gen Genet* 250: 305–315
22 Sassaman DM, Dombroski BA, Moran JV, Kimberland ML, Naas TP, DeBerardinis RJ, Gabriel A, Swergold GD, Kazazian J (1997) Many human L1 elements are capable of retrotransposition. *Nat Genet* 16: 37–43
23 Leeton PRJ, Smyth DR (1993) An abundant LINE-like element amplified in the genome of *Lilium speciosum*. *Mol Gen Genet* 237: 97–104
24 Goubely C, Arnaud P, Tatout C, Harrison G, Heslop-Harrison JS, Deragon J-M (1999) S1 SINE retroelements are methylated at symmetrical and non-symmetrical positions in *Brassica napus*: identification of a preffered target site for asymmetrical methylation. *Plant Mol Biol* 39: 243–255
25 Brandes A, Heslop-Harrison JS, Kamm A, Kubis S, Doudrick RL, Schmidt T (1997) Comparative analysis of the chromosomal and genomic organization of *Ty1-copia*-like retrotransposons in pteridophytes, gymnosperms and angiosperms. *Plant Mol Biol* 33: 11–21
26 Heslop-Harrison JS, Brandes A, Taketa S, Schmidt T, Vershinin AV, Alkhimova EG, Kamm A, Doudrick RL, Schwarzacher T, Katsiotis A et al (1997) The chromosomal distributions of Ty1-*copia* group retrotransposable elements in higher plants and their implications for genome evolution. *Genetica* 100: 197–204
27 Bureau TE, Ronald PC, Wessler SR (1996) A computer-based systematic survey reveals the predominance of small inverted-repeat elements in wild-type rice genomes. *Proc Natl Acad Sci USA*

93: 8524–8529

28 Flavell AJ, Dunbar E, Anderson R, Pearce SR, Hartley R, Kumar A (1992) Ty1-copia group retrotransposons are ubiquitous and heterogeneous in higher plants. *Nucl Acid Res* 20: 3639–3644

29 Hirochika H, Fukuchi A, Kikuchi F (1992) Retrotransposon families in rice. *Mol Gen Genet* 233: 209–216

30 Katsiotis A, Schmidt T, Heslop-Harrison JS (1996) Chromosomal and genomic organization of Ty1-copia-like retrotransposon sequences in the genus *Avena*. *Genome* 39: 410–417

31 Leggett JM, Markhand GS (1995) The genomic structure of *Avena* revealed by GISH. *In*: PE Brandham, MD Bennett (eds): *Kew Chromosome Conference IV*, UK, HMSO, 133–139

32 Katsiotis A, Hagidimitriou M, Heslop-Harrison JS (1997) The close relationship between the A and B genomes in *Avena* L. determined by molecular cytogenetic analysis of total genomic, tandemly and dispersed repetitive DNA sequences. *Ann Bot* 79: 103–109

33 Pearce SR, Harrison G, Heslop-Harrison JS, Flavell AJ, Kumar A (1997) Characterisation, genomic organisation of Ty1-copia group retrotransposons in rye (*Secale cereale*). *Genome* 40: 617–625

34 Wessler SR, Bureau TE, White SE (1995) LTR-retrotransposons and MITEs: important players in the evolution of plant genomes. *Curr Opin Genet Develop* 5: 814–821

35 Moran JV, DeBerardinis RJ, Kasasian HH (1999) Exon reshuffling by L1 retrotransposition. *Science* 283: 1530–1534

36 Schaechter M, von Freiesleben U (1993) The equivalent of mitosis in bacteria. *In*: JS Heslop-Harrison, RB Flavell (eds): *The Chromosome*. Bios, Oxford, 61–73

37 Sherratt DJ, Blakely G, Burke M, Colloms S, Leslie N, McCulloch R, May G, Roberts J (1993) Site-specific recombination and the partition of bacterial chromosomes. *In*: JS Heslop-Harrison, RB Flavell (eds): *The Chromosome*. Bios, Oxford, 25–41

38 Saitoh S, Takahashi K, Yanagida M (1997) Mis6, a fission yeast inner centromere protein, acts during G1/S and forms specialized chromatin required for equal segregation. *Cell* 90: 131–143

39 Uzawa S, Yanagida M (1992) Visualization of centromeric and nucleolar DNA in fission yeast by fluorescence *in situ* hybridization. *J Cell Sci* 101: 267–275

40 Hegemann JH, Fleig UN (1993) The centromere of budding yeast. *Bioessays* 15: 451–460

41 Pluta AF, Cooke CA, Earnshaw WC (1990) Structure of the human centromere at metaphase. *Trends Biochem Sci* 15: 181–185

42 Lee C, Wevrick R, Fisher RB, Ferguson-Smith MA, Lin CC (1997) Human centromeric DNAs. *Hum Genet* 100: 291–304

43 Kipling D, Warburton PE (1997) Centromeres, CENP-B and Tigger too. *Trends Genet* 13: 141–145

44 Tyler-Smith C, Corish P, Burns E (1998) Neocentromeres, the Y chromosome and centromere evolution. *Chromosome Res* 6: 65–71

45 Harrington JJ, Van Bokkelen G, Mays RW, Gustashaw K, Willard HF (1997) Formation of *de novo* centromeres and construction of first-generation human artificial microchromosomes. *Nat Genet* 15: 345–355

46 Goldberg IG, Sawhney H, Pluta AF, Warburton PE, Earnshaw WC (1996) Surprising deficiency of CENP-B binding-sites in African green monkey alpha-satellite DNA – implications for CENP-B function at centromeres. *Mol Cell Biol* 16: 5156–5168

47 Harrison GE, Heslop-Harrison JS (1995) Centromeric repetitive DNA in the genus *Brassica*. *Theor Appl Genet* 90: 157–165

48 Brandes A, Thompson H, Dean C, Heslop-Harrison JS (1997) Multiple repetitive DNA sequences in the paracentromeric regions of *Arabidopsis thaliana* L. *Chromosome Res* 5: 238–246

49 Galasso I, Schmidt T, Pignone D, Heslop-Harrison JS (1995) The molecular cytogenetics of *Vigna unguiculata* (L.) Walp: the physical organization and characterization of 18S-5.8S-25S rRNA genes, 5S rRNA genes, telomere-like sequences, and a family of centromeric repetitive DNA sequences. *Theor Appl Genet* 91: 928–935

50 Leach CR, Donald TM, Franks TK, Spiniello SS, Hanrahan CF, Timmis JN (1995) Organization and origin of a B chromosome centromeric sequence from *Brachycome dichromosomatica*. *Chromosoma* 103: 708–714

51 Aragon-Alcaide L, Miller T, Schwarzacher T, Reader S, Moore G (1996) A cereal centromeric sequence. *Chromosoma* 105: 261–268

52 Jiang J, Nasuda S, Dong F, Scherrer CW, Woo S-S, Wing RA, Gill BS, Ward DC (1996) A conserved repetitive DNA element located in the centromeres of cereal chromosomes. *Proc Natl Acad Sci USA* 93: 14210–14213

53 Nagaki K, Tsujimoto H, Sasakuma T (1998) A novel repetitive sequence of sugar cane, SCEN family, locating on centromeric regions. *Chromosome Res* 6: 295–302
54 Maluszynska J, Heslop-Harrison JS (1991) Localization of tandemly repeated DNA sequences in *Arabidopsis thaliana*. *Plant J* 1: 159–166
55 Murata M, Ogura Y, Motoyoshi F (1994) Centromeric repetitive sequences in *Arabidopsis thaliana*. *Jpn J Genet* 69: 361–370
56 Pelissier T, Tutois S, Tourmente S, Deragon JM, Picard G (1996) DNA regions flanking the major *Arabidopsis thaliana* satellite are principally enriched in Athila retroelement sequences. *Genetica* 97: 141–151
57 Kamm A, Galasso I, Schmidt T, Heslop-Harrison JS (1995) Analysis of a repetitive DNA family from *Arabidopsis arenosa* and relationships between *Arabidopsis* species. *Plant Mol Biol* 27: 853–862
58 Presting GG, Malysheva L, Fuchs J, Schubert I (1998) A TY3/GYPSY retrotransposon-like sequence localizes to the centromeric regions of cereal chromosomes. *Plant J* 16: 721–728
59 Finch JT, Klug A (1976) Solenoid model for superstructure in chromatin. *Proc Natl Acad Sci USA* 73: 1897–1901
60 Vershinin AV, Heslop-Harrison JS (1998) Comparative analysis of the nucleosomal structure of rye, wheat and their relatives. *Plant Mol Biol* 36: 149–161
61 Meyne J, Ratliff RL, Buckingham JM, Jones MD, Wilson JS, Moyzis RK (1990) The human telomere. *Chromosome Today* 10: 75–80
62 Biessmann H, Walter MF, Kurenova E, Mason JM (1993) Retrotransposons at *Drosophila* telomeres and terminal chromosome deficiencies. *Chromosome Today* 12: 104–121
63 Richards EJ, Vongs A, Walsh M, Yang J, Chao S (1993) Substructure of telomere repeat arrays. *In*: JS Heslop-Harrison, RB Flavell (eds): *The Chromosome*. Bios, Oxford, 103–114
64 Fuchs J, Brandes A, Schubert I (1995) Telomere sequence localization and karyotype evolution in higher plants. *Plant Syst Evol* 196: 227–241
65 Greider CW, Autexier C, Avilion AA, Collins K, Harrington LA, Mantell LL, Prowse KR, Smith SK, Allsopp RC, Counter CM et al (1993) Telomeres and telomerase in mortal and immortal cells. *In*: JS Heslop-Harrison, RB Flavell (eds): *The Chromosome*. Bios, Oxford, 115–125
66 Schmidt T, Heslop-Harrison JS (1996) The physical and genomic organization of microsatellites in sugar beet. *Proc Natl Acad Sci USA* 93: 8761–8765
67 Cuadrado A, Schwarzacher T (1999) The chromosomal organization of simple sequence repeats in wheat and rye genomes. *Chromosoma* 107: 587–594
68 Schmidt T, Heslop-Harrison JS (1998) Genomes, genes and junk: the large scale organization of plant chromosomes. *Trends Plant Sci* 3: 195–199

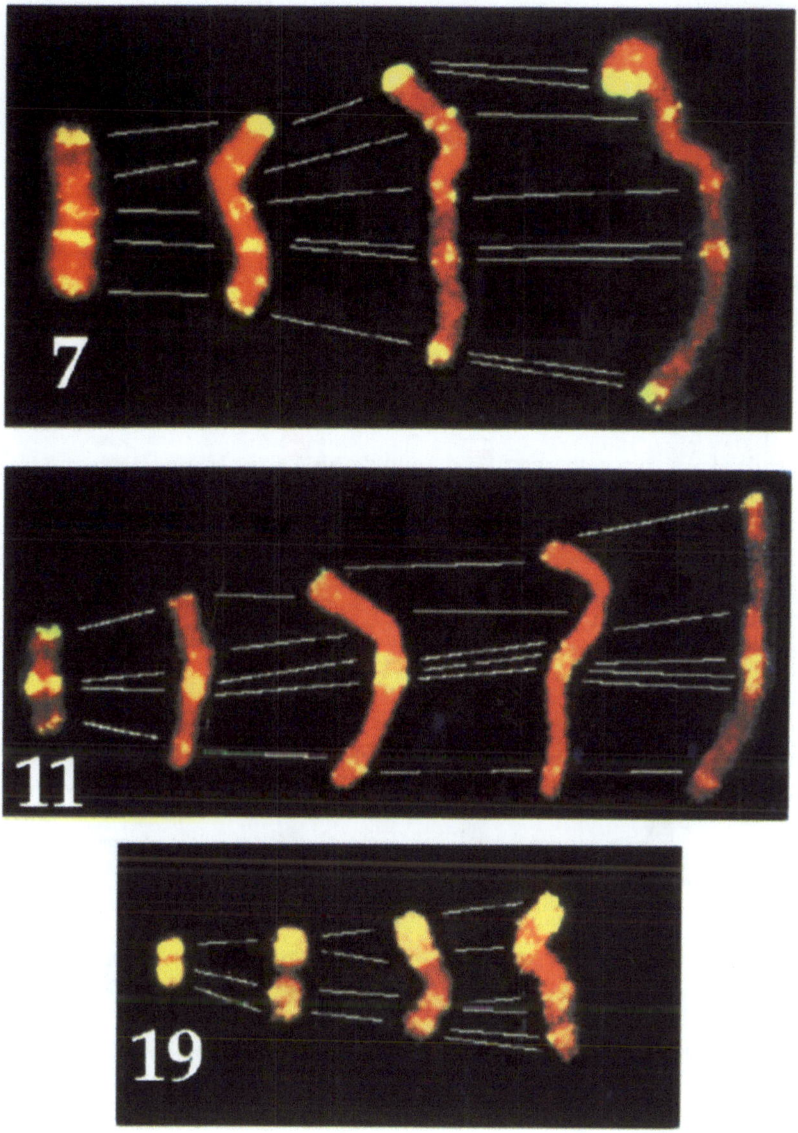

Figure 1. Legend see p. 26

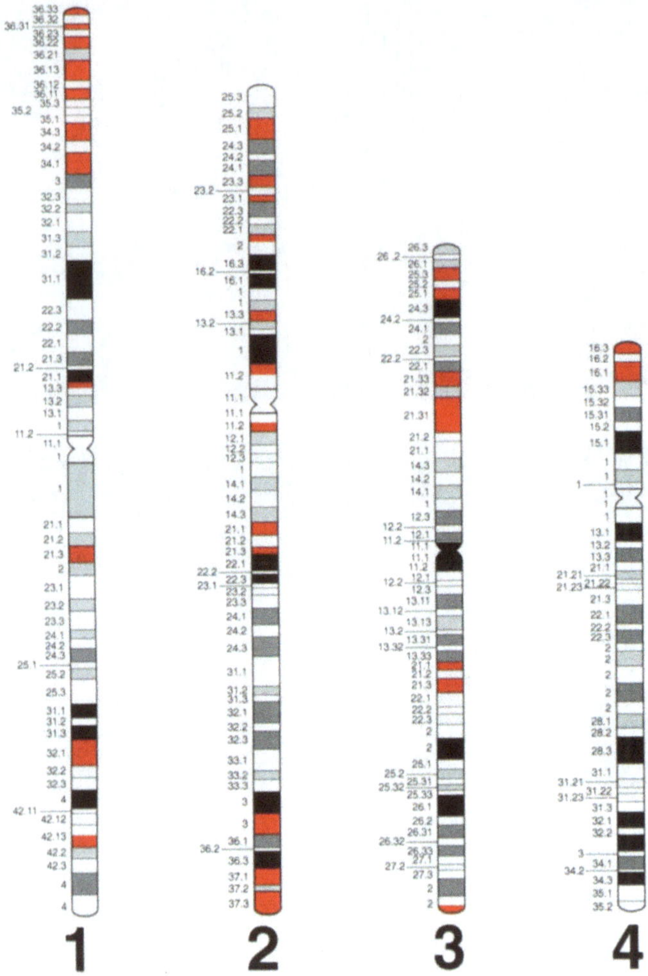

Figure 2. Legend see p. 26

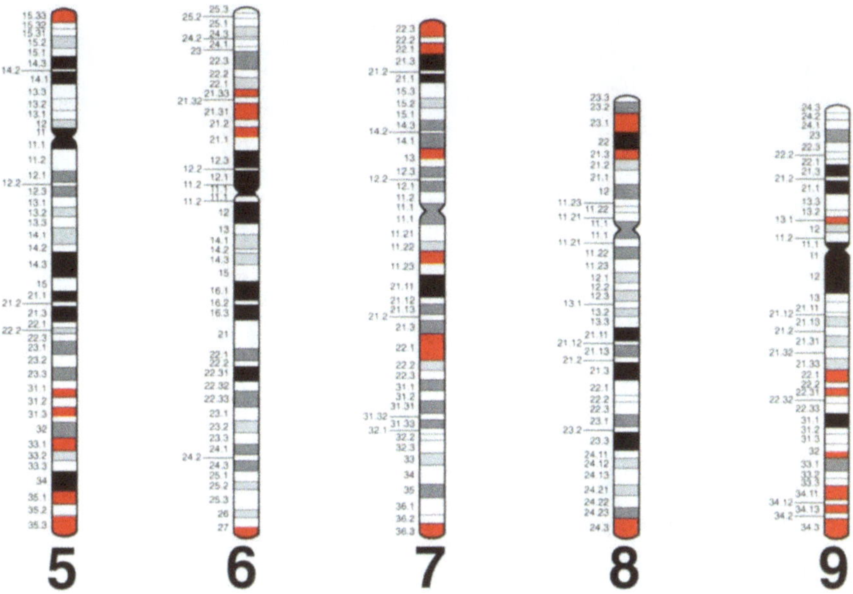

Figure 3. Legend see p. 26

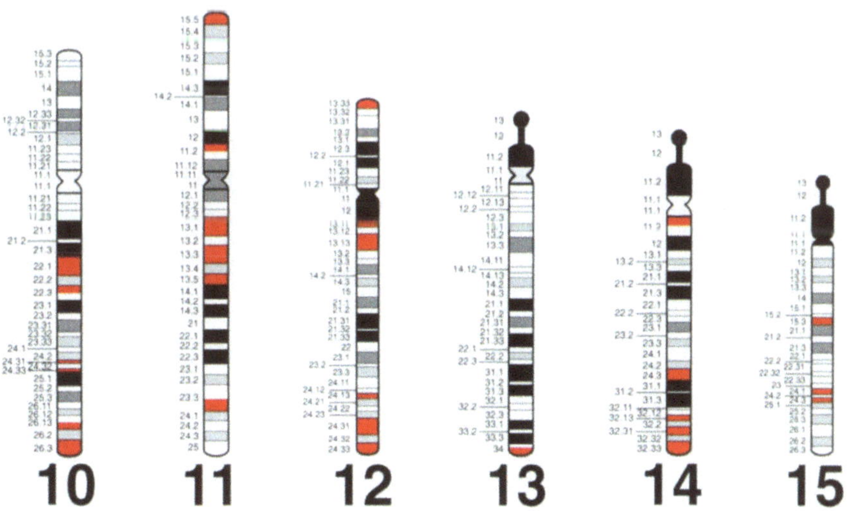

Figure 4. Legend see p. 26

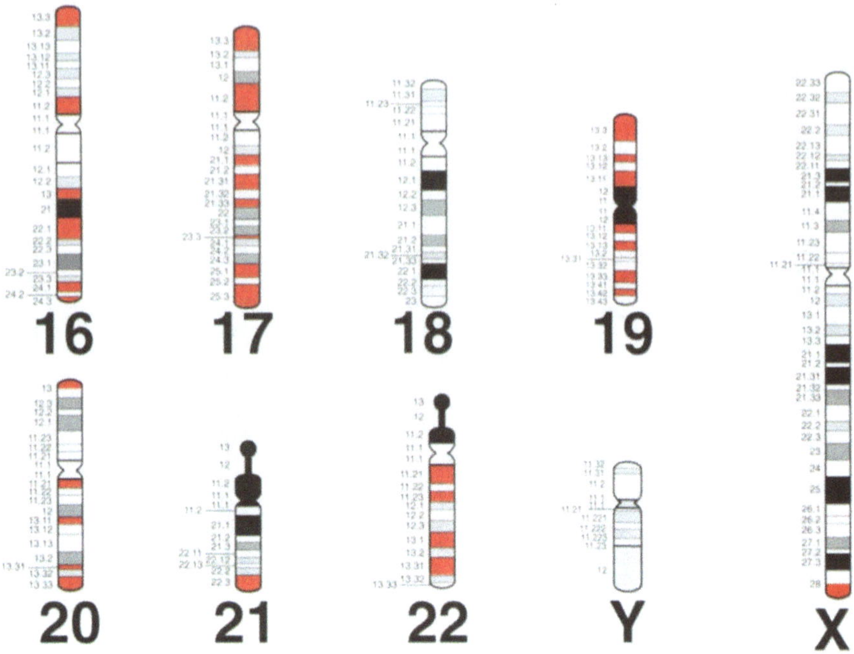

Figure 5. Legend see p. 26

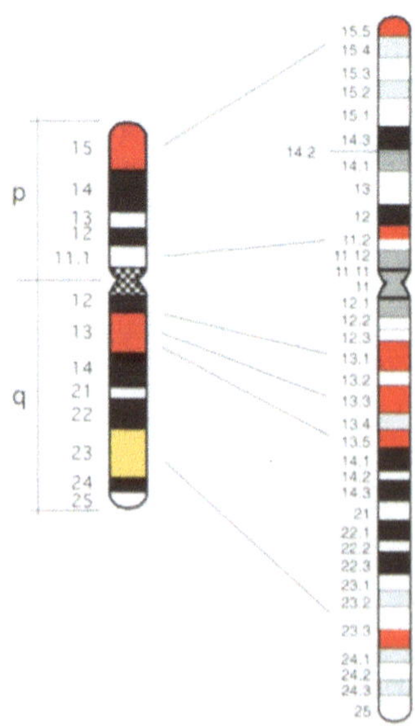

Figure 6. Legend see p. 26

Chr 11

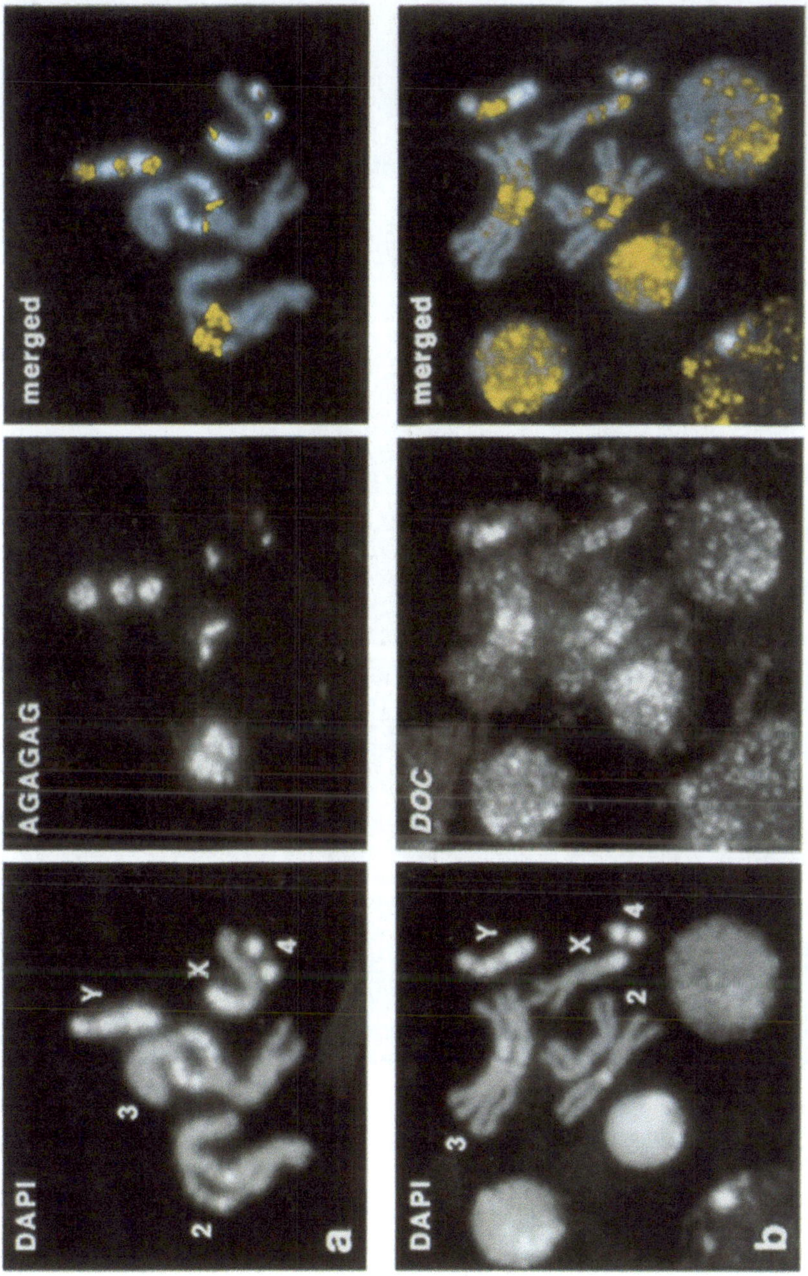

Figure 2. Legend see p. 39

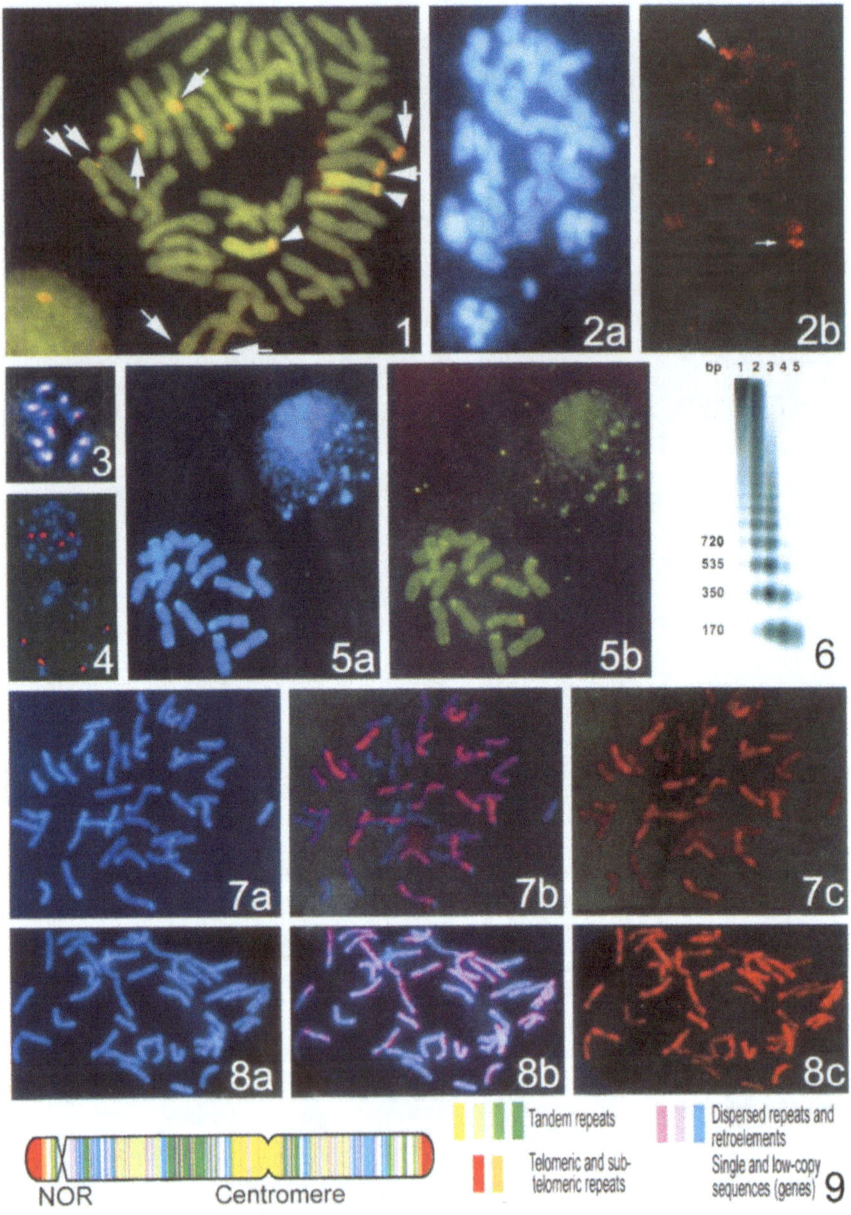

Figures 1–4. Legends see p. 47.
Figures 5–9. Legends see p. 50.

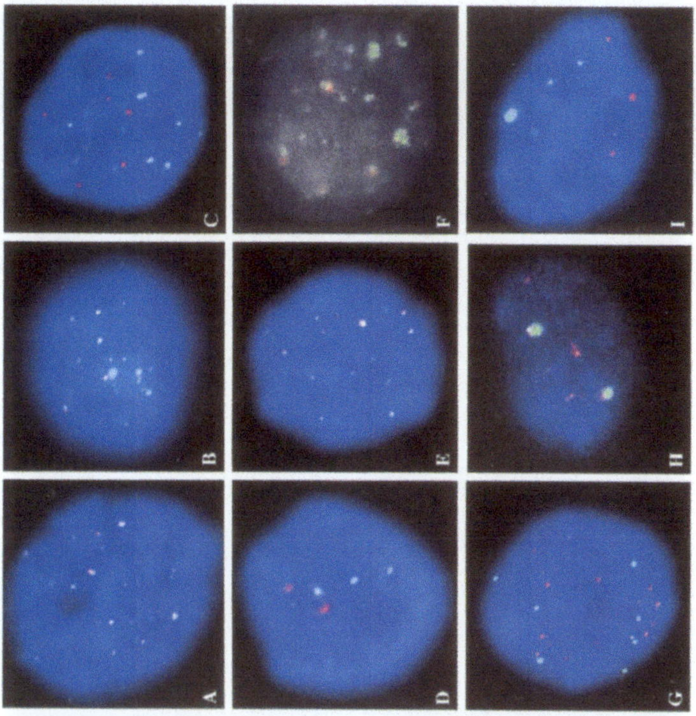

Figure 3. Legend see p. 63

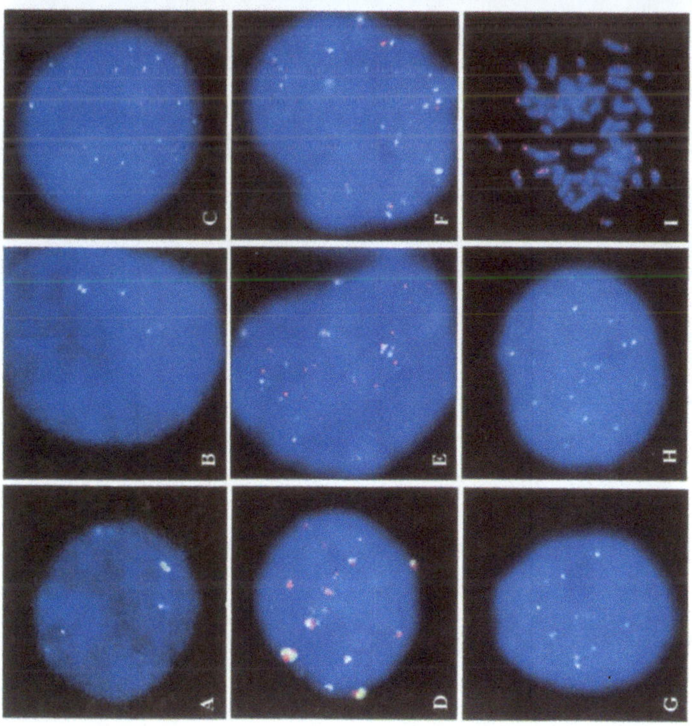

Figure 2. Legend see p. 61

Chromosomes Today, Vol. 13
ed. by E. Olmo and C.A. Redi

Non-B DNA structures spatially and sequence-specifically associated with individual centromeres in the human interphase nucleus

Mizuki Ohno[1, 2], Toyoaki Tenzen[1], Yoshihisa Watanabe[1], Tetsushi Yamagata[1], Shigehiko Kanaya[3] and Toshimichi Ikemura[1]

[1] *Department of Evolutionary Genetics, National Institute of Genetics, The Graduate University for Advanced Studies, Mishima, Shizuoka-ken 411-8540, Japan*
[2] *Institute of Genetic Information, Kyushu University, Fukuoka 812-0054, Japan*
[3] *Faculty of Engineering, Yamagata University, Yonezawa, Yamagata-ken 992-8510, Japan*

Introduction

Chromosome domains such as centromeres and telomeres occupy specific nuclear compartments supporting spatial organization of chromosomal DNA in the interphase nucleus. Telomere repeat-sequences can form a non-B quadruplex structure, presumably to support telomere-telomere association [1]. Several sequences abundant in human centromeres are also known to have the potential to form non-B structures, that are thought to support the ordered arrangement of centromeric repeat sequences and also chromosomal DNA. In the human genome there exists many polypurine/polypyrimidine tract sequences (pur/pyr tracts), each several hundred bp in length and scattered along the genome [2]. Interest in pur/pyr tracts has increased due to the *in vitro* observation that these sequences can form a triple-helix (triplex) structure even under physiological conditions [3, 4]. Triplex formation is thought to be important as a molecular mechanism determining spatial organization in an interphase nucleus [5–7]. In triplex formation, single-stranded DNA is left unpaired, and it can hybridize with other single-stranded DNA and RNA with sufficient complementarity under physiological conditions. This enables distantly-spaced DNA regions along the genome to associate with each other in the nucleus. It is also conceivable that distantly-spaced DNAs can form trans-molecular triplexes organizing themselves into an ordered array. The purpose of this paper is to show that non-B structures such as triplexes are present in the human interphase nucleus and introduce methods for their investigation. One of the pur/pyr tracts of interest is that found in the DNA-replication switch region described below.

Pur/Pyr sequences in DNA-replication switch region

Chromosome band zones have been associated with various genome characteristics such as DNA replication timing during S phase and long-range G+C% (GC%) mosaic structures [8–13]; therefore, band boundaries are thought to correspond to switching points of DNA replication timing (i.e., from early to late replication) and of GC% mosaic boundaries. Early- and late-replicating domains are highly compartmentalized within the nucleus, instructing individual band zones to replicate in spatially separated nuclear domains [14–17]. Band boundaries with control signals for replication timing may reside at characteristic positions, functioning as spatial landmarks in nuclear organization. The human major histocompatibility complex (MHC) spans about 4 Mb and is composed of plural bands at a high-resolution level. We previously found a boundary for megabase (Mb)-sized GC% mosaic domains at the junction between MHC classes II and III, and proposed this boundary as a chromosome band boundary [18, 19]. Because chromosome bands are structures observed under the microscope, precise location of their boundaries might seem meaningless. However, considering genome behaviors connected with chromosome bands, there exists a possibility that band boundaries are precisely assignable by identifying the early-to-late switch point for replication timing. We measured the replication timing for a 500 kb region harboring the MHC class II and III junction, for which we completed bidirectional chromosome walking and a large portion has been sequenced [20–22]. Many sets of PCR primers closely spaced (e.g., even within one kb) could be designed for clarifying the switch of replication timing fairly precisely, and the replication timing was found to change precisely at the GC% boundary with a two hour time difference between the GC- and AT-rich sides (Fig. 1 A and B)[23]. At exactly the midpoint of the replication switch zone a 209 bp pur/pyr tract with triplex-forming potential exists. Its purine strand can be expressed mostly as palindromic sequences composed of two tetranucleotides, GGAA and AGAA. Under proper superhelical strain, this type of palindromic pur/pyr sequence (termed H-palindrome) undergoes a distinct structural transition involving an intrastrand disproportionation in which half of the mirror repeat sequence dissociates into single strands, and one of the two single strands folds back on the other half of the mirror repeat duplex to form an intramolecular triplex (known as H-DNA)[24]. Figure 1*C* shows examples of such H-DNA triplexes formed

Figure 1. The 209 bp pur/pyr tract found at the exact midpoint of the switching region of replication timing during S phase. Replication timing (A) and GC% distribution (B) in the junction between MHC classes II and III is from Tenzen et al. [23]. (A) Y-axis means the replication timing; the start time of S-phase was taken as 0, and replication timings for individual loci were plotted as reported by Tenzen et al. [23]. (B) For the GC% distribution in this and other MHC regions, see Fukagawa et al. [19]. (C) An example of H-DNA triplexes for the 209 bp pur/pyr tract leaving the longest single-strand, when only the perfectly consecutive Hoogsteen-pairing longer than 10 bps was assumed. The 209 bp sequence corresponds to that from nucleotide 5301 to 5510 in GenBank Locus HSMHC3A5 (Accession No. U89335), and its location in A, B is indicated by a vertical arrowhead line.

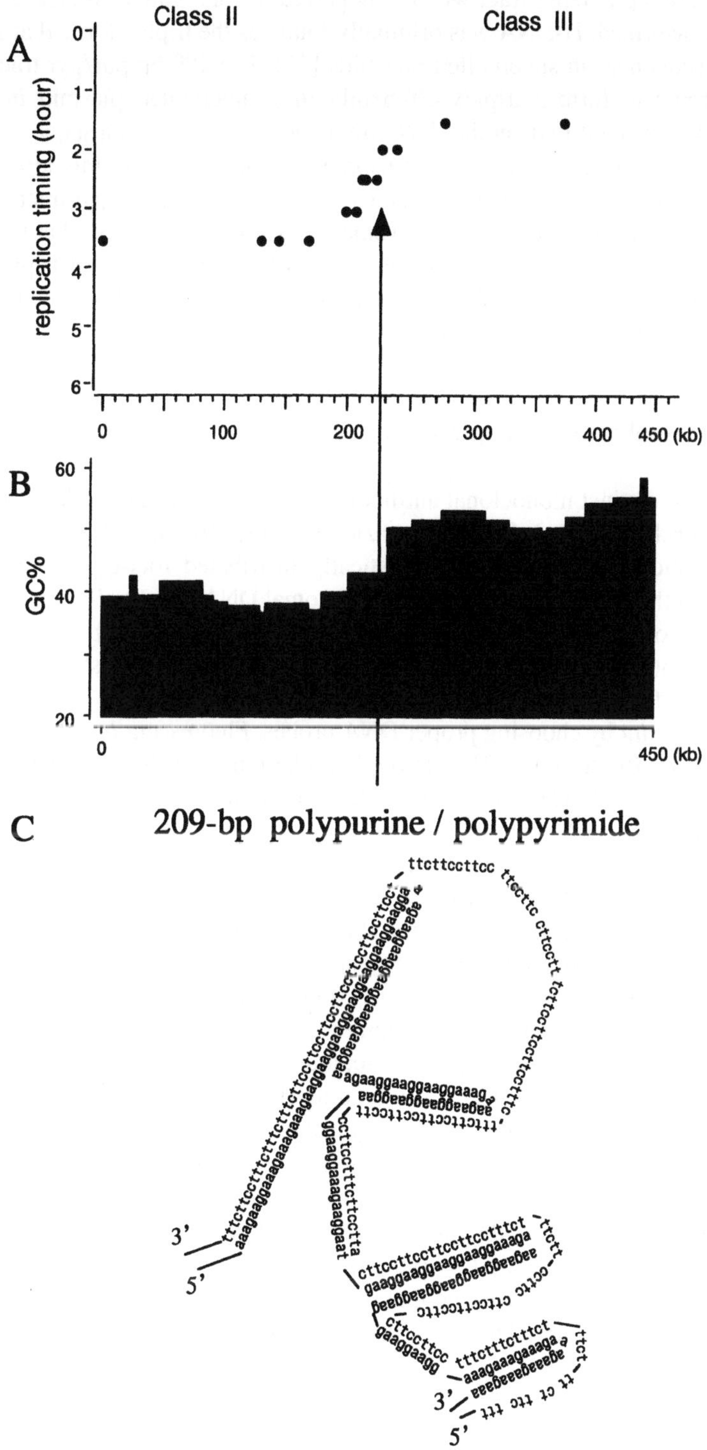
A
Class II
Class III
replication timing (hour)
0
1
2
3
4
5
6
0
100
200
300
400
450 (kb)
B
GC%
60
40
20
0
450 (kb)
C
209-bp polypurine / polypyrimide
3'
5'
3'
5'

for the 209 bp pur/pyr tract when only perfectly consecutive Hoogsteen-pairing was assumed. H-DNA was originally found as the triplex formed at mirror repeat sequences in supercoiled plasmids [25]. The 209 bp pur/pyr tract here was shown to form a triplex efficiently in a supercoiled plasmid *in vitro*. Recently, we identified another DNA-replication switching at a sequence level in the XIST locus on the X chromosome, and predicted that this also corresponds to a chromosome band boundary. There also exists pur/pyr tracts in this switch region. Replication fork movement is known to be arrested at triplexes formed in pur/pyr tracts *in vitro* [26, 27]. While the function of triplexes *in vivo* has not yet been clarified, it has become clear that triplex structures are formed in the interphase nucleus.

In situ visualization of triplexes in human interphase nuclei

Using two distinct monoclonal antibodies specific for triplex DNAs, Lee et al. [28], Burkholder et al. [6], and Agazie et al. [7, 29] showed that the mammalian nucleus revealed characteristically distributed foci-type signals and proposed that a certain portion of chromosomal DNA is naturally organized in a triplex configuration. If individual signals stained by the antibody contained a sufficient amount of triplex DNA, and the resultant single-stranded DNA, sequence-dependent foci-type signals could be visualized in the non-denatured nucleus *in situ* by choosing proper DNA probes. Fluorescent *in situ* hybridization (FISH) for detecting RNA in the interphase nucleus has been successfully developed [30–32]; for detecting RNA in the presence of double-stranded genomic DNA, non-denatured nuclei were used because RNAs of interest are mostly present in a single-stranded form. An analogous method may detect the single-stranded DNA displaced by triplex-formation and the genomic DNA with which probe DNAs can form a stable transmolecular triplex under physiological conditions. When involvement of probe DNAs in triplex-formation (and other non-B structure formations) is considered, non-Watson-Crick pairing such as Hoogsteen pairing becomes prominent. In such cases, the term "hybridization" may not be appropriate; thus "binding" is used. We refer to the *in situ* assay as fluorescence *in situ* binding (FISB) instead of FISH, although the experimental procedure itself is equivalent. The 209 bp pur/pyr tract gave clear foci-type signals (Fig. 2A). The pattern was also obtained when analyzing nuclei pre-treated by RNase A, demonstrating that the respective binding sites are not RNAs. A wide range of DNA probes with triplex-forming potential gave sequence-specific foci-type signals and a few examples are presented below. Recently, we have found that these signals exactly overlap with triplexes immunolocalized by known antitriplex antibodies [28, 29].

Figure 2. (See color plate VII between pages 56 and 57.) FISB (A–F), lipofection (G, H) or FISH (I) assay using DNAs with non-B forming potential. Human monocytic leukemia THP-1 cells ([46], XY) were analyzed. While the nuclei fixed with acetic acid-methanol (1:3) [53] were mainly used for FISB, the paraformaldehyde Triton X-100 treated nuclei [54] gave similar results, demonstrating that the triplex formation is not due to the acid treatment of nuclei. To obtain highly-fluorescent DNA probes, PCR with fluorescent deoxynucleotides was done according to the method of [55]. The DNA probe mix for FISB was composed of 50% formamide in 2x SSC, 5% dextran sulfate, 100 µg/ml sheared salmon sperm DNA, and 0.1–1 µg/ml fluorescent DNA probe. The reaction solution can be replaced by Hybrisol (ONCOR). After 5 min denaturation at 70 °C followed by cooling on ice, 30 µl of the mixture was applied to each slide covered with a glass coverslip and incubated for 4 h at 37 °C in a humid chamber. The slides were then washed three times for 5 min each in 2x SSC containing 0.1% Tween 20 at room temperature, once at 37 °C for 10 min in 2x SSC, and once for 5 min in 2x SSC containing 0.1% Tween 20 at room temperature. Probes were labeled by either Cy3- (red) or FITC- (green) deoxynucleotide, and nuclei were counterstained by DAPI (blue). For the lipofection assay (G, H), cells were grown until they were 50% confluent. For each lipofection, 0.1–1 µg of DIG-labeled DNA was diluted into 100 µl serum-free medium and mixed with LipofectAmine Reagent (Gibco) according to the manufacturer's protocol. The lipofected cells were incubated at 37 °C in a CO_2 incubator, and an equal amount of serum-containing medium was added after 1 h. After an additional 5 h incubation, cells were washed three times in PBS, then cytocentrifuged onto a microscope glass slide, and the slide was washed in PBD (0.1% Tween 20, 0.1% Triton X-100, and 0.1% NP40 in PBS) three times for 5 min each. After an ethanol series (70%, 80%, 95%, 100%), bound DIG was detected by antidigoxigenin antibody. (A) FISB with the FITC-labeled 209 bp pur/pyr tract. RNase-treated nuclei gave analogous patterns. (B) FISB with FITC (TGGAA/TTCCA)n. RNase-treated nuclei were analyzed. (C) FISB with FITC pancentromeric α satellite probe. (D) FISB with FITC pancentromeric α satellite; centromeres were immunolocalized by the centromere-specific antisera, and visualized by rhodamine-conjugated human IgGs (red). (E) FISB with FITC pancentromeric α satellite and Cy3 850-bp pur/pyr tract. The 850 bp pur/pyr tract near PAB (pseudoautosomal boundary) on human sex chromosomes corresponds to that from nucleotides 26 081 to 26 931 in GenBank Locus HSCAMF3X1 (Accession No. X96421). (F) FISB with FITC pancentromeric α satellite and Cy3 209-bp pur/pyr tract. (G, H) Lipofection assay with (GAA/TTC)n or pancentromeric α satellite probe, respectively. (I) DNA-FISH on metaphase chromosomes with Cy3 (GAA/TTC)n according to a standard FISH method.

FISB assay with centromere repeat sequences

Several sequences abundant in human centromeres are know to have the potential to form non-B structures *in vitro*. One sequence is (TGGAA/TTCCA)n, which is similar to the telomere sequence. This repeat is present in the human satellite III element found in a restricted class of centromeres [33, 34]. The (TGGAA)n sequence can form a stable "self-complementary" antiparallel duplex *in vitro* [35, 36], indicating that the other strand is displaced as a single-stranded form. Actually, FISB with the (TGGAA/TTCCA)n probe gave foci-type signals for RNase-treated nuclei (Fig. 2B). Significant portions of α satellite sequences are composed of alternate purine and pyrimidine clusters of about 10 bp, that are relatively rich in GAA and GGA motifs known to form non-B structures *in vitro* [36–39]. Alternate purine and pyrimidine clusters can form "alternate-strand triplexes" [5]. If the hypothesized non-B structures were formed even in a very restricted portion of α satellite sequences, the structures may be seen *in situ* because of the huge multiplication of the repeat sequences. A pancentromeric α satellite probe has been designed to stain all centromeres of human chromosomes by a standard DNA-FISH [40], where nuclear DNA has to be denatured. This probe gave several tens of foci-type FISB signals in non-denatured nuclei (Fig. 2C). To examine whether binding sites for this α satellite probe actually correspond to subnuclear zones occupied by individual centromere sequences in the interphase nucleus ("centromere territories"), centromeres were detected by anticentromere antibody visualized with rhodamine-conjugated human IgGs (Fig. 2D). Almost all FISB signals (FITC, green) overlapped or were associated with anticentromere antibody signals (rhodamine, red), showing that the binding sites for the α satellite probe represent centromere territories. The centromeres visualized by anticentromere antibody after FISB reactions were often larger than those visualized by the same antibody without the FISB-treatment; this will be discussed later.

Differentiated centromere territories with respect to pur/pyr binding activity

That centromeres can be visualized with FISB implies that locations of other FISB signals relative to centromeres can be clarified in nondenatured nuclei. Elucidation of this location should give valuable information for characterizing binding sites for individual probes. During the course of FISB studies with a wide range of triplex-forming DNA probes, we found the 850 bp pur/pyr tract near the pseudoautosomal boundary (PAB) of human sex chromosomes (refer to Fig. 2E legend) to give foci-type signals that overlapped or were associated with centromeres visualized by pancentromeric α satellite probe (Fig. 2E). In the case of the 209 bp tract (Fig. 2F, red), signal number was reduced significantly and the signals were closely associated with a subset of centromeres. The signal number of (GAA/TTC)n was also less than centromere number, and again the signals were closely associated with a subset of centromeres. The finding that the latter two pur/pyr probes gave signals associated with only a fraction of centromeres suggests the possibility of a differential binding ability associated with individual centromeres. If so, it may provide a tool to elucidate differentiated characteristics and roles of individual centromeres in organizing chromosomal DNA in the interphase nucleus.

Human α satellite sequences are known to differ significantly between different chromosomes, and based on this difference, chromosome-specific centromere probes for a standard DNA-FISH have been developed [40, 41]. Even for FISB, individual centromeres might be differentially marked or at least classified into certain groups using chromosome-specific probes. Commercially available chromosome-specific α satellite probes were tested with FISB. Since the condition for post-washings in FISB was less stringent (refer to Fig. 2 legend) than that recommended by the manufacturer for FISH detection of a unique chromosome centromere, FISB signals may not necessarily be restricted to the chromosome specified with DNA-FISH. Actually, the

Figure 3. (See color plate VII between pages 56 and 57.) FISB assay with chromosome (Chr.) -specific α satellite probe. (A) FITC-labeled pancentromeric α satellite and Cy3 Chr. 15-specific α satellite probe. (B) Cy3 Chr. 3 α satellite probe; centromeres were immunolocalized by centromere-specific antisera, and visualized by FITC-conjugated human IgGs (green). (C) FITC Chr. 3 and Cy3 Chrs. 13/21 probes. (D) FITC Chr. 8 and Cy3 Chr. 7 probes. (E) FITC Chr. 15 and Cy3 Chr. 6 probes. (F) FITC 209 bp tract and Cy3 Chr. 6 probes; DAPI color was omitted. Chr. 15 probe showed similar association. (G) FITC (GAA/TTC)n and Cy3 Chr. 3. (H) FITC (TGGAA/TTCCA)n and Cy3 Chrs. 13/21 probes. (I) FITC (TGGAA/TTCCA)n and Cy3 Chrs. 1/5/19 probes. As a separate experiment, correlation of FISB signals visualized by the biotin-labeled Chrs. 1/5/19 probe with the centromeres specified by a standard DNA-FISH using the Cy3-labeled Chrs. 1/5/19 probe was shown.

signal number for FISB usually ranged from a few to more than ten (Fig. 3). FISB signals visualized by a chromosome-specific probes were found to overlap with a subset of centromeres visualized by a pancentromeric α satellite probe: Fig. 3A shows an example for a chromosome 15 specific centromere probe (Chr. 15 probe, red). Colocalization of FISB signals with centromeres was also confirmed using the anticentromere antibody; Fig. 3B shows an example for a Chr. 3 probe.

To investigate the differential binding characteristics for individual centromeres, different chromosome-specific probes, distinctly labeled either with Cy3 or FITC, were tested with FISB for various pairings of specific probes. For example, centromeres visualized by a Chr. 3 probe were almost completely separated from, but characteristically arranged with, those visualized by a set of Chrs. 13/21 (Fig. 3C). Chr. 7 and 8 probes also visualized distinct centromeres with reduced numbers (Fig. 3D). In contrast, those visualized by a Chr. 6 probe mostly overlapped those visualized by a Chr. 15 probe (Fig. 3E). Based on this differential marking for centromeres with FISB, relative locations of binding sites for various pur/pyr probes to distinct centromeres could be investigated. FISB signals visualized by a Chr. 6 probe (Cy3, red) were closely associated with those of the 209 bp pur/pyr tract (Fig. 3F). Chr. 3 probe visualized more than ten foci-type signals, that were widely-separated from, but characteristically arranged with, those for (GAA/TTC)n (Fig. 3G). FISB signals visualized by a Chrs. 13/21 probe overlapped or were closely associated with those of (TGGAA/TTCCA)n (Fig. 3H), which is predominantly present in satellite III, but those for a Chrs. 1/5/19 probe were widely separated from (TGGAA/TTCCA)n signals (Fig. 3I). This is consistent with the data that chromosomes 13 and 21, but not chromosomes 1, 5, and 19, have satellite III in the respective centromeres [34].

Lipofection assay

Using DNA probes with non-B forming potential, FISB visualizes sequence-specific foci-type signals in non-denatured nuclei. Investigation of such probe-binding sites in living cell nuclei is of particular interest for understanding their biological significance. The investigation was conducted according to antigene strategy. In antisense and antigene technologies, single-stranded DNAs introduced into living cells are expected to associate with target RNAs or DNAs in a sequence-specific manner and to exert designated biological functions. In the antigene strategy, to modulate transcription, the introduced DNA is designed to associate with the target genome site and to work as an artificial gene repressor [42, 43]. The expected association should be based on molecular mechanisms operating under physiological conditions, and therefore the most probable DNA segment is a series of potentially-triplex sequences [44–48]. Digoxigenin (DIG)-labeled DNA probes that gave clear FISB signals were introduced into living cells by lipofection according to the antigene method, and lipofected cells were cultured for up to 12 h. The cells were deposited onto a microscope slide by low speed cytocentrifugation and washed to remove unbound probes trapped either in the nucleus or cytoplasm. The residual bound probes were then detected with FITC-conjugated antidigoxigenin antibody. Figure 2G shows that (GAA/TTC)n gave strong foci-type signals. The 850 and 209 bp pur/pyr tracts also gave foci-type signals, and general signal characteristics of the lipofection assay were fairly similar to those of FISB. Pancentromeric α satellite repeats gave several tens of clear foci-type signals (Fig. 2H), and colocalization of the signals with centromeres was confirmed using anticentromere antibody. While the main concern of this study was to investigate the molecular mechanisms supporting the highly ordered nuclear organization, the results appear to provide basic information for developing antisense and antigene technologies. There have been several reports describing the use of triplex-forming oligonucleotides to inhibit expression of endogenous genes [42, 43]. Oligonucleotides are designed to bind to double-stranded DNA at oligopurine/oligopyrimidine sites where they form a local triplex structure [49]. While this appears to provide a fairly general approach to sequence-dependent targeting for a specific genome portion, the actual probe targeting in the nucleus is still largely not understood. Recent reports indicate that in both antigene or antisense strategies, certain introduced DNAs appear to bind to nuclear structures at sites other than the target sites but still in a sequence-specific manner exerting some biological activities [50–52]. Some of the nuclear structures with such binding activities may correspond to the sequence-dependent binding sites explained here.

Characteristics and usefulness of the FISB method

Either methanol, acetic acid fixed [53] or paraformaldehyde, Triton-X100 treated [54] nuclei gave similar FISB signals, demonstrating that the triplex formation is not due to the acid treatment of nuclei. According to the FISH protocol for detecting RNA, probe-binding was performed in 50% formamide containing denatured salmon sperm DNA at a neutral pH (refer to Fig. 2 legend). In common with the RNA-FISH, the inclusion of formamide is thought not only to ensure nucleotide-sequence specificity but also to presumably suppress protein interaction with DNA probes. In the 50% formamide solution, special sequences might be preferentially melted at 37 °C, even during a several-hour incubation, and selectively detected without prior nuclear denaturation. This melting, if present, should not be responsible for the binding observed here because analogous binding was observed in the lipofection assay. Furthermore, as a separate FISB experiment, clear strand-specificity was observed for a complementary pair of single-stranded DNA probes.

Centromere visualization in interphase nuclei without a stringent denaturation process provides a valuable tool for investigating centromere arrangements in a fairly intact nucleus. Furthermore, when using chromosome-specific α satellite probes, differential binding characteristics among centromeres were disclosed. This differential marking cannot be accomplished by immunocytogenetic detection using anticentromere antibodies. The lack of a need for nuclei denaturation will allow us to use centromeres as landmark positions for locating other nuclear structures detected by their biological activities. While DNA-FISH can assign a unique centromere by a chromosome-specific probe, FISB presently visualizes a group of centromeres with similar binding characteristics. By combining a chromosome-specific probe with specific repeat-sequence probes, a unique centromere may be assignable, further clarifying differential characteristics of individual centromere territories and supplying the data required to understand their differential roles.

Discussion

Judging by signal intensities and probe sizes (e.g., the 209 bp pur/pyr tract), FISB signals of interest may not represent a single copy DNA. When GenBank was searched with the 209 bp sequence using the FASTA program, more than 50 sequences had greater than 80% nucleotide identity over the 200 bp stretch, and their locations are scattered and away from centromeres. This was also the case for the 850 bp tract and (GAA/TTC)n. A standard DNA-FISH assay on metaphase chromosomes showed (GAA/TTC)n signals were scattered among and within chromosomes (Fig. 2I). Therefore, the close association of triplexes and centromeres should reflect a spatial association in the nucleus rather than in the genome sequence. Individual FISB signals should visualize a supramolecular structure composed of clustered DNAs with particular repetitive sequences such as those with triplex-forming potential. As noted in the Figure 2D experiment, centromeres visualized by anticentromere antibody after FISB were often larger than those without FISB-treatment. Furthermore, in FISB using the pur/pyr and pancentromeric α satellite probes, individual centromere territories often appeared to be subdivided, giving complex signals (Figs 2E and 3F, H). We hypothesize that a certain portion of the potential triplex-forming genomic DNAs may get into centromere territories and participate in forming the large-scale supramolecular structures. If this is the case, binding of probes may dissect the supramolecular structures and provide information about the internal architecture of individual centromere territories.

Conclusion

When ordered arrangements between very distantly spaced DNAs along the genome or belonging to different chromosomes is considered, subnuclear structures other than nuclear scaffold/matrix-associated regions (SARs/MARs) should be requisite. We propose the formation of non-B structures as triplexes to be responsible for this spatial association and the probe-binding sites of interest to represent the respective structures. There should exist a wide variety of sequence-specific binding sites in the nucleus, which may afford molecular mechanisms for the highly-ordered nuclear organization.

DNA probes with triplex-forming potential gave sequence-dependent foci-type signals closely associated with centromeres. Differential characteristics among centromeres may indicate differential roles of individual centromeres in supporting the highly organized arrangement of chromosomal DNA in the interphase nucleus. Centromeres are undoubtedly key structures for chromosomal DNA organization and a wide range of important cellular functions. The close association between centromere territories and DNA-binding sites is attractive not only for the optimal nuclear folding design of chromosomal DNA but also for the coordinately-regulated and interactive functions executed by DNAs distantly spaced along the genome. Together with centromeres and telomeres, chromosomal band boundaries, in which DNA replication-timing switches, may constitute important landmark positions for nuclear organization. Interestingly, the 209 bp pur/pyr tract found at the midpoint of the switch region gave binding signals closely associated with the centromeres visualized by the centromere probes specific for Chrs. 6 and 15 (Fig. 3F), and the signals of the two centromere probes mostly overlap each other (Fig. 3E). These findings show that the binding characteristics of the two centromere territories and their surrounding regions are similar to each other. Centromere regions replicate very late during S phase and fulfill crucial roles in genome duplication and chromosome segregation. Spatial association of binding sites for the 209 bp pur/pyr in MHC (on 6p21.3) with the Chr. 6 centromere is of particular interest in relation to their interactively-regulated function and duplication.

Acknowledgments

The authors are very grateful to Dr. S. Mizuno for valuable suggestions and encouragements; to Mrs. K. Suzuki, Y. Miyauchi, and K. Kita for technical assistance. This work was supported by Grant-in-Aid of Scientific Research from the Ministry of Education, Science and Culture of Japan.

References

1 Rhodes D, Giraldo R (1995) Telomere structure and function. *Curr Opin Struct Biol* 5: 311–322
2 Siedlaczck I, Epplen C, Rieb O, Epplen JT (1993) Simple repetitive (GAA)n loci in the human genome. *Electrophoresis* 14: 973–977
3 Hoffman-Liebermann B, Liebermann D, Troutt A, Kedes LH, Cohen SN (1986) Human homologs of TU transposon sequences: polypurine/polypyrimidine sequence elements that can alter DNA conformation *in vitro* and *in vivo*. *Mol Cell Biol* 6: 3632–3642
4 Hampel KJ, Crosson P, Lee JS (1991) Polyamines favor DNA triplex formation at neutral pH. *Biochemistry* 30: 4455–4459
5 Frank-Kamenetskii MD, Mirkin SM (1995) Triplex DNA structures. *Annu Rev Biochem* 64: 65–95
6 Burkholder GD, Latimer LJP, Lee JS (1988) Immunofluorescent staining of mammalian nuclei and chromosomes with a monoclonal antibody to triplex DNA. *Chromosoma* 97: 185–192
7 Agazie YM, Burkholder GD, Lee JS (1996) Triplex DNA in the nucleus: direct binding of triplex-specific antibodies and their effect on transcription, replication and cell growth. *Biochem J* 316: 461–466
8 Bernardi G, Olofsson B, Filipski J, Zerial M, Salinas J, Cuny G, Meunier-Rotival M, Rodier F (1985) The mosaic genome of warm-blooded vertebrates. *Science* 228: 953–958
9 Ikemura T (1985) Codon usage and tRNA content in unicellular and multicellular organisms. *Mol Biol Evol* 2: 13–34
10 Ikemura T, Aota S (1988) Global variation in G+C content along vertebrate genome DNA: Possible correlation with chromosome band structures. *J Mol Biol* 203: 1–13
11 Ikemura T, Wada K (1991) Evident diversity of codon usage patterns of human genes with respect to chromosome banding patterns and chromosome numbers; relation between nucleotide sequence data and cytogenetic data. *Nucl Acid Res* 19: 4333–4339
12 Holmquist GP (1992) Chromosome bands, their chromatin flavors, and their functional features. *Amer J Hum Genet* 51: 17–37
13 Craig JM, Bickmore WA (1994) The distribution of CpG islands in mammalian chromosomes. *Nat Genet* 7: 376–382
14 O'Keefe RT, Henderson SC, Spector DL (1992) Dynamic organization of DNA replication in mammalian cell nuclei: Spatially and temporally defined replication of chromosome-specific α-satellite DNA sequences. *J Cell Biol* 116: 1095–1110
15 Hozak P, Hassan AB, Jackson DA, Cook PR (1993) Visualization of replication factories attached to a nucleoskeleton. *Cell* 73: 361–373
16 Hassan AB, Errington RJ, White NS, Jackson DA, Cook PR (1994) Replication and transcription sites are colocalized in human cells. *J Cell Sci* 107: 425–434
17 Hassan AB, Cook PR (1993) Visualization of replication sites in unfixed human cells. *J Cell Sci* 105: 541–550
18 Ikemura T, Wada K, Aota S (1990) Giant G+C% mosaic structures of the human genome found by arrangement of GenBank human DNA sequences according to genetic positions. *Genomics* 2: 207–216
19 Fukagawa T, Sugaya K, Matsumoto K, Okumura K, Ando A, Inoko H, Ikemura T (1995) A boundary of long-range G+C% mosaic domains in the human MHC locus: pseudoautosomal boundary-like sequence exists near the boundary. *Genomics* 25: 184–191
20 Matsumoto K, Arai M, Ishihara N, Ando A, Inoko H, Ikemura T (1992) Cluster of fibronectin type III repeats found in the human major histocompatibility complex class III region shows the highest homology with the repeats in an extracellular matrix protein, Tenascin. *Genomics* 12: 485–491
21 Sugaya K, Fukagawa T, Matsumoto K, Mita K, Takahashi E, Ando A, Inoko H, Ikemura T (1994) Three genes in the human MHC class III region near the junction with the class II gene for receptor of advanced glycosylation end products, PBX2 gene and a Notch homolog, human counterpart of mouse mammary tumor gene int-3. *Genomics* 23: 408–419
22 Sugaya K, Sasanuma S, Nohata J, Kimura T, Fukagawa T, Nakamura Y, Ando A, Inoko H, Ikemura T, Mita K (1997) Gene organization of human NOTCH4 and (CTG)n polymorphism in this human counterpart gene of mouse proto-oncogene Int3. *Gene* 189: 235–244
23 Tenzen T, Yamagata T, Fukagawa T, Sugaya K, Ando A, Inoko H, Gojobori T, Fujiyama A, Okumura K, Ikemura T (1997) Precise switching of DNA replication timing in the GC content transition area in the human major histocompatibility complex. *Mol Cell Biol* 17: 4043–4050
24 Mirkin SM, Frank-Kamenetskii MD (1994) H-DNA and related structures. *Annu Rev Biophys*

Biomol Struct 23: 541–576
25 Wells RD, Collier DA, Hanvey JC, Shimizu M, Wohlrab F (1988) The chemistry and biology of unusual DNA structures adopted by oligopurine•oligopyrimidine sequences. *FASEB J* 2: 2939–2949
26 Baran N, Lapidot A, Manor H (1991) Formation of DNA triplexes accounts for arrests of DNA synthesis at d(TC)n and d(GA)n tracts. *Proc Natl Acad Sci USA* 88: 507–511
27 Brinton BT, Caddle MS, Heintz NH (1991) Position and orientation-dependent effects of a eukaryotic Z-triplex DNA motif on episomal DNA replication in COS-7 cells. *J Biol Chem* 266: 5153–5161
28 Lee JS, Burkholder GD, Latimer LJP, Haug BL, Braun RP (1987) A monoclonal antibody to triplex DNA binds to eucaryotic chromosomes. *Nucl Acid Res* 15: 1047–1061
29 Agazie YM, Lee JS, Burkholder GD (1994) Characterization of a new monoclonal antibody to triplex DNA and immunofluorescent staining of mammalian chromosomes. *J Biol Chem* 269: 7019–7023
30 Lawrence JB, Singer RH, Marselle LM (1989) Highly localized tracts of specific transcripts within interphase nuclei visualized by *in situ* hybridization. *Cell* 57: 493–502
31 Johnson CV, Singer RH, Lawrence JB (1991) Fluorescent detection of nuclear RNA and DNA: implications for genome organization. *Meth Cell Biol* 35: 73–99
32 Dirks RW, Van de Rijke FM, Fujishita S, Van der Ploeg M, Raap AK (1993) Methodologies for specific intron and exon RNA localization in cultured cells by haptenized and fluorochromized probes. *J Cell Sci* 104: 1187–1197
33 Grady DL, Ratliff RL, Robinson DL, McCanlies EC, Meyne J, Moyzis RK (1992) Highly conserved repetitive DNA sequences are present at human centromeres. *Proc Natl Acad Sci USA* 89: 1695–1699
34 Lee C, Wevrick R, Fisher RB, Ferguson-Smith MA, Lin CC (1997) Human centromeric DNAs. *Hum Genet* 100: 291–304
35 Luderus MEE, Steensel B, Chong L, Sibon OCM, Cremers FFM, Lange T (1996) Structure, subnuclear distribution, and nuclear matrix association of the mammalian telomeric complex. *J Cell Biol* 135: 867–881
36 Zhu L, Chou S-H, Reid BR (1995) The structure of a novel DNA duplex formed by human centromere d(TGGAA) repeats with possible implications for chromosome attachment during mitosis. *J Mol Biol* 254: 623–637
37 Chou SH, Zhu L, Reid BR (1994) The unusual structure of the human centromere $(GGA)_2$ motif. *J Mol Biol* 244: 259–268
38 Chou SH, Zhu L, Reid BR (1996) On the relative ability of centromeric GNA triplets to form hairpins versus self-paired duplexes. *J Mol Biol* 259: 445–457
39 Catasti P, Gupta G, Garcia AE, Ratliff R, Hong L, Yau P, Moyzis RK, Bradbury EM (1994) Unusual structure of the tandem repetitive DNA sequences located at human centromeres. *Biochemistry* 33: 3819–3830
40 Matera AG, Ward DC (1992) Oligonucleotide probes for the analysis of specific repetitive DNA sequences by fluorescence *in situ* hybridization. *Hum Mol Genet* 1: 535–539
41 Willard HF, Waye JS (1987) Hierarchical order in chromosome-specific human alpha satellite DNA. *TIG* 3: 192–198
42 Cooney M, Czernuszewicz G, Postel EH, Flint SJ, Hogan ME (1988) Site-specific oligonucleotide binding represses transcription of the human c-myc gene *in vitro*. *Science* 241: 456–459
43 Maher LJ (1996) Prospects for the therapeutic use of antigene oligonucleotides. *Cancer Invest* 14: 66–82
44 Moser HE, Dervan PB (1987) Sequence-specific cleavage of double helical DNA by triple helix formation. *Science* 238: 645–650
45 Wilson WD, Tanious FA, MIzan S, Yao S, Kiselyov AS, Zon G, Strekowski L (1993) DNA triple-helix specific intercalators as antigene enhancers: unfused aromatic cations. *Biochemistry* 32: 10614–10621
46 Noonberg SB, Scott GK, Garovoy MR, Benz CC, Hunt CA (1994) *In vivo* generation of highly abundant sequence-specific oligonucleotides for antisense and triplex gene regulation. *Nucl Acid Res* 22: 2830–2836
47 Xodo L, Alunni-Fabbroni M, Manzini G, Quadrifoglio F (1994) Pyrimidine phosphorothioate oligonucleotides form triple-stranded helices and promote transcription inhibition. *Nucl Acid Res* 22: 3322–3330

48 Kandimalla ER, Agrawal S (1995) Single strand targeted triplex-formation. Destabilization of guanine quadruplex structures by foldback triplex-forming oligonucleotides. *Nucl Acid Res* 23: 1068–1074
49 Giovannangeli C, Diviacco S, Labrousse V, Gryaznov S, Charneau P, Helene C (1997) Accessibility of nuclear DNA to triplex-forming oligonucleotides: the integrated HIV-1 provirus as a target. *Proc Natl Acad Sci USA* 94: 79–84
50 Burgess TL, Fisher EF, Ross SL, Bready JV, Qian YX, Bayewitch LA, Cohen AM, Herrera CJ, Hu SS-F, Kramer TB et al (1995) The antiproliferative activity of c-myb and c-myc antisense oligonucleotides in smooth muscle cells is caused by a nonantisense mechanism. *Proc Natl Acad Sci USA* 92: 4051–4055
51 Stein CA (1995) Does antisense exist? *Nat Med* 1: 1119–1121
52 Bishop JS, Guy-Caffey JK, Ojwang JO, Smith SR, Hogan ME, Cossum PA, Rando RF, Chaudhary N (1996) Intramolecular G-quartet motifs confer nuclease resistance to a potent anti-HIV oligonucleotide. *J Biol Chem* 271: 5698–5703
53 Barch MJ (ed) (1991) *The ACT cytogenetics laboratory manual, second edition.* Raven Press Ltd, New York
54 Wiegant J, Kalle W, Mullenders L, Brookes S, Hoovers JMN, Dauwerse JG, Van Ommen GJB, Raap AK (1992) High-resolution *in situ* hybridization using DNA halo preparations. *Hum Mol Genet* 1: 587–591
55 Michel D, Chatelain G, Herault Y, Harper F, Brun G (1993) H-DNA can act as a transcriptional insulator. *Cell Mol Biol Res* 39: 131–140

Part two
Chromosome function and behaviour

R. Bruce Nicklas

Maria J. Puertas

Wolf Reik

Chromosomes Today, Vol. 13
ed. by E. Olmo and C.A. Redi

Chromosomes in mitosis: chance and checkpoint

R. Bruce Nicklas

Departments of Zoology and Cell Biology, Duke University, Durham, NC 27708, USA

Errors in chromosome distribution in meiosis and mitosis result in cells with missing or extra chromosomes, which can cause birth defects in children and cancer in adults. Chromosome distribution is determined by chromosome attachment to the mitotic spindle (reviewed in [1]). Proper attachment of daughter chromosomes to opposite spindle poles leads to movement to opposite poles in anaphase, with the fortunate consequence that each daughter cell contains one copy of each chromosome. Attachment begins with chance events, in which growing microtubules accidentally encounter a kinetochore, the microtubule attachment site on each chromosome (Fig. 1). Reliance on

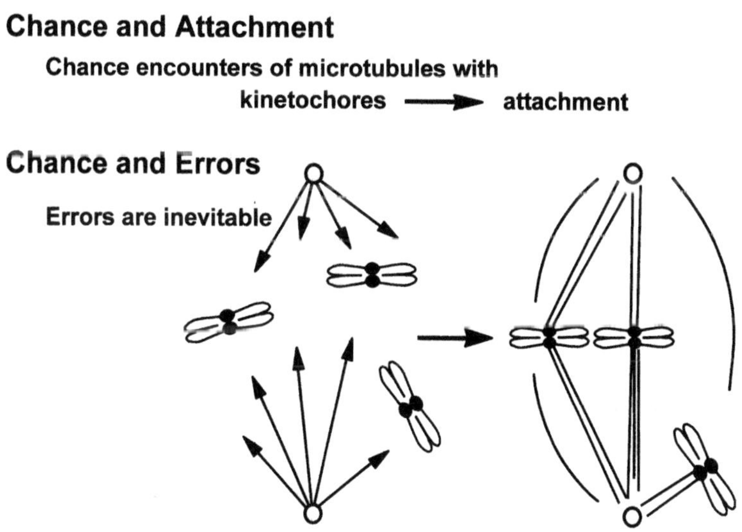

Figure 1. Chromosome attachment to the spindle depends on chance encounters. The diagram on the left shows long protein filaments, microtubules, depicted as arrows. Microtubules grow outward from each spindle pole (circles). If they happen, by chance, to encounter a kinetochore (black ovals) they may be captured, and the chromosome is attached to the spindle. The diagram on the right shows the consequence: two chromosomes on the left are properly attached, with one sister kinetochore attached to one pole and the other to the opposite pole, while the chromosome on the right is improperly attached, since one sister kinetochore is unattached. The errant kinetochore just happened to lie so that it faced away from both poles, in a position that growing microtubules from either pole would only rarely contact.

chance makes errors in initial attachment inevitable (Fig. 1). A quality-control checkpoint allows the correction of most errors (Fig. 2) [2]. The checkpoint detects the presence of an unattached or improperly attached chromosome and delays the onset of anaphase, allowing time for error correction. In the example shown in Figure 2, the delay allows additional time for the capture of microtubules by a kinetochore that lies in an unfavorable position (microtubules growing from the upper pole in Figure 2 will rarely grow at the right angle to touch the unattached kinetochore). Thus, the checkpoint is a cellular adaptation to chance, a matter of buying enough time that rare events are likely to occur by chance. The checkpoint works by inhibiting the machine that drives the exit from cell division, the anaphase-promoting complex.

The fundamental problem for the cell is to convert a mechanical state, the presence or absence of a mechanical attachment to the spindle, into a chemical signal to the anaphase-promoting complex (Fig. 2). What is it about a missing attachment that is detected? The answer, as first proposed by McIntosh [3], is the absence of tension from spindle forces, or some consequence of the absence of tension. We showed this by micromanipulating insect spermatocytes, using a fine glass needle to pull on chromosomes [4]. By pulling, we imposed tension on an improperly attached chromosome. This mimics the ten-

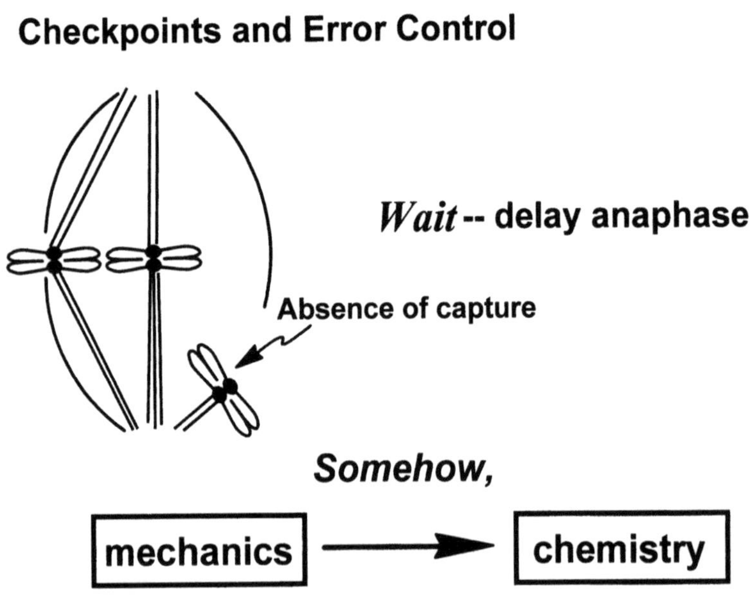

Figure 2. A checkpoint reduces errors in mitosis. The spindle checkpoint detects the unattached kinetochore indicated by the arrow and acts to delay the start of anaphase, allowing more time for attachment to occur. The checkpoint must somehow convert a mechanical state, the absence of a mechanical connection to the spindle, into a chemical signal that halts the cell cycle machine that drives the entry into anaphase – the anaphase-promoting complex.

sion produced by the normal spindle forces when chromosomes are properly attached to opposite poles and experience forces toward opposite poles. The tension we imposed on one chromosome caused the cell to enter anaphase hours ahead of control cells with improperly attached chromosomes. Pulling on a chromosome caused the cell to enter anaphase even when this resulted in daughter cells with unbalanced numbers of chromosomes. But how does tension result in a chemical signal? The probable answer is that tension changes kinetochore chemistry [5]. Certain kinetochore proteins are phosphorylated before chromosomes become attached to the spindle and they become dephosphorylated when chromosomes are properly attached and the kinetochores are under tension from mitotic forces. Tension from a micromanipulation needle causes the dephosphorylation of improperly attached kinetochores [5]. Rich correlations, but not direct proof, connect kinetochore phosphorylation with the "wait" signal from the checkpoint (reviewed in [1, 6]). We do not know the mechanism of tension-sensitive phosphorylation. It may be that tension affects phosphorylation directly, for instance, by deforming and inactivating a protein kinase. Alternatively, tension may have an indirect effect – tension may lead to an increase in the number of kinetochore microtubules, which could inactivate the phosphorylation system [7, 8].

Whatever the mechanism of tension-sensitive phosphorylation may be, phosphorylation very likely is how the checkpoint detects an errant chromosome [1, 6]. We also know how the checkpoint inhibits the onset of anaphase. Central players are Mad2, a highly conserved checkpoint component first discovered in yeast [9], and Cdc20 (p55/fizzy), a protein that normally activates the anaphase-promoting complex [10, 11]. When the checkpoint is aroused by an errant chromosome, Mad2 forms a complex with Cdc20 that prevents activation of the anaphase-promoting complex [11–16]. Mad2 shows a striking localization that correlates with the attachment status of kinetochores: it is found at kinetochores that are not attached to the spindle but it is absent from kinetochores after they become properly attached to the spindle and accumulate microtubules [17–19]. Cdc20 is also found at the kinetochores of unattached chromosomes [11], suggesting that the Mad2-Cdc20 complex that delays the onset of anaphase may be formed at unattached kinetochores, the site of the error.

Checkpoint operation involves two steps, *detection* followed by *action* (Fig. 3). We understand, at least in outline, how errant attachments are detected and how the checkpoint acts to delay the start of anaphase. But how are detection and inhibition linked together? In particular, how is the formation of the Mad2/Cdc20 complex linked to tension-sensitive kinetochore phosphorylation? An unexpectedly easy experiment provides a likely answer. Dephosphorylated kinetochores in lysed, washed cells retain a kinase at the kinetochore, and the kinetochores can be rephosphorylated simply by incubating the preparation in ATP plus a phosphatase inhibitor [20, 21]. Recombinant Mad2 protein can be added to preparations containing either dephosphorylat-

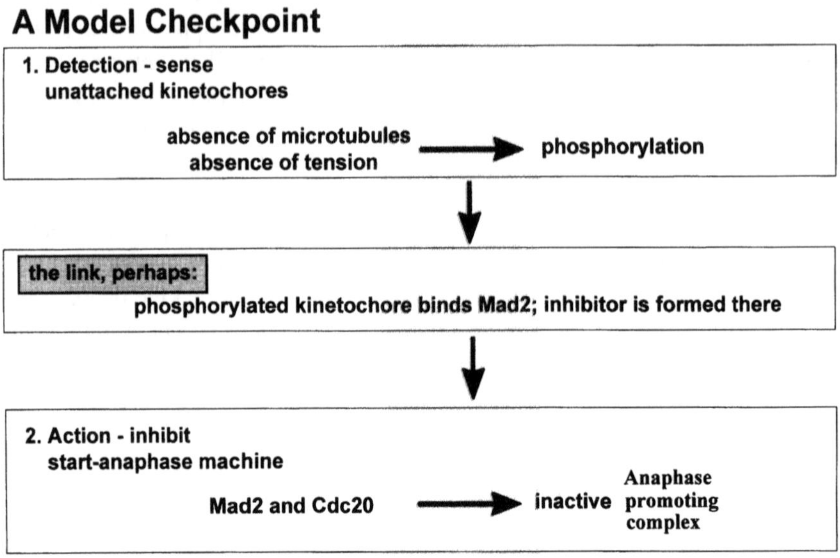

Figure 3. Outline of a model for the spindle checkpoint. The checkpoint detects unattached kinetochores by tension-sensitive phosphorylation. It then acts to delay anaphase by the formation of a complex between Mad2 and Cdc20 that prevents activation of the anaphase-promoting complex. The binding of Mad2 to phosphorylated kinetochores may provide the link between detection and action – the link that connects the start of anaphase to the attachment status of all the kinetochores in the cell.

ed or phosphorylated kinetochores. We found that Mad2 bound to phosphorylated but not to unphosphorylated kinetochores [22]. This observation provides a plausible link between error detection and checkpoint action inhibition, giving a complete pathway for error control in mitosis (Fig. 3).

Together, these observations suggest a model for the checkpoint [22]. Unattached kinetochores are phosphorylated, which promotes Mad2 binding. Mad2 at the kinetochores forms a complex with Cdc20 that prevents activation of the anaphase-promoting complex [11–16]. The inhibitory complex is continually being formed at and released from unattached kinetochores. As an unattached kinetochore becomes attached to the spindle, kinetochore microtubules accumulate, which promotes the loss of Mad2 [19]. Proper attachment to opposite spindle poles leads to tension and to kinetochore dephosphorylation [5], which reduces the binding of new Mad2 by the kinetochore. The Mad2-Cdc20 complex is no longer being formed and its concentration begins to fall; at some point the anaphase-promoting complex is no longer being inhibited, so the cell progresses to anaphase and divides.

This model provides a first view of checkpoint operation. It satisfies the cardinal principle of the spindle checkpoint: the production of the inhibitor that delays anaphase is linked to the attachment status of the chromosomes.

Anaphase onset is delayed when, and only when, one or more unattached chromosomes are present in the cell. Even if correct in principle, the model is obviously incomplete [22]. Among the unanswered questions are the roles of other known players, such as the other Mads, particularly Mad1 [23], the budding unhibited by benzimidazole (Bubs) [24, 25] and mitogen-activated protein (MAP) kinase [26].

In conclusion, chromosome distribution in cell division depends upon chance – chance events whose outcome is monitored by a quality-control checkpoint. We and our children depend mightily on the proper operation of that half-understood checkpoint. Much still remains to be learned but we are on the way.

Acknowledgments

I thank my collaborators, who make this endeavor a pleasure – Mike Campbell, Rey Chen, Gary Gorbsky, Xiaotong Li, Andrew Murray, Ted Salmon, Suzanne Ward and Jennifer Waters. Work in my group is supported in part by grant GM-13745 from the Institute of General Medical Sciences, National Institutes of Health.

References

1 Nicklas RB (1997) How cells get the right chromosomes. *Science* 275: 632–637
2 Hartwell LH, Weinert TA (1989) Checkpoints: controls that ensure the order of cell cycle events. *Science* 246: 629–634
3 McIntosh JR (1991) Structural and mechanical control of mitotic progression. *Cold Spring Harbour Symp Quant Biol* 56: 613–619
4 Li X, Nicklas RB (1995) Mitotic forces control a cell-cycle checkpoint. *Nature* 373: 630–632
5 Nicklas RB, Ward SC, Gorbsky GJ (1995) Kinetochore chemistry is sensitive to tension and may link mitotic forces to a cell cycle checkpoint. *J Cell Biol* 130: 929–939
6 Gorbsky GJ (1997) Cell cycle checkpoints: arresting progress in mitosis. *BioEssays* 19: 193–197
7 Rieder CL, Cole RW, Khodjakov A, Sluder G (1995) The checkpoint delaying anaphase in response to chromosome monoorientation is mediated by an inhibitory signal produced by unattached kinetochores. *J Cell Biol* 130: 941–948
8 Wells WAE (1996) The spindle-assembly checkpoint: aiming for a perfect mitosis, every time. *Trends Cell Biol* 6: 228–234
9 Li R, Murray AW (1991) Feedback control of mitosis in budding yeast. *Cell* 66: 519–531
10 Fang G, Yu H, Kirschner MW (1998) Direct binding of CDC20 protein family members activates the anaphase-promoting complex in mitosis and G1. *Mol Cell* 2: 163–171
11 Kallio M, Weinstein J, Daum JR, Burke DJ, Gorbsky GJ (1998) Mammalian p55CDC mediates association of the spindle checkpoint protein Mad2 with the cyclosome/anaphase-promoting complex, and is involved in regulating anaphase onset and late mitotic events. *J Cell Biol* 141: 1393–1406
12 Hwang LH, Lau LF, Smith DL, Mistrot CA, Hardwick KG, Hwang ES, Amon A, Murray AW (1998) Budding yeast Cdc20: a target of the spindle checkpoint. *Science* 279: 1041–1044
13 Kim SH, Lin DP, Matsumoto S, Kitazono A, Matsumoto T (1998) Fission yeast Slp1: an effector of the Mad2-dependent spindle checkpoint. *Science* 279: 1045–1047
14 Fang G, Yu H, Kirschner MW (1998) The checkpoint protein MAD2 and the mitotic regulator CDC20 form a ternary complex with the anaphase-promoting complex to control anaphase initiation. *Gene Develop* 12: 1871–1883
15 Li Y, Gorbea C, Mahaffey D, Rechsteiner M, Benezra R (1997) MAD2 associates with the cyclo-

some/anaphase-promoting complex and inhibits its activity. *Proc Natl Acad Sci USA* 94: 12431–12436
16 Wassmann K, Benezra R (1998) Mad2 transiently associates with an APC/p55Cdc complex during mitosis. *Proc Natl Acad Sci USA* 95: 11193–11198
17 Chen RH, Waters JC, Salmon ED, Murray AW (1996) Association of spindle assembly checkpoint component XMAD2 with unattached kinetochores. *Science* 274: 242–246
18 Li Y, Benezra R (1996) Identification of a human mitotic checkpoint gene: hsMAD2. *Science* 274: 246–248
19 Waters JC, Chen R-H, Murray AW, Salmon ED (1998) Localization of Mad2 to kinetochores depends on microtubule attachment, not tension. *J Cell Biol* 141: 1181–1191
20 Bousbaa H, Correia L, Gorbsky GJ, Sunkel CE (1997) Mitotic phosphoepitopes are expressed in Kc cells, neuroblasts and isolated chromosomes of *Drosophila melanogaster*. *J Cell Sci* 110: 1979–1988
21 Shapiro PS, Vaisberg E, Hunt AJ, Tolwinski NS, Whalen AM, McIntosh JR, Ahn NG (1998) Activation of the MKK/ERK pathway during somatic cell mitosis: direct interactions of active ERK with kinetochores and regulation of the mitotic 3F3/2 phosphoantigen. *J Cell Biol* 142: 1533–1545
22 Waters JC, Chen R-H, Murray AW, Gorbsky GJ, Salmon ED, Nicklas RB (1999) Phosphorylated kinetochores bind Mad2 *in vitro*: A link between error detection and checkpoint action in mitosis. *Curr Biol* 9: 649–652
23 Chen RH, Shevchenko A, Mann M, Murray AW (1998) Spindle checkpoint protein xmad1 recruits xmad2 to unattached kinetochores. *J Cell Biol* 143: 283–295
24 Taylor SS, Ha E, McKeon F (1998) The human homologue of Bub3 is required for kinetochore localization of Bub1 and a Mad3/Bub1-related protein kinase. *J Cell Biol* 142: 1–11
25 Farr KA, Hoyt MA (1998) Bub1p kinase activates the *Saccharomyces cerevisiae* spindle assembly checkpoint. *Mol Cell Biol* 18: 2738–2747
26 Zecevic M, Catling AD, Eblen ST, Renzi L, Hittle JC, Yen TJ, Gorbsky GJ, Weber MJ (1998) Active MAP kinase in mitosis: localization at kinetochores and association with the motor protein CENP-E. *J Cell Biol* 142: 1547–1558

Chromosomes Today, Vol. 13
ed. by E. Olmo and C.A. Redi

Genetic control of B chromosome transmission in maize and rye

María J. Puertas[1], Guillermo Jiménez [1], Silvia Manzanero[1], A. Mauricio Chiavarino[2,3], Marcela Rosato[2,3], Carlos A. Naranjo[3], Lidia Poggio[2,3]

[1] *Departamento de Genética, Facultad de Biología, Universidad Complutense, 28040 Madrid, Spain*
[2] *Departamento de Biología, Facultad de Ciencias Exactas y Naturales, Universidad de Buenos Aires, 1428 Buenos Aires, Argentina*
[3] *Instituto Fitotécnico Santa Catalina (FCAF, UNLP)-Centro de Investigaciones Genéticas (UNLP-CONICET-CIC), C.C.4 1836 Llavallol, Buenos Aires, Argentina*

The B chromosome system in rye and maize

B chromosomes (Bs) are dispensable genetic elements which lack homology with any members of the basic set and have non-Mendelian modes of inheritance. They are widespread in many animal and plant species and our main interest is to understand the genetic mechanisms underlying their polymorphisms.

Our experiments to investigate the genetic control of B chromosome transmission in maize and rye were basically the same. It is interesting to compare both species because we can see whether the Bs have a common strategy for establishing themselves as a particular part of the genome, or if the evolutionary story of the Bs is different in each species.

In rye and maize the number of Bs varies from plant to plant but it is constant in all somatic cells within a plant. We carried out fluorescence *in situ* hybridization (FISH) with the D1100 rye B specific probe [1, 2] in sections of various tissues, both in seedlings and mature plants, and all nuclei showed a similar signal corresponding to the Bs. It seems that the Bs are repressed in the somatic tissues because there is no significant effect of rye Bs on the phenotype at the somatic level [3, 4]. The stability of maize Bs in the sporophyte was documented following a B-A translocation [5].

Rye Bs display a complicated behaviour during male meiosis and gametogenesis. At pachytene they form bivalents or multivalents [6]. At metaphase I, they can form univalents, bivalents or multivalents; the univalents may be lost as micronuclei. This instability creates a permanent variation in B number which results in a polymorphism at the population level. The maize Bs are more stable than the Bs of rye. When the plant has only one B chromosome it is frequently included in one pole at anaphase I and so the loss of the B univalent is prevented. When there are more Bs, they usually pair and the microspores show few Bs lost as micronuclei.

A main event occurs during postmeiotic mitosis. In rye, at anaphase of the first pollen grain mitosis, the B chromosome undergoes non-disjunction and migrates to the generative pole from which the gametes will form. The second mitosis is normal and both sperm nuclei carry two Bs. In this way the Bs are accumulated and driven to the next generation. In maize, non-disjunction typically occurs during the second pollen mitosis. If the microspore begins with a single B chromosome, non-disjunction will result in one sperm nucleus possessing two B chromosomes and the other containing no Bs. Few data are available on the behaviour of the Bs on the female side, but the progeny of the appropriate crosses show that the Bs of rye have a mechanism of accumulation; actually, rye is the only known plant species where the Bs are accumulated on both sides. In maize the inheritance through the female parent is conventional (reviews in [7, 8]).

All these peculiar events do not occur in all cells, but there is a variation within and between plants. Figure 1 summarizes B chromosome behaviour outlining the moments where variation of the processes may result in variation of B transmission rate. The main purpose of our work is to determine the genetic control of these sources of variation. This is important to understand the evolution of B polymorphisms.

The first source of variation is the proportion of bivalents to univalents at metaphase I. In rye, the loss of univalents is the most important event in determining B transmission rate. Univalent conservation depends on their ability to divide reductionally since all equationally dividing univalents will be lost as micronuclei. At male meiosis the maize Bs are more stable, showing normal meiosis in general terms.

The second source of variation is the proportion of non-disjunction. In rye, directed non-disjunction to the generative pole is almost fixed [3]. In maize, non-disjunction frequency varies among lines [8]. In a line carrying a B/A translocation, it has been shown using a B specific probe [9, 10], that it occurs at about 60% of the pollen grains and it can occasionally also occur during the first microspore division. However, in the native maize population studied in this paper, non-disjunction is almost fixed. Non-disjunction would not accumulate the Bs if it were not coupled with preferential distribution of the Bs towards the nuclei forming or acting as gametes. In rye the non-disjoined B chromosome is included in the generative nucleus which will divide once more giving rise to the sperm nuclei. We have never observed the reverse distribution, so it seems that this process does not vary.

In maize, the two sperm nuclei differ in B chromosome content, in this situation transmission rate depends on the frequency with which the sperm containing B chromosomes fertilizes the egg. We will see below that this is the main source of variation in determining B chromosome transmission rate in this species.

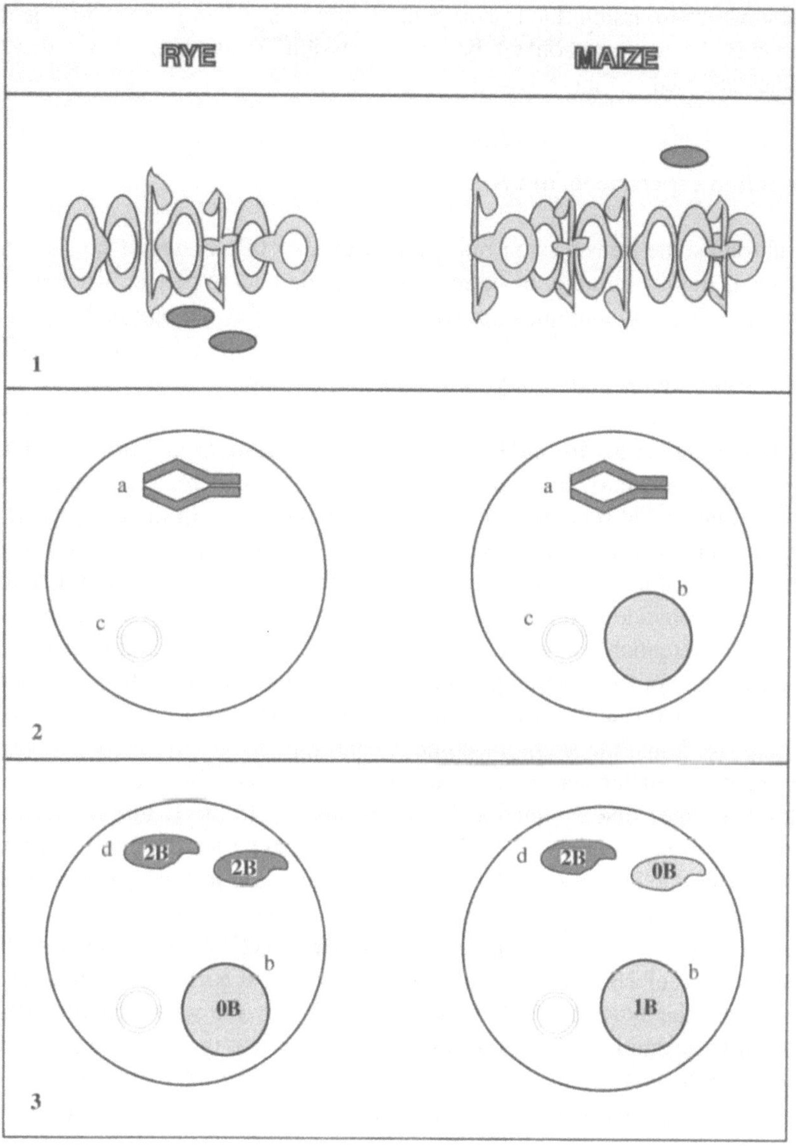

Figure 1. Sources of variation affecting B transmission rate (B TR) in rye and maize.
1. At first meiotic division B univalents may not be included in the poles. In rye the proportion of univalents/bivalents at metaphase I is the main factor determining B TR.
2. The B standard chromosome undergoes non-disjunction at pollen grain mitosis in nearly 100% of the cases. In rye, directed non-disjunction towards the generative pole occurs at the 1st mitosis. In maize non-disjunction occurs at the 2nd mitosis. a = B chromosome undergoing non-disjunction; b = tube nucleus; c = pollen grain pore
3. In rye the sperm nuclei (d) coming from the generative pole contain a doubled B number. In maize the two sperm nuclei differ in B content and B TR depends on the frequency with which the sperm containing B chromosomes fertilizes the egg.

It should be noted that there are other sources of variation affecting B chromosome transmission due to competition between pollen grains carrying different B chromosome content, but our discussion in this paper is limited to the stages outlined in Figure 1.

Selection experiments in rye

B chromosome transmission rate (B TR) in 2B × 0B crosses of non-selected plants is highly variable. In some crosses the Bs tend to be lost, whereas others show B accumulation. In previous papers [11, 12] we reported the selection of genotypes which significantly differ in their high (H) or low (L) B TR, demonstrating the heritability of the H or L character.

We have ample evidence that the difference between the lines is not related to a differential non-disjunction ability, firstly because the proportion of plants with odd numbers of Bs is always low; that is, the difference between the H and L lines is the proportion of 0B *vs.* 2B or 4B plants in the progeny of all types of crosses. Secondly, because we have observed pollen mitosis directly, and in both lines the frequency of Bs undergoing non-disjunction at pollen anaphase coincides with the frequency of Bs present at pollen metaphase.

The differential behaviour of B chromosomes at meiosis and gametogenesis accounts for the differences in B TR in the H and L lines. 2B plants of the L line form bivalents in only 20% of the pollen mother cells (PMCs) at metaphase I, and the B chromosome is conserved in 45% of the pollen grains at first pollen metaphase, whereas the remaining Bs are eliminated as micronuclei when they divide equationally at anaphase I. In the H line the Bs form bivalents in nearly 90% of the PMCs at metaphase I and they are present in 85% of the pollen grains. The different transmission in the two lines results from their differential ability to form uni- or bivalents at metaphase I, which determines their loss or conservation in the gametes (Fig. 2) [13]. On the other hand, the Bs of 2B plants regularly form a bivalent at pachytene in all cases as shown by the synaptonemal complexes studied by the whole mount surface spreading technique. This allows us to conclude that the difference between the Bs of the two lines is the ability to have bound arms at metaphase I.

We made 2B × 0B crosses between rye plants of both lines to determine the location of the genes controlling B chromosome transmission rate. Our results show that the transmission of the Bs depends on the H or L source of the Bs whereas the 0B parent has no significant effect on the progeny. B TR is low when the Bs come from the L line, and high when the Bs come from the H line, irrespective of the 0B parent. This demonstrates that the genetic control of B TR is located on the Bs [14]. Statistical analyses of the observed revealed the significance of other factors during transmission, suggesting that the A chromosomes also affect B transmission, but to a relatively small extent.

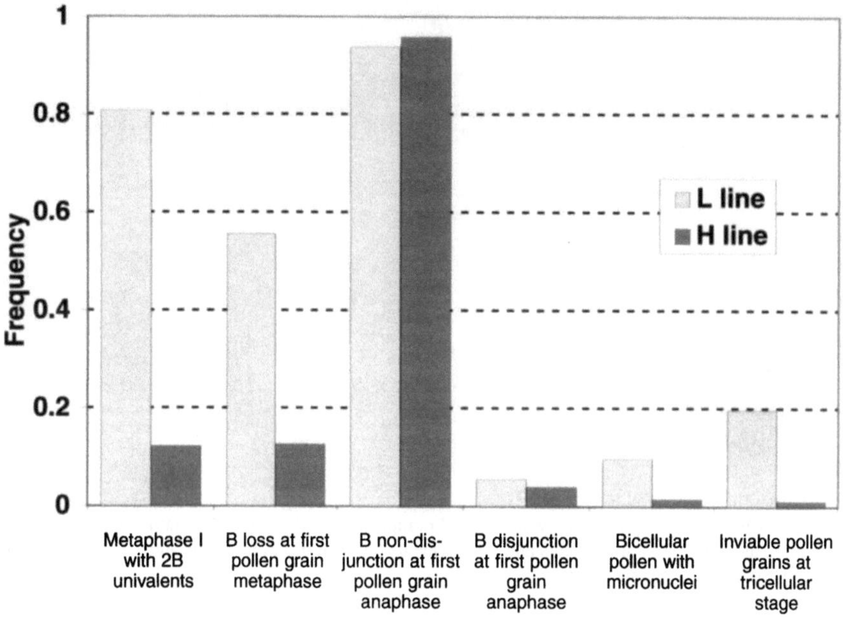

Figure 2. Differences in 2B chromosome behaviour in the L and H B transmission rate rye lines in relevant stages of male meiosis and gametogenesis. The main difference between the two lines is the frequency of cells showing two B univalents at metaphase I. Many of these univalents are lost as micronuclei which are degraded during pollen maturation. The frequency of non-disjunction in the pollen grains with the B is nearly the same in both lines. The frequency of inviable pollen grains is higher in the L line.

According to all our data, the simplest assumption is that the genes for high and low transmission rate selected in our lines actually are sites for chiasma formation, or at least B homologous arm binding sites, located on the B chromosomes.

It is interesting to know the behaviour of the H and L Bs under heterozygous conditions, to study their interaction when they are together in the same cell. In rye, it is not possible to obtain 2B plants heterozygous for the Bs because the two Bs always come from non-disjunction. Therefore, we made 2B × 2B crosses to obtain 4B plants in all possible combinations: HxH, LxL where the four Bs are of the same type, and HxL and LxH where there is heterozygosity for the Bs and crosses to obtain 4B plants were reciprocal.

We studied the behaviour of the four classes at pachytene and metaphase I. At pachytene, the four Bs may form either two bivalents or one quadrivalent (Fig. 3); the frequency of each type is shown in Table 1. In contrast to the 2B plants, where the Bs always form a bivalent at pachytene irrespective of their H or L status, 4B plants differ in synapsis behaviour. According to Sved's model [15], when four homologues are present, one quadrivalent is expected

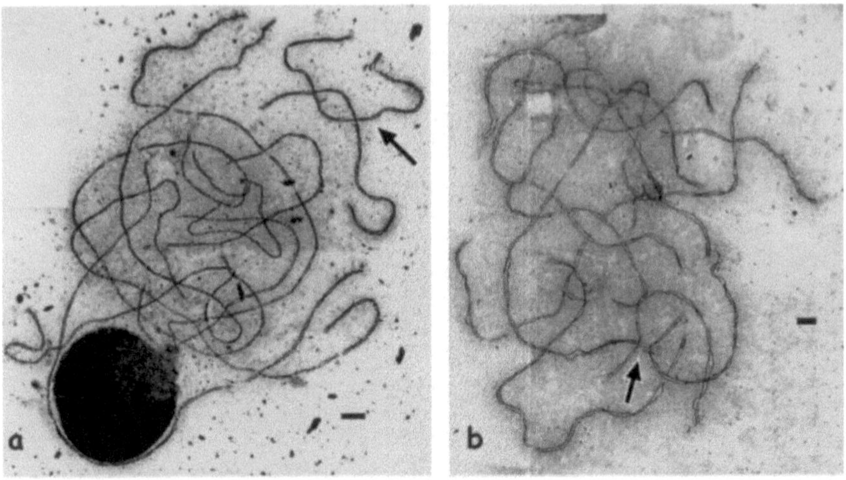

Figure 3. Electron micrographs of synaptonemal complexes in 4B rye nuclei obtained by the surface spreading technique. a. 7 A bivalents and 2 B bivalents (arrow); b. 7 A bivalents and 1 B quadrivalent (arrow). Scale bar represents 2 μm.

Table 1. Meiotic configurations of 4B chromosomes at pachytene (Pch) and metaphase I (MI)

Type of cross		2 II	1 IV	4 I	1 II + 2 I	1 III + 1 I
2BH × 2BH	Pch	21	9			
		(70.00)	*(30.00)*			
	MI	137	6	10	66	0
		(62.55)	*(2.7)*	*(4.56)*	*(30.11)*	
2BL × 2BL	Pch	17	21			
		(44.73)	*(55.26)*			
	MI	60	11	69	149	6
		(20.34)	*(3.73)*	*(23.39)*	*(50.51)*	*(2.03)*
2BH × 2BL	Pch	17	14			
		(54.84)	*(45.16)*			
	MI	167	17	50	130	4
		(45.38)	*(4.62)*	*(13.59)*	*(35.33)*	*(1.08)*
2BL × 2BH	Pch	13	14			
		(48.15)	*(51.85)*			
	MI	57	2	22	129	5
		(26.51)	*(0.93)*	*(10.23)*	*(60.00)*	*(2.33)*

in 2/3 of the cells and two bivalents in the other 1/3. This prediction is based on the assumption that there are two independent initiation sites at the telomeres and that the beginning of synapsis is at random. In our case, it was found that both LL and LH 4B plants fitted the 2/3, 1/3 distribution, indicating random pairing. However, HL and especially HH, significantly differed since the number of bivalents was higher than expected. It seems that these two lines have restricted pairing towards the formation of bivalents. We do not know how such restriction takes place, but we can hypothesize that there is a maternal effect since H B chromosomes coming from a H female parent form more bivalents than the other types.

The results at metaphase I (Tab. 1) show, firstly, that the proportion of cells with 2 II is correlated with those at pachytene showing also 2II, indicating that synapsis at pachytene influences further associations of the Bs at metaphase I in 4B plants. Metaphase I data of the HL and LH classes are of special interest because they are significantly different. This indicates a lack of complementation in the hybrid between the H and L Bs at this level. This result is in agreement with the hypothesis that chiasma formation depends on the B chromosome itself, because if a product produced by the normal A chromosomes determined chiasma formation in the Bs, the HL and LH would have a similar behaviour. On the other hand, since LH has the same distribution of bivalents to non-bivalents as LL, and HL is closer to HH, a maternal effect on chiasma formation seems evident.

We also studied the correlation between B transmission rate and meiotic configurations at metaphase I. The only significant correlation ($r = 0.993$, $P = 0.003$) is between the frequency of PMCs with 2II and transmission rate. Therefore, our final conclusion is that both in 2B or 4B plants the main feature which determines the B transmission rate is the ability to form bivalents at metaphase I.

Selection experiments in maize

Similar experiments on selection for the character H or L B transmission rate were carried out in maize. The parental population was the native race Pisingallo from northwest Argentina, with a frequency of plants with Bs of 44% [16].

To study B TR on the male side, twenty 0B × 1B crosses were made. They produced progeny with either 0B or 2B due to non-disjunction, which occurred regularly in nearly all cases. The variation of B TR was large, from 0.17 to 0.98, the mean being Mendelian (0.5). The progenies showing the highest and nearly the lowest B TR were selected (Fig. 4). In the H line the mean B TR was 0.65 with a proportion of 2B plants significantly higher than the Mendelian expectation. The mean B TR was higher than 0.5 in all crosses. In the L line the mean B TR was 0.4 with a proportion of 2B plants and a B TR significantly lower than the Mendelian expectation. B TR was significantly different in the L and H lines, indicating a selection gain and therefore a genetic component of the variation [17]. Selection was repeated in four generations. The results were similar for the H line, whereas in the L line B TR resulted in a Mendelian distribution.

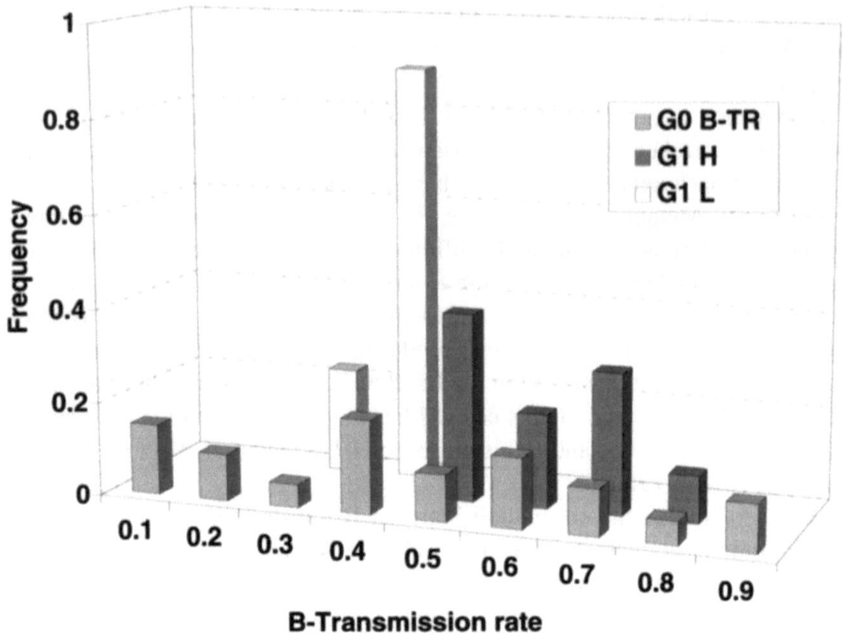

Figure 4. Male B transmission rate (B TR) distribution in 0B × 1B crosses of the native race Pisingallo of maize (G0) and in 0B × 2B crosses (G1) selected for high (H) or low (L) male B TR.

The same type of crosses were made to select the Bs transmitted on the female side [17]. In this case, the distribution of B TR in twenty 1B × 0B crosses in the non-selected population was less variable (0.3 to 0.6) than in the former reciprocal cross 0B × 1B. The average mean was 0.47, again Mendelian (Fig. 5). In these crosses the progeny had either 1B or 0B due to the lack of a mechanism of accumulation on the female side. In this case, significant selection gain was obtained only in the line selected for low B TR, with a mean of 0.4 whereas in the line selected for high B TR the mean remained Mendelian. Similar results were found in further generations of selection. It is possible that the B univalent tends to be lost or to be preferentially directed to the non-functional megaspore, or both.

We studied male meiosis in the selected maize plants to see if there is a correlation between meiotic behaviour and B TR as occurs in rye. In 2B plants, the Bs formed bivalents in about 90% of the metaphase I cells, irrespective of their H or L status and the generation of selection studied. On the other hand, the number of micronuclei at the tetrad stage was lower than 5% in all cases. It was concluded that the differential transmission of the Bs on the male side is not due to a differential B loss at meiosis [18]. Therefore, rye and maize are different in this respect.

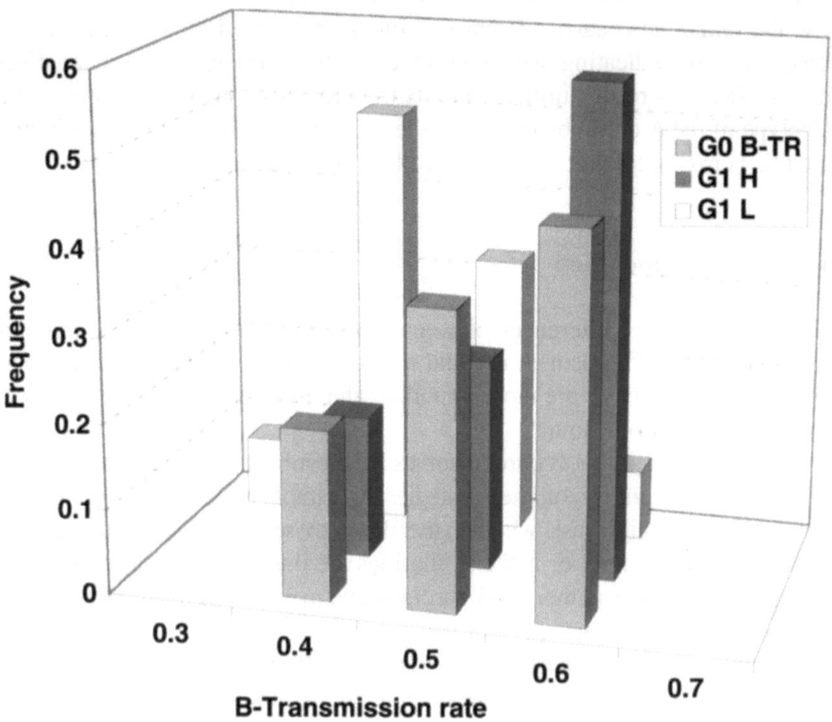

Figure 5. Female B transmission rate (B TR) distribution in 1B × 0B crosses of the native race Pisingallo of maize (G0) and in 1B × 0B crosses (G1) selected for high (H) or low (L) male B TR.

0B × 2B crosses were made between and within the H and L lines in the third generation of selection to determine the location of genes controlling male B TR in maize [19]. We observed, firstly, that during the three generations of selection the H line transmitted the Bs with a rate of about 0.7 whereas the L line transmits the Bs at about a Mendelian rate. Secondly, that B transmission rate depends on the 0B female parent, because the transmission rate is low when the 0B parent is L, and high when the 0B female parent is H, irrespective of the Bs. Therefore, the genes controlling B TR are located on the A chromosomes.

Bearing in mind that the sperm nuclei carry 0B or 2B by means of non-disjunction, our results indicate that the genotype of the 0B female determines the rate at which the egg cell is fertilized by the sperm nucleus carrying the Bs. This means that a 0B female parent of the H genotype selects the 2B sperm to fertilize the egg; in contrast, the 0B L female plant does not select any sperm nucleus because fertilization is random in this case, resulting in a Mendelian B TR.

In the next generation of selection we made 2B × 2B crosses, that is, in this case the female also carries B chromosomes. B TR in these crosses was the same as before, indicating that the presence of Bs in the egg cell does not affect male B TR. This result further supports the previous conclusion that it is the genotype of the A chromosomes in the female which controls male B TR, and that this is irrespective of the presence of Bs in the egg cell.

B chromosome evolution

In our opinion, the differences between the genetic systems for the maintenance of B polymorphism in rye and maize are only secondarily important, whereas the similarities are more important and may help us to understand the evolution of B chromosomes.

We should accept that B chromosomes have evolved separately in both species, but the similarity of the strategies to establish the polymorphism is remarkable. We might ask whether the Bs of these two species have a very ancient common origin, or if the similarities are due to an obligated common strategy for any supernumerary B chromosome to establish itself in the genetic system of the gramineae.

It is difficult to hypothesize on the origin of Bs in maize and rye because both are cultivated plants and therefore human migration has had a great influence on their distribution. However, it seems that the origin of Bs is ancient in both cases because the distribution of populations with Bs is widespread and they are present in native races. In addition, the mechanism of non-disjunction at pollen mitosis is common to all related species [7]. It is necessary to perform a molecular study to compare B DNA sequences in these species, although the answer will not be easy due to the high mutability of B chromosomes because of their negligible effects on the phenotype.

It has to be emphasized that the regularity of meiotic segregation is strongly related to B transmission rate. In the varieties of maize reported to date, meiosis is mostly regular and the Bs at male meiosis are conserved even if they are present as univalents. The same is true for the H line of rye, whereas the L line tends to lose the Bs because they are not bound at metaphase I.

The importance of a regular meiosis is further supported because structural variants of the Bs undergoing non-disjunction, but having irregular meiosis, soon disappear after they are formed *de novo,* and only the standard type is maintained at the population level. We can conclude that regular meiotic behaviour is essential for a B chromosome to be maintained in the long term.

The classical hypothesis for parasitic Bs assumed that B chromosome polymorphisms are stable due to antagonistic forces, the equilibrium being the result of B accumulation, and the detrimental effects on the fitness of B carriers [20–22]. A more recent non-equilibrium hypothesis of B chromosome evolution has been proposed by Camacho et al. [23–25]. They propose that the long term evolution of the Bs can be considered as the outcome of an arms race between the host genome that attempts to eliminate the Bs or suppress their parasitic effects and the Bs themselves that try to escape through generating new selfish variants. This hypothesis views B chromosome evolution as a continuous conflict between parts of the genome with different interests so that B influences may shift back and forth between parasitic to, neutral and possibly beneficial effects.

Our results fit this new hypothesis only partially. We have found H and L B variants in rye, but the variant with the higher transmission rate is not the most parasitic. In fact just the opposite, since in 2B plants the higher transmission corresponds to a regular meiosis and proper pollen development, whereas the presence of univalents in the variant with low transmission influences pollen development thus decreasing its viability. 4 Bs in rye are very parasitic, producing a strong reduction in fertility but, even in this case, the H line is linked to a more regular meiosis and transmission; therefore, it seems that the Bs are protecting themselves by linking higher transmission to less parasitic effects.

On the other hand, the "arms race" hypothesis predicts that the A genome will be the counterpart that will tend to eliminate the Bs. Therefore, only "anti B" genes and not "pro B" genes are expected to exist in the A genome. It is therefore difficult to fit into the hypothesis the existence of alleles that increase B transmission rate in maize. We have found "anti B" genes in the L line of maize which tend to eliminate the Bs during female gametogenesis. This is in agreement with the hypothesis, but we have also found that the female parent of the H line preferentially chooses the sperm carrying B chromosomes to fertilize herself, and this is a "pro B" action of the A chromosomes because this behaviour tends to increase B transmission rate.

Nevertheless, we found the arms race hypothesis very good to explain that the Bs of maize are more evolved than the Bs of rye. In maize, B effects are neutral at the B frequencies normally found in populations and all Bs form chiasmata and segregate normally. In this more evolved situation, the overall B transmission rate is Mendelian.

Our result show that in rye, B transmission rate depends on the Bs themselves, whereas in maize, B TR depends on the normal complement of A. Why the difference? If we assume that rye is a younger system where the A chromosomes have not developed alleles against the Bs, the variation of B transmission rate will depend on the Bs themselves. If we assume that the Bs of maize have reached a more evolved status, the effect of the A chromosomes becomes apparent. Probably the As have developed alleles for B tolerance and the Bs can be maintained at a Mendelian rate although displaying a complicated behaviour as a relict of their previous, more parasitic nature.

We would like to finish by saying that the results reported here correspond to one race of maize and one of rye. We have, however, preliminary results showing that in other races the B polymorphism is established at a different level. Figure 6 shows the distribution of B TR in three different populations:

Puyo corresponding to the experiments reported in the present paper, Paldang, where the average mean B TR is lower than Mendelian, the H Bs do not exist

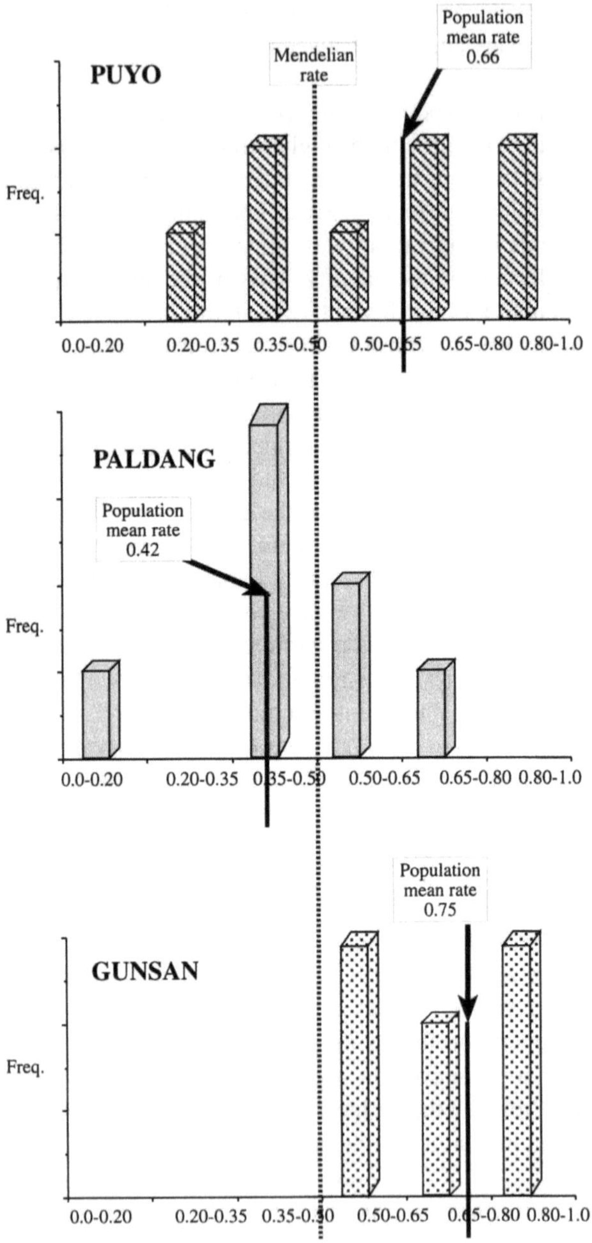

Figure 6. Distribution of B transmission rate in 2B × 0B crosses in three different rye populations.

and B frequency is low (20% of individuals carry Bs), and Gunsan where it seems that L Bs do not exist, but to our surprise B frequency in the population is as low as in Paldang. Fortunately, we have lots of B variants providing interesting data to study B evolution in the future.

Acknowledgements

This research was supported by the grant PB98-0678 from the Spanish MEC, and grants of the CONICET, Argentina, Universidad de Buenos Aires and Universidad Nacional de La Plata. M.R. and A.M.C. were supported by scholarships from CONICET

References

1 Sandery MJ, Forster JW, Blunden R, Jones RN (1990) Identification of a family of repeated sequences on the rye B chromosome. *Genome* 33: 908–913
2 Blunden R, Wilkes TJ, Forster JW, Jiménez MM, Sandery MJ, Karp A, Jones RN (1993) Identification of the E3900 family, a 2nd family of rye chromosome-B specific repeated sequences. *Genome* 36: 706–711
3 Romera F, Vega JM, Diez M, Puertas MJ (1989) B chromosome polymorphism in Korean rye populations. *Heredity* 62: 117–121
4 Jiménez MM, Romera F, Puertas MJ, Jones RN (1994) B chromosomes in inbred lines of rye *(Secale cereale* L.) I. Vigour and fertility. *Genetica* 92: 149–154
5 Alfenito MR, Birchler JA (1990) Studies on B chromosome stability during development. *Maydica* 35: 359–366
6 Santos JL, Jiménez MM, Díez M (1993) Synaptic patterns of rye B chromosomes. I: The standard type. *Chromosome Res* 1: 145–152
7 Jones RN, Rees H (1982) *B chromosomes*. Academic Press, New York
8 Carlson WR (1986) The B chromosome of maize. *Crit Rev Plant Sci* 3: 201–226
9 Shi L, Zhu T, Mogensen HL, Keim P (1996) Sperm identification in maize by fluorescence *in situ* hybridization. *Plant Cell* 8: 815–821
10 Rusche ML, Mogensen HL, Shi L, Keim P, Rougier M, Chaboud A, Dumas C (1997) B chromosome behaviour in maize pollen as determined by a molecular probe. *Genetics* 147: 1915–1921
11 Romera F, Jiménez MM, Puertas MJ (1991) Genetic control of the rate of transmission of rye B chromosomes I. Effects in 2B × 0B crosses. *Heredity* 66: 61–65
12 Jiménez MM, Romera F, Gallego A, Puertas MJ (1995) Genetic control of the rate of transmission of rye B chromosomes II. 0B × 2B crosses. *Heredity* 74: 518–523
13 Jiménez MM, Romera F, González-Sánchez M, Puertas MJ (1997) Genetic control of the rate of transmission of rye B chromosomes III. Male meiosis and gametogenesis. *Heredity* 78: 636–644
14 Puertas MJ, González-Sánchez M, Manzanero S, Romera F, Jiménez MM (1998) Genetic control of the rate of transmission of rye B chromosomes IV. Localization of the genes controlling B transmission rate. *Heredity* 80: 209–213
15 Sved JA (1966) Telomere attachment of chromosomes. Some genetical and cytological consequences. *Genetics* 53: 747–756
16 Chiavarino AM, Rosato M, Naranjo CA, Cámara-Hernández J, Poggio L (1995) B chromosome polymorphism in N. Argentine populations of maize. *Maize Genet Coop Newslett USA* 69: 94
17 Rosato M, Chiavarino AM, Naranjo CA, Puertas MJ, Poggio L (1996) Genetic control of B chromosome transmission rate in *Zea mays* ssp. *mays* (*Poaceae*). *Amer J Bot* 83: 1107–1112
18 Chiavarino AM, Rosato M, Rosi P, Naranjo CA, Poggio L (1997) B chromosomes in native races of maize from Argentina. III. Selection of H and L transmission genotypes and meiotic behaviour of Bs. *Maize Genet Coop Newslett USA* 71: 48–49
19 Chiavarino AM, Rosato M, Rosi P, Poggio L, Naranjo CA (1998) Localization of the genes controlling B chromosome transmission rate in maize *Zea mays* ssp. *mays* (*Poaceae*). *Amer J Bot* 85:

1581–1585
20 Jones RN (1985) Are B chromosomes selfish? *In*: T Cavalier-Smith (ed.): *The evolution of genome size*. Wiley, London, 397–425
21 Shaw MW, Hewitt GM (1990) B chromosomes, selfish DNA and theoretical models: where next? *In*: D Futuyuma, J Antonovics (eds): *Oxford surveys in evolutionary biology*, vol 7. Oxford University Press, Oxford, 197–223
22 Beukeboom LW (1994) Bewildering Bs: an impression of the 1st B-chromosome conference. *Heredity* 73: 328–336
23 Camacho JPM, Cabrero J, López-León MD, Shaw MW (1997) Evolution of a near-neutral B chromosome. *In*: N Henriques-Gil, JS Parker, MJ Puertas (eds): *Chromosomes Today*, vol. 12. Chapman and Hall, London, 301–318
24 Camacho JPM, Shaw MW, López-León MD, Pardo MC, Cabrero J (1997) Population dynamics of a selfish B chromosome neutralized by the standard genome in the grasshopper *Eyprepocnemis plorans*. *Amer Naturalist* 149: 1030–1050
25 Zurita S, Cabrero J, López-León MD, Camacho JPM (1998) Polymorphism regeneration for a neutralized selfish B chromosome. *Evolution* 52: 274–277

Chromosomes Today, Vol. 13
ed. by E. Olmo and C.A. Redi

Igf2 imprinting in development and disease

W. Reik[1], M. Constancia[1], W. Dean[1], K. Davies[1], L. Bowden[1], A. Murrell[1], R. Feil[1], J. Walter[2], G. Kelsey[1]

[1] *Laboratory of Developmental Genetics and Imprinting, Developmental Genetics Programme, The Babraham Institute, Cambridge CB2 4AT, UK*
[2] *Max-Planck Institute of Molecular Genetics, Berlin, Germany*

Introduction

Insuline-like growth factor 2 gene (*Igf2*) was one of the first imprinted genes to be discovered and occupies a centre stage in the study of imprinting. This is because it has dramatic effects on the control of fetal growth, it is involved in growth disorders and in cancer, it interacts with products of other imprinted genes, and its imprinting status is under complex regulation in a cluster of tightly linked imprinted genes. Here we review briefly the key features of *Igf2* imprinting in normal development and in disease, and hope to show what a fascinating subject of study this gene and its biology provides.

Role of *Igf2* in placenta and fetus

The mouse (and other mammalian) *Igf2* genes encode a single polypeptide (Fig. 1). This polypeptide is thought to have autocrine (same cell), paracrine (surrounding cells), and endocrine (circulation) actions (Efstratiadis, 1998, see below). The IGF-II ligand signals through the IGF1 receptor (IGF1R) and the insulin receptor, and is also bound by the IGF2 receptor (which does not transduce any signal). The biological action of IGF-II is further modulated by IGF-binding proteins in serum and tissues. Various effects of IGF-II *in vitro* on cell proliferation and apoptosis are documented, however, its precise roles in growth *in vivo* are not fully established (Efstratiadis, 1998). A recent analysis suggests effects on both dry weight and wet weight, but less on DNA content, suggesting roles in cell and tissue reorganisation as well as cell proliferation (Gardner et al., 1999).

Igf2 transcripts are primarily produced in mesodermal, endodermal, and extraembryonic tissues, and different enhancers are largely responsible for this tissue specificity (see below). The three promoters P1–P3 are transcribed in all these tissues whereas promoter PO is active specifically in the placenta (Moore et al., 1997). Although there is some potential for translational differences

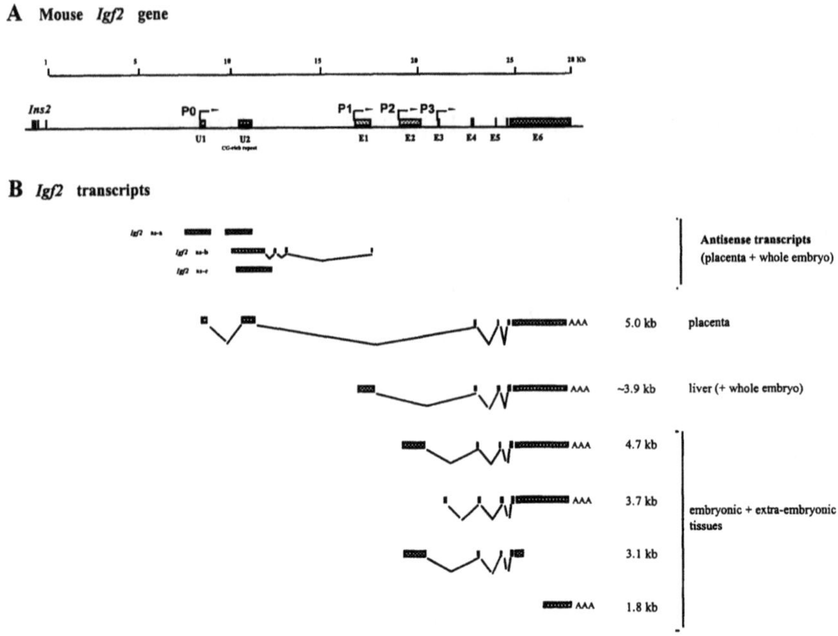

Figure 1. (A) Genomic map of the mouse *Igf2* gene. The *Igf2* gene maps to distal chromosome 7 and is flanked 5' by the insulin II gene (*Ins2*). Multiple transcripts are produced from alternative promoters (horizontal arrows): P0, P1, P2 and P3. *Igf2* exons (filled boxes): U1, U2, E1-E6. The coding sequences are located in exons 4-6 and are shown as open boxes. A CG rich repeat is found in the U2 exon. (B) *Igf2* transcripts. The various *Igf2* transcripts are shown below the structure of the *Igf2* gene. Extensive antisense transcription occurs through the *Igf2* upstream region (*Igf2*as-a,b,c). The function of these untranslated transcripts remains unknown. All the different spliced, sense transcripts contain exons 4-6 that code for the 67 amino acid IGF-II protein. The P0 transcript is placenta-specific. P1 transcripts are predominantly expressed in fetal and neonatal liver. P3 and P4 transcripts are the major embryonic and fetal promoters. The 3' untranslated region (UTR) sequences vary between transcripts because of alternative polyadenylation (AAA).

between these transcripts, they are all thought to be translated (Nielsen et al., 1999). While *Igf2* begins to be transcribed shortly after implantation, first in extraembryonic tissues and then throughout mesodermal and endodermal tissues in postimplantation embryos (Lee et *a*l., 1999), the action of gene expression on growth is only seen from E13 onward (Efstratiadis, 1998). Since this is also the case for some of the other components of the insuline-like growth factor/insuline (IGF/INS) pathway (IGF-1, IGF1R), it is likely that critical components of the signalling pathways are only assembled after this point. Both under and overexpression of *Igf2* has dramatic and apparently dose dependent effects. Fetuses (and their placentae) completely lacking IGF-II are 40% growth retarded at birth (60% of normal size) and are apparently normally proportioned dwarfs (De Chiara et al., 1990). Fetuses lacking *Igf2* expression in endoderm (liver, gut etc.) through knockout (KO) of the endo-

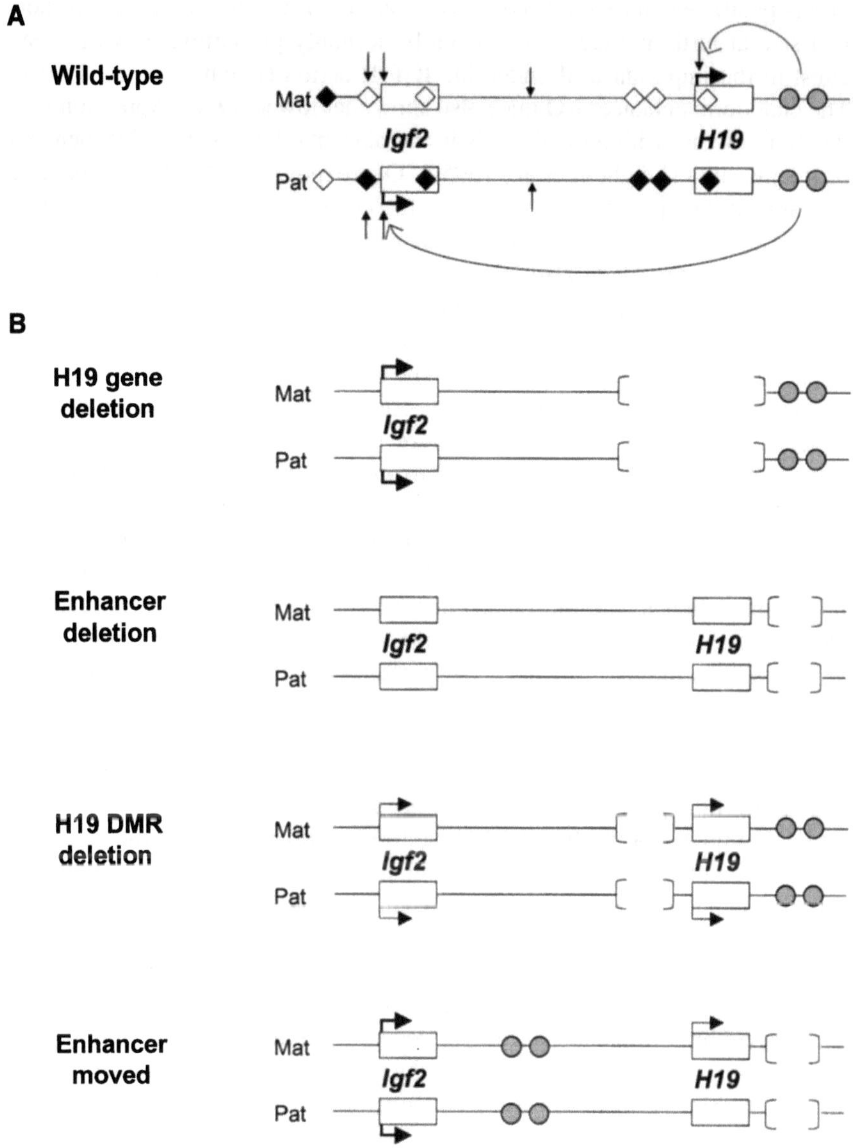

Figure 2. Epigenetic modifications in the *Igf2 -H19* region and their functional testing. (A) shows the wildtype situation on both maternal and paternal chromosomes. Open diamonds, unmethylated DMRs, filled diamonds, methylated DMRs. *Igf2* has DMRs 0-2, *H19* has an upstream DMR which may be a boundary element, and the promoter and gene is also differentially methylated. Arrows, DNase I hypersensitive sites in chromatin. Circles, endoderm specific enhancers. (B) shows knockout experiments that address the function of some of the elements. Knockout of the *H19* gene and its upstream region results in biallelic expression of *Igf2*. Deletion of the endoderm enhancers results in reduced expression of *H19* and *Igf2* in endodermal tissues, thus showing that the enhancers are used by both genes. Deletion of the *H19* DMR results in biallelic expression of both genes, suggesting a boundary element has been removed. Moving the enhancers results in biallelic expression of *Igf2* but *H19* is only expressed from the unmethylated allele. Other regions have not yet been tested.

derm-specific enhancers located downstream of *H19* (Fig. 2) are 70% of normal size at birth and are also apparently normally proportioned, which may attest to the importance of endocrine IGF-II action (Leighton et al., 1995a). The endoderm enhancer KO mice also show that lowered *Igf2* expression limited to the fetus (with normal levels in the placenta) has an effect independent of the placenta (Leighton et al., 1995a). Overexpression of *Igf2* can increase size at birth by up to 160% (Sun et al., 1997), and size at E17 by up to 200% (Eggenschwiler et al., 1997), in a manner dependent on IGF-II dosage. High levels of overexpression also lead to various malformations and intrauterine death (Sun et al., 1997, Eggenschwiler et al., 1997, see below). Individual organs can be enlarged in proportion to their *Igf2* levels, suggesting autocrine or paracrine control (Sun et al., 1997). The issue of paracrine or endocrine control is not fully resolved, but new genetic models including the endoderm enhancer KO and a mutant that we are studying in collaboration with B. Cattanach (Harwell) which lacks *Igf2* expression in mesoderm (K. Davies et al., unpublished observations), may allow further clarification.

While IGF-II action can affect the growth of the fetus (through signalling via IGF1R and insuline receptor (INSR)), it has a particularly unique and important action in the placenta. Thus, while *Igf1* and *Igf1r* KOs result in small fetuses and normal sized placentae, all manipulations involving *Igf2* or *Igf2r* result in altered growth of the placenta (Efstratiadis, 1998). IGF-II is thought to signal through a separate receptor in the placenta which remains unknown (Efstratiadis, 1998). Therefore, placental IGF-II action and its effect on the fetus could thus far not be examined specifically. However, since a placenta-specific *Igf2* transcript has been recently identified (Moore et al., 1997), which is transcribed in the labyrinthine trophoblast where maternal and fetal circulation interface (M. Constancia et al., unpublished observations), the question can now be addressed. Significantly, a knockout of the placenta-specific transcript results in smaller size of the placenta and fetus from E14, and fetal weight is 70% of normal at birth (M. Constancia et al., unpublished observations). Therefore, reduction of size of the placenta or dysfunction of the labyrinthine trophoblast leads to severe intrauterine growth retardation. The importance of placental IGF-II production is confirmed in another system, in which chimeric conceptuses were constructed that lacked IGF-II in trophectoderm and hence labyrinthine and spongiotrophoblast of the placenta, but the fetus was of a normal genotype (Gardner et al., 1999). The effects on fetal and placental weights on E16.5 were identical to the ones recorded in the placenta-specific KO (M. Constancia et al., unpublished observations; Gardner et al., 1999).

The observation that genes for fetal growth factors which are particularly important for placental growth and development are imprinted (*Igf2, Igf2r),* whereas those whose roles are limited to the fetus are not (*Igf1, Igf1r, Insr*), raises the question of whether the primary selection for imprinting comes from the action of growth factors in the placenta with a direct effect on nutrient transfer from mother to offspring. It also raises the possibility that there might be differences in mechanistic aspects of imprinting in placenta and fetus.

IGF-II interacts with the products of other imprinted genes (*Igf2r, H19,* probably *Grb10,* Miyoshi et al., 1998) and it will be interesting to see whether other imprinted, growth-related genes also interact with the IGF pathway, in particular perhaps in the placenta.

Mechanisms of *Igf2* imprinting

The *Igf2* gene is part of a cluster of imprinted genes on distal chromosome seven in the mouse, with a homologous cluster on the syntenic region on chromosome 11p15.5 in the human (Paulsen et al., 1998, Fig. 3). A number of

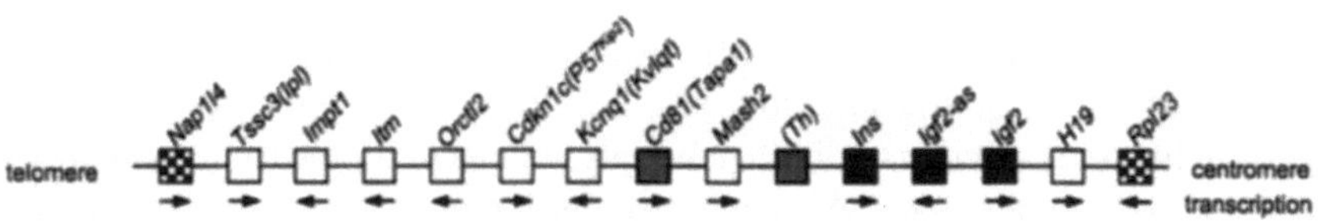

Figure 3. Imprinting cluster on distal chromosome 7. Schematic physical map of the distal chromosome 7 imprinting cluster in mouse showing imprinting status and transcriptional orientation. Maternally expressed genes are shown in white, paternally expressed genes in black, non-imprinted genes hatched and genes for which the imprinting status is unknown in grey. *Th* has so far not been mapped in mouse hence its position (based on human data) is shown in parentheses. For details see http://www.mgu.har.mrc.ac.uk/imprinting/implink.html.

sequences that are important for *Igf2* imprinting and expression have now been defined. *Igf2* imprinting is under regional control being particularly influenced by the neighbouring maternally expressed *H19* gene (Leighton et al., 1995b; Ainscough et al., 1997; Brannan and Bartolomei, 1999). The influence of the *H19* control sequences apparently also extend to *Ins2*, but probably not further to other genes in the cluster (Caspary et al., 1998). Whether other genes or control sequences in the cluster have an influence on *Igf2* or *H19* imprinting or expression is currently uncertain.

Two sequences have been identified so far with clearly defined roles *in vivo* in imprinting and expression of *Igf2/H19*. The first is a set of enhancers called the endoderm enhancers which are located 3' of the *H19* gene (Fig. 2). Knockout of these enhancers abolishes expression of *H19* (on the maternal chromosome) in mainly endodermal tissues such as liver and gut, and of *Igf2* (on the paternal chromosome) in the same tissues (Leighton et al., 1995a). This establishes that these enhancers are shared by both genes. So how is the action of the enhancers limited to either gene on different parental chromosomes? Upstream of *H19* is located a differentially methylated region (DMR) with paternal methylation and maternal undermethylation (the paternal methylation also extends into the *H19* promoter and gene body and is thought to silence the paternal gene copy). When this region is deleted, *H19* is no longer methylated on the paternal chromosome, and is now expressed. Similarly, the maternal copy of *Igf2* is expressed from a chromosome with the deletion (Thorvaldsen et al., 1998). The currently favoured interpretation is that this DMR represents some sort of chromatin insulator or boundary element which, when methylated, allows access of the *Igf2* promoters to the *H19* enhancer (Thorvaldsen et al., 1998; Webber et al., 1998). In its unmethylated state, the element represents a closed boundary which prevents activation of the *Igf2* promoters by their enhancers. Interestingly, a specialised non-histone type chromatin structure has recently been shown to exist on the maternal (unmethylated) DMR region (Hark et al., 1998; Szabo et al., 1998; Khosla et al., 1999). The DMR consists of various elements, including a direct repeat structure and a potential silencer (Lycko et al., 1997), whose detailed functions *in vivo* are unclear at present. The knockout of the DMR also results in a reduced level of *Igf2* expression on a chromosome that also expresses *H19* (and *vice versa*) suggesting that the promoters of both genes also compete for the shared enhancers (Thorvaldsen et al., 1998). As mentioned earlier, enhancers for lineages other than endoderm have so far not been found, but we are studying a radiation-induced mutant in which *Igf2* expression is abolished in mesodermal tissues (K. Davies et al., unpublished observations).

How the *Igf2* promoters access the distant enhancers (or are prevented from doing so presumably by an unmethylated DMR) is unclear. It is remarkable that strong DNase I hypersensitive sites are present in *Igf2* promoters and upstream on both parental chromosomes (Sasaki et al., 1992; Feil et al., 1995). The *Igf2* gene itself also has three DMRs (Fig. 2) but whether these have functions independent of *H19* and its DMR/enhancers or in conjunction with these is currently not known (Sasaki et al., 1992, Feil et al., 1994, Moore et al., 1997). Knockouts in DMR1 and DMR2 have been prepared, however, and are under study, so that some of these answers should soon emerge.

The *H19* DMR seems to carry a 'germline methylation imprint' in the sense that the sperm copy is methylated, the oocyte copy is not, and these methylation patterns are inherited through all stages of development (Olek and Walter, 1997; Tremblay et al., 1997), except in germ cells where they are switched as appropriate. DMR1 in *Igf2* (which was the first DMR to be discovered, Sasaki et al., 1992) is apparently not differentially methylated in germ cells but becomes so soon after fertilisation (Shemer et al., 1996). DMR2 is differentially methylated in germ cells, and loses this methylation in the early preimplantation embryo, which then becomes re-established later on (Oswald et al., 2000). Thus, whether these sequences carry germline imprints is not totally clear, however, it should be borne in mind that there could be epigenetic memory systems other than methylation that interact with the methylation system (Pickard et al., unpublished observations). The methylation patterns at *Igf2*

DMRs 0-2 are not independent of the *H19* DMR/gene, but whether this is because of altered expression of *Igf2* as a consequence of deletion of the *H19* DMR/gene, or whether there are elements that are responsible for regional establishment and spreading of the 'epigenotype', is currently not clear (Forné et al., 1997; Moore et al., 1997).

Although other genes in the cluster also have DMRs, and the cluster as a whole displays asynchronous DNA replication during the cell cycle as a global feature (Kitsberg et al., 1993), other regional or local control elements have so far not been defined (but see below for some hints). In any event, the *H19–Igf2* system of genes is an excellent model system for the study of germline imprints and regional control of imprinting. The mechanistic details of this system are under intense scrutiny and discussion (Reik and Walter, 1998; Constancia et al., 1998; Surani 1998; Tilghman 1999; Brannan and Bartolomei, 1999), but presentation of these details goes beyond the scope of the current review.

Altered *Igf2* expression in disease

There are now a number of disease situations in which levels of *IGF2* expression are altered (Ward 1997; Feinberg 1999; Morison et al., 1998). *IGF2* expression is often increased and this is associated with enhanced growth or increased cell proliferation. Increases are most frequently caused by loss of imprinting (LOI), *i.e.* expression of the otherwise repressed maternal allele in addition to the paternal one. The *IGF2* gene is part of a cluster of imprinted genes on distal chromosome 7 (in mouse) and 11p15.5 (in human) as discussed (Fig. 3), and imprinting mechanisms are thought to be regional, so potentially a variety of mechanisms could lead to altered imprinting/expression of *IGF2* (Reik and Maher, 1997; Feinberg 1999).

The dependence of *Igf2* imprinting on *H19* has already been described. Further towards telomeric (Fig. 3), lies the *Mash2* gene which has an important role in placenta development, where imprinting controls have not been precisely defined. Remarkably though, it is possible that imprinting of *Mash2* can proceed independently of DNA methylation, which would further support the notion of epigenetic memory other than methylation (Caspary et al., 1998). Close to *Mash2* there is a region of transcripts including *Tapa1* (*Cd81*) which may show some bias in gene expression towards one of the parental alleles, but may not show clear cut imprinting as do other genes in the cluster (Caspary et al., 1998). Further centromeric are *Kvlqt1/Kcnq1* (potassium channel), *p57*^{Kip2}*/Cdkn1c* (negative cell cycle regulator), *Orctl2* (organic cation transporter), *Hm* (transmembrane domain), *Impt1* (transporter-like), and *Tssc3/Ipl,* all *bona fide* imprinted genes with maternal expression. *Kvlqt1/Kcnq1* is overlapped by an antisense transcript with paternal expression (Lee et al., 1999; Smilinich et al., 1999), and this system may therefore be in 'expression competition' as are *H19- Igf2* (Brannan and Bartolomei, 1999), *Igf2r–Igf2r*-antisense (Wutz et al., 1997) and possibly others. Although various DMRs have been described in these genes, the controls have not been elucidated in functional assays.

Loss of imprinting (LOI) of *Igf2* has been described as an isolated somatic event, largely associated with various pediatric and adult cancers, or an embryonic or germline event associated with the overgrowth and cancer syndrome Beckwith-Wiedemann syndrome (BWS) (see below). Many cancers (but with notable exceptions) show LOI of *IGF2* (Feinberg, 1999). Such LOI could either be causally involved in tumour initiation or growth, or be an epi-phenomenon because of altered maintenance of DNA methylation and potentially imprinting in tumours. To our knowledge, a large scale analysis of the stability or otherwise of imprints in tumours has not yet been carried out. However, there is evidence that overexpression of *IGF2* can indeed lead to tumour initiation or progression (Christofori et al., 1994; Bates et al., 1995).

In terms of mechanisms, the best studied system is Wilms' tumour, a childhood tumour of the kidney. A large proportion of tumours with *IGF2* LOI also have maternal methylation of the *H19* gene (Moulton et al., 1994; Steenman et al., 1994). This is consistent with the model described above in which methylation of the *H19* DMR would allow access of *IGF2* promoters to the shared enhancers. This hypermethylation of *H19* can sometimes be observed in adjacent healthy tissues, suggesting that it is an early, or even predisposing event, in tumourigenesis (Moulton et al., 1994; Okamotomo et al., 1997). However, no mutational mechanisms leading to this epigenetic switch have been identified, and to our knowledge this type of event has not been observed in rare familial cases of the disease. Hence, it is conceivable that this epigenetic switch could be the result of 'epimutation'. Although this could arise randomly, a proposal has been made for a mechanism of 'methylation transfer' from the methylated paternal to the unmethylated maternal *H19* DMR (Bestor and Tycko, 1996). Confirmation of this mechanism awaits analysis in a genetic model.

Germline or early embryonic alterations in *IGF2* imprinting could lead to LOI in many cells of the developing organism (Dean et al., 1998). This seems to be the case in the fetal overgrowth syndrome, Beckwith-Wiedemann syndrome, in which the majority of patients show apparent LOI at least in fibroblasts cultured from their skin (Weksberg et al., 1993; Reik et al., 1995; Reik and Maher, 1997). In mouse models, the majority of the symptoms of the disease can indeed be attributed to overexpression of *Igf2* (Sun et al., *1997;* Eggenschwilder et al., 1997*)*. The mechanisms by which LOI arises, and the molecular pathogenesis in BWS patients in which there is no LOI of *IGF2*, are apparently complex but all molecular abnormalities detected so far are linked to 11p15.5 (Fig. 4, Reik and Maher, 1997). The majority of BWS patients are sporadic, and familial cases are rare. A number of sporadic cases show maternal methylation of *H19*, with associated LOI of *IGF2* (Fig. 4, Reik et al., 1995). Surprisingly, other patients with LOI of *IGF2* show normal imprinting

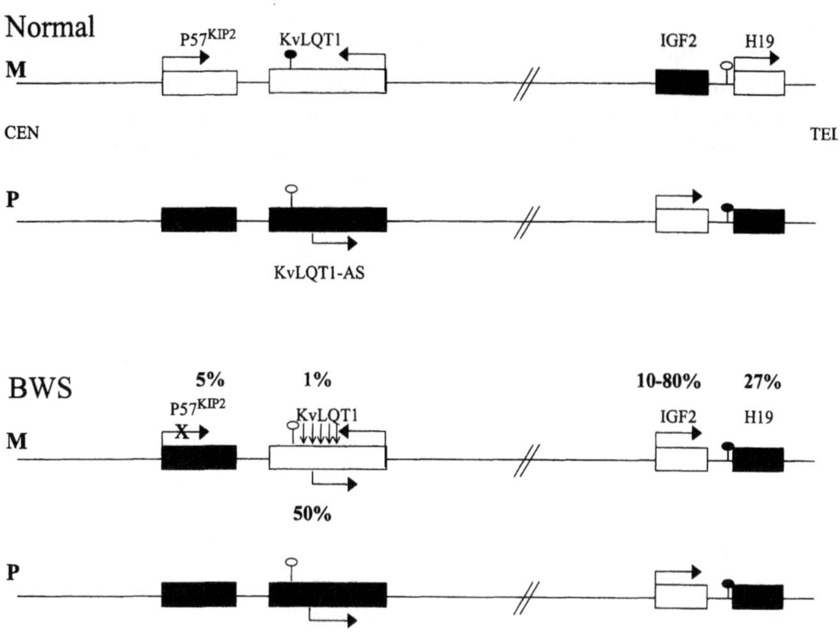

Figure 4. Altered imprinting of chr11p15.5 is involved in Beckwith-Wiedemann syndrome (BWS). Mutations and epimutations in the $p57^{KIP2}$, *KvLQT1, IGF2* and *H19* genes on chromosome 11p15.5 have so far been shown to be involved in the aetiology of BWS. White boxes, expressed genes; black boxes, silent genes; X, point mutations; arrows in *KvLQT1*, breakpoint cluster for translocations; filled/open circles, methylated/demethylated CpGs; arrows above the boxes, direction of sense transcription; arrows below boxes, antisense transcription. The frequency of genetic and epigenetic mutations in BWS patients is also indicated. These are very approximate. The frequency of LOI of *IGF2* varies between studies (*e.g.* compare Reik and Maher (1997) and Lee et al. (1999)).

of *H19* (Joyce et al., 1997). A minority of BWS patients have maternally inherited translocations in 11p15.5, with several breakpoints in *KvLQT1* (Lee et al., 1997), or regions further towards the centromere and outside of the present clusters (Mannens et al., 1996). In two families in which this could be analysed, translocation was associated with LOI of *IGF2* (Brown et al., 1996; Smilinich et al., 1999), but the mechanism of this is unknown. In another series of sporadic patients, LOI of the *KvLQT1* antisense gene occurred at a high frequency (Fig. 4, Lee et al., 1999; Smilinich et al., 1999) which in some cases was associated with LOI of *IGF2* but in others not. Hence, currently it is unclear whether mutational (translocation) or likely epigenetic (LOI of the antisense gene) events in the *KvLQT1* region interfere with *IGF2* imprinting/expression. If they do not, this would reinforce the notion of a bipartite structure of the cluster, with potentially independent imprinting controls (Caspary et al., 1998). However, strikingly, it would also lend support to the previous suggestion that alterations in linked imprinted genes in the cluster could result in similar phenotypes, perhaps because these genes interact in common physiological pathways (Reik and Maher, 1997). Indeed, the only point mutations discovered so far in BWS patients have been found in the cell cycle inhibitor *p57*^{KIP2}*/CDKN1C* (Fig. 4, Hatada et al., 1996; Lee et al., 1997; O'Keefe et al., 1997; Lam et al., 1999), and these do not seem to be associated with LOI of *IGF2*. These mutations are more frequent in familial cases than in sporadic ones (Lam et al., 1999), whereas the epigenetic alterations in *H19* or *KvLQT1* antisense are exclusive to sporadic patients. The suspicion is therefore that the majority of BWS patients have epimutations – in either *H19* or *KvLQT1* antisense – rather than genetic mutations.

Conclusions

The study of imprinting and phenotypic effects of the *Igf2* gene, its imprinted neighbours in the cluster, and its interacting components encoded elsewhere in the genome, will continue to be a fascinating subject of imprinting research. Since growth and cell proliferation are such central themes in imprinting, progress in understanding the regulation of *Igf2* will continue to lead to key insights into the biology and pathology of growth and cell proliferation.

This article has first been published in *Int J Dev Biol* 44: 145–150, Reik et al. (2000).

References

Ainscough JF, Koide T, Tada M, Barton S, Surani MA (1997) Imprinting of *Igf2* and *H19* from a 130 kb YAC transgene. *Development* 124: 3621–3632

Bartolomei MS, Tilghman SM (1997) Genomic imprinting in mammals. *Annu Rev Genet* 31: 493–525

Bates P, Fisher R, Ward A, Richardson L, Hill DJ, Graham CF (1995) Mammary cancer in transgenic mice expressing insulin-like growth factor II (*IGF-II*). *Brit J Cancer* 72: 1189–1193

Bestor TH, Tycko B (1996) Creation of genomic methylation patterns. *Nat Genet* 12: 363–367

Brannan CI, Bartolomei MS (1999) Mechanisms of genomic imprinting. *Curr Opin Genet Develop* 9: 164–170

Brown KW, Villar AJ, Bickmore W, Clayton-Smith J, Catchpoole D, Maher ER, Reik W (1996) Imprinting mutation in the Beckwith-Wiedemann syndrome leads to biallelic *IGF2* expression through an *H19*-independent pathway. *Hum Mol Genet* 5: 2027–2032

Caspary T, Cleary MA, Baker CC, Guan XJ, Tilghman SM (1998) Multiple mechanisms regulate imprinting of the mouse distal chromosome 7 gene cluster. *Mol Cell Biol* 18: 3466–3474

Christofori G, Naik P, Hanahan D (1994) A second signal supplied by insulin-like growth factor II in oncogene-induced tumorigenesis. *Nature* 369: 414–418

Constancia M, Pickard B, Kelsey G, Reik W (1998) Imprinting mechanisms. *Genome* 8: 881–900

Dean W, Bowden L, Aitchison A, Klose J, Moore T, Meneses JJ, Reik W, Feil R (1998) Altered imprinted gene methylation and expression in completely ES cell-derived mouse fetuses: association with aberrant phenotypes. *Development* 125: 2273–2282

De Chiara TM, Efstratiadis A, Robertson EJ (1990) A growth-deficiency phenotype in heterozygous mice carrying an insulin-like growth factor II gene disrupted by targeting. *Nature* 345: 78–80

Efstratiadis A (1998) Genetics of mouse growth. *Int J Develop Biol* 42: 955–976

Eggenschwiler J, Ludwig T, Fisher P, Leighton PA, Tilghman SM, Efstratiadis A (1997) Mouse mutant embryos overexpressing *IGF-II* exhibit phenotypic features of the Beckwith-Wiedemann and Simpson-Golabi-Behmel syndromes. *Gene Develop* 11: 3128–3142

Feil R, Handel MA, Allen ND, Reik W (1995) Chromatin structure and imprinting: developmental control of DNase-I sensitivity in the mouse insulin-like growth factor 2 gene. *Develop Genet* 17: 240–252

Feil R, Walter J, Allen ND, Reik W (1994) Developmental control of allelic methylation in the imprinted mouse *Igf2* and *H19* genes. *Development* 120: 2933–2943

Feinberg AP (1999) Imprinting of a genomic domain of 11p15 and loss of imprinting in cancer: an introduction. *Cancer Res* 59: 1743–1746

Forné T, Oswald J, Dean W, Saam JR, Bailleul B, Dandolo L, Tilghman SM, Walter J, Reik W (1997) Loss of the maternal *H19* gene induces changes in *Igf2* methylation in both cis and trans. *Proc Natl Acad Sci USA* 94: 10243–10248

Gardner RL, Squire S, Zaina S, Hills S, Graham CF (1999) Insulin-like growth factor-2 regulation of conceptus composition: effects of the trophectoderm and inner cell mass genotypes in the mouse. *Biol Reprod* 60: 190–195

Hark AT, Tilghman SM (1998) Chromatin conformation of the *H19* epigenetic mark. *Hum Mol Genet* 7: 1979–1985

Hatada I, Ohashi H, Fukushima Y, Kaneoko Y, Inoue M, Komoto Y, Okada A, Ohishi S, Nabetani A, Morisaki H et al (1996) An imprinted gene $p57^{Kip2}$ is mutated in Beckwith-Wiedemann syndrome. *Nat Genet* 14: 171–173

Joyce JA, Lam WK, Catchpoole DJ, Jenks P, Reik W, Maher ER, Schofield PN (1997) Imprinting of *IGF2* and *H19:* lack of reciprocity in sporadic Beckwith-Wiedemann syndrome. *Hum Mol Genet* 6: 1543–1548

Khosla S, Aitchison A, Gregory R, Allen ND, Feil R (1999) Parental allele-specific chromatin configuration in a boundary-imprinting-control element upstream of the mouse *H19* gene. *Mol Cell Biol* 19: 2556–2566

Kitsberg D, Selig S, Brandeis M, Simon I, Keshet I, Driscoll DJ, Nicholls RD, Cedar H (1993) Allele-specific replication timing of imprinted gene regions. *Nature* 364: 459–463

Lam WW, Hatada I, Ohishi S, Mukai T, Joyce JA, Cole JA, Cole TR, Donnai D, Reik W, Schofield PN, Maher ER (1999) Analysis of germline CDKN1C (p57KIP2) mutations in familial and sporadic Beckwith-Wiedermann Syndrome (BWS) provides a novel genotype-phenotype correlation. *J Med Genet* 36: 518–523

Lee JE, Pintar J, Efstratiadis A (1990) Pattern of the insulin-like growth factor II gene expression during early mouse embryogenesis. *Development* 110: 151–159

Lee MP, Debaun M, Randhawa G, Reichard BA, Elledge SJ, Feinberg AP (1997) Low frequency of $p57^{Kip2}$ mutation in Beckwith-Wiedemann syndrome. *Amer J Hum Genet* 61: 304–309

Lee MP, Debaun MR, Mitsuya K, Galonek HL, Brandenburg S, Oshimura M, Feinberg AP (1999) Loss of imprinting of a paternally expressed transcript, with antisense orientation to *KvLQT1*, occurs frequently in Beckwith-Wiedemann syndrome and is independent of insulin-like growth factor II imprinting. *Proc Natl Acad Sci USA* 96: 5203–5208

Lee MP, Hu RJ, Johnson LA, Feinberg AP (1997) Human *KvLQT1* gene shows tissue-specific imprinting and encompasses Beckwith-Wiedemann syndrome chromosomal rearrangements. *Nat Genet* 15 181–185

Leighton PA, Ingram RS, Eggenschwiler J, Efstratiadis A, Tilghman SM (1995a) Disruption of imprinting causes by deletion of the *H19* gene region in mice. *Nature* 375: 34–39

Leighton PA, Saam JR, Ingram RS, Stewart CL, Tilghman SM (1995b) An enhancer deletion affects both *H19* and *Igf2* expression. *Gene Develop* 9: 2079–2089

Lyko F, Brenton JD, Surani MA, Paro R (1997) An imprinting element from the mouse *H19* locus functions as a silencer in *Drosophila. Nat Genet* 16: 171–173

Mannens M, Alders M, Redeker B, Bliek J, Steenman M, Wiesmeyer C, de Meulemeester M, Ryan A, Kalikin L, Voute et al (1996) Positional cloning of genes involved in the Beckwith-Wiedemann syndrome, hemihypertrophy and associated childhood tumors. *Med Pediat Oncol* 27: 490–494

Miyoshi N, Kuroiwa Y, Kohda T, Shitara H, Yonekawa H, Kawabe T, Hasegawa H, Barton SC, Surani MA, Kaneko-Ishino T et al (1998) Identification of the Meg1/Grb10 imprinted gene on mouse proximal chromosome 11, a candidate for the Silver-Russell syndrome gene. *Proc Natl Acad Sci USA* 95: 1102–1107

Moore T, Constancia M, Zubair M, Bailleul B, Feil R, Sasaki H, Reik W (1997) Multiple imprinted sense and antisense transcripts, differential methylation and tandem repeats in a putative imprinting control region upstream of mouse *Igf2*. *Proc Natl Acad Sci USA* 94: 12509–12514

Morison IM, Reeve AE (1998) Insulin-like growth factor 2 and overgrowth: molecular biology and

clinical implications. *Mol Med Today* 4: 110–115
Moulton T, Crenshaw T, Hao Y, Moosikasuwan J, Lin N, Dembitzer F, Hensle T, Weiss L, Mcmorrow L, Loew T et al (1994) Epigenetic lesions at the *H19* locus in Wilms' tumour patients. *Nat Genet* 7: 440–447
Nielsen J, Christiansen J, Lykke-Andersen J, Johnsen AH, Wewer UM, Nielsen FC (1999) A family of insulin-like growth factor II mRNA-binding proteins represses translation in late development. *Mol Cell Biol* 19: 1262–1270
O'Keefe D, Dao D, Zhao L, Sanderson R, Warburton D, Weiss L, Anyane-Yeboa K, Tycko B (1997) Coding mutations in $p57^{Kip2}$ are present in some cases of Beckwith-Wiedemann syndrome but are rare or absent in Wilms' tumors. *Amer J Hum Genet* 61: 295–303
Okamotomo K, Morison IM, Taniguchi T, Reeve AE (1997) Epigenetic changes at the insulin-like growth factor *II/H19* locus in developing kidney is an early event in Wilms' tumorigenesis. *Proc Natl Acad Sci USA* 94: 5367–5371
Olek A, Walter J (1997) The preimplantation ontogeny of the *H19* methylation imprinting. *Nat Genet* 17: 275–276
Oswald J, Engemann S, Lane M, Mayer W, Olek A, Fundele R, Dean W, Reik W, Walter J (2000) Active demethylation of the paternal genome in the mouse zygote. *Curr Biol; in press*
Paulsen M, Davies KR, Bowden LM, Villar AJ, Franck O, Fuermann M, Dean WL, Moore TF, Rodrigues N, Davies KE et al (1998) Syntenic organization of the mouse distal chromosome 7 imprinting cluster and the Beckwith-Wiedemann syndrome region in chromosome 11p15.5. *Hum Mol Genet* 7: 1149–1159
Reik W, Walter J (1998) Imprinting mechanisms in mammals. *Curr Opin Genet Develop* 8: 154–164
Reik W, Maher ER (1997) Imprinting in clusters: lessons from Beckwith-Wiedemann syndrome. *Trends Genet* 13: 330–334
Reik W, Brown KW, Schneid H, Le Bouc Y, Bickmore W, Maher ER (1995) Imprinting mutations in the Beckwith-Wiedemann syndrome suggested by altered imprinting pattern in the *IGF2-H19* domain. *Hum Mol Genet* 4: 2379–2385
Sasaki H, Jones PA, Chaillet JR, Ferguson-Smith AC, Barton SC, Reik W, Surani MA (1992) Parental imprinting: potentially active chromatin of the repressed maternal allele of the mouse insulin-like growth factor (*Igf2*) gene. *Gene Develop* 6: 1843–1856
Shemer R, Birger Y, Dean WL, Reik W, Riggs AD, Razin A (1996) Dynamic methylation adjustment and counting as part of imprinting mechanisms. *Proc Natl Acad Sci USA* 93: 6371–6376
Smilinich NJ, Day CD, Fitzpatrick GV, Caldwell GM, Lossie AC, Cooper PR, Smallwood AC, Joyce JA, Schofield PN, Reik W et al (1999) A maternally methylated CpG-island in *KvLQT1* is associated with an antisense paternal transcript and loss of imprinting in Beckwith-Wiedemann syndrome. *Proc Natl Acad Sci USA* 96: 8064–8069
Steenman MJ, Rainier S, Dobry CJ, Grundy P, Horon IL, Feinberg AP (1994) Loss of imprinting of *IGF2* is linked to reduced expression and abnormal methylation of *H19* in Wilms' tumour. *Nat Genet* 7: 433–439
Sun FL, Dean WL, Kelsey G, Allen ND, Reik W (1997) Transactivation of *Igf2* in a mouse model of Beckwith-Wiedemann syndrome. *Nature* 389: 809–815
Surani MA (1998) Imprinting and the initiation of gene silencing in the germ line. *Cell* 93: 309–312
Szabo PE, Pfeifer GP, Mann JR (1998) Characterization of novel parent-specific epigenetic modifications upstream of the imprinted mouse *H19* gene. *Mol Cell Biol* 18: 6767–6776
Thorvaldsen JL, Duran KL, Bartolomei MS (1998) Deletion of the *H19* differentially methylated domain results in loss of imprinted expression of *H19* and *Igf2*. *Gene Develop* 12: 3693–3702
Tilghman SM (1999) The sins of the fathers and mothers: genomic imprinting in mammalian development. *Cell* 96: 185–193
Tremblay KD, Duran KL, Bartolomei MS (1997) A 5' 2-kilobase-pair region of the imprinted mouse *H19* gene exhibits exclusive paternal methylation throughout development. *Mol Cell Biol* 17: 4322–4329
Ward A (1997) Beckwith-Wiedemann syndrome and Wilms' tumour. *Mol Hum Reprod* 3: 157–168
Webber AL, Ingram RS, Levorse JM, Tilghman SM (1998) Location of enhancers is essential for the imprinting of *H19* and *Igf2* genes. *Nature* 391: 711–715
Weksberg R, Shen DR, Fei YL, Song QL, Squire J (1993) Disruption of insulin-like growth factor 2 imprinting in Beckwith-Wiedemann syndrome. *Nat Genet* 5: 143–150
Wutz A, Smrzkda OW, Schweifer N, Schellander K, Wagner EF, Barlow DP (1997) Imprinted expression of the *Igf2r* gene depends on an intronic CpG island. *Nature* 389: 745–749

Part three
Meiosis

Nancy Kleckner

Josef Loidl

Bruce D. McKee

Chromosomes Today, Vol. 13
ed. by E. Olmo and C.A. Redi

Seeing meiotic chromosomes without seeing them

Nancy Kleckner

Department of Molecular and Cellular Biology, Harvard University, Cambridge, MA 02138, USA

Introduction

Cytological visualization of meiotic chromosomes has revealed the fundamental basis for Mendelian inheritance and for the exceptions to Mendel's rules provided by genetic linkage and recombination as well as revealing and describing many complex aspects of chromosome morphogenesis (for recent review, see [1]). As a complement to cytological approaches, our laboratory has developed assays which provide other ways of probing chromosome organization and behavior.

Homolog juxtaposition

By the end of meiotic prophase, homologs are coaligned along their lengths, without any appreciable number of entanglements between non-homologous chromosomes. Three distinct processes contribute to this juxtaposition, at least as defined thus far in yeast: (i) double-strand break (DSB)-independent pairing, (ii) DSB-initiated strand invasion and exchange of the recombination process and (iii) synaptonemal complex (SC) formation. The three types of interactions likely occur approximately in succession, though perhaps with significant temporal overlap on a per-nucleus, per-chromosome and/or per-region basis.

Assays for "chromosome pairing". Classical cytological tools can be used to describe the coming-together of homologs only at stages where chromosomes are either individualized enough to be seen by light microscopy and/or contain full length axial elements traceable by electron microscopy. Various lines of evidence from such studies raised the possibility that a substantial degree of homolog juxtaposition occurs during meiosis before the chromosomes can be seen by either of these approaches [2]. One recent complement to classical studies has been the use of chromosome painting (see, e.g. [3–5]). Our laboratory has developed three other types of assays and applied them to analysis of homolog disposition in yeast:

Fluorescent in situ *hybridization (FISH) analysis of dispersed chromosomes with locus-specific probes*

Highly dispersed chromosomes are analyzed by FISH with 5–30 kb probes [6–8]. This approach permits the detection of specific contacts between homologs because, when performed in such a way that the nuclear contents are fully disrupted, spatial localization effects resulting from factors other than direct contact are eliminated (see [7] for discussion). Furthermore, such an analysis can be performed at nuclear stages in which the chromosomes are completely diffuse (e.g. premeiosis, " meiotic interphase" and, in yeast, all stages of the mitotic cycle).

Using this approach it has been possible to infer that homologs are connected during early meiotic prophase via "multiple interstitial pairing contacts" which are independent of recombination initiation by formation of DSBs ([6]; Fig. 1). Moreover, such contacts are present during premeiosis, lost during meiotic S-phase, and restored in early prophase. Such contacts have also been found to be present during the mitotic cell cycle, implying that yeast has "somatic pairing". Pairing is present in the G1 and G2 phases but not during the intervening S-phase; the status during M-phase is still ambiguous, though significant homolog pairing may persist then as well [7]. Interestingly, also, homolog pairing is disrupted in certain conditions of G2 or G2/M checkpoint arrest, which may imply direct cell cycle regulation of pairing status. Importantly, pairing in both somatic and premeiotic cells is independent of recA homologs, further emphasizing the mechanistic distinction between this "pairing process" and DNA repair or recombination. Finally, these studies imply that the mechanism of homolog pairing is strongly similar in vegetative, premeiotic and early meiotic cells, in support of early speculations (e.g. [9]).

Identification of "trans" effects on chromatin structure and DSB formation

In a diploid nucleus, the nature or behavior of one chromosome is sometimes influenced by the status of its homolog. Such "*trans* " effects might involve direct physical contact between the partners or might occur indirectly, e.g. because one chromosome titrates proteins away from the other. We have long favored the idea that DSB-independent homolog pairing, including that in vegetative and premeiotic cells, might involve direct DNA/DNA contacts between homologous regions [2]. If so, it seemed likely that such contacts might occur in regions of the DNA which were free of nucleosomes (identifiable as "nuclease hypersensitive sites"), rather than in nucleosome-coated segments. We therefore investigated the possibility that the genetic status of a particular nucleosome-free region might influence the chromatin structure of the corresponding region on the homolog. In particular, it might be expected that the

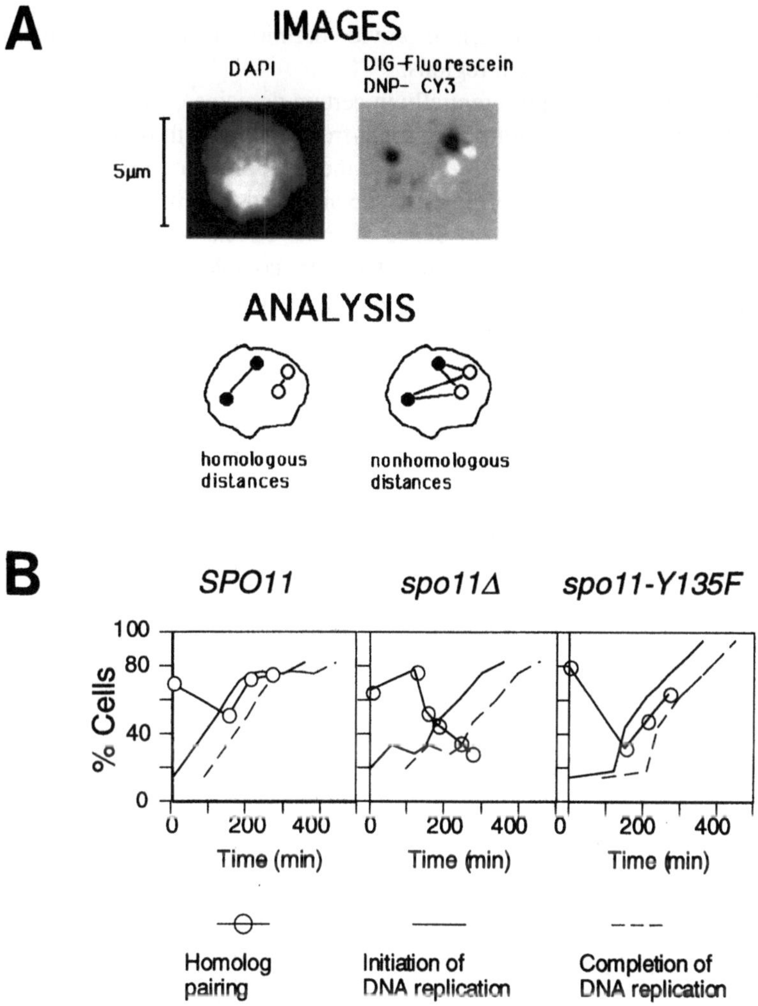

Figure 1. (A) Analysis of homologous chromosome pairing. Contents of a yeast nucleus are spread and subjected to FISH with differentially labeled probes to loci on two different chromosomes. (Images) (Left) Staining of nuclear DNA with DAPI. (Right) Fluorescein (green signal) detection of a 20 kb segment on chromosome III (white spots; probe m) and CY3 (red signal) detection of a 40 kb segment on chromosome XI (black spots; probe f). Images were obtained, stored, analyzed, and presented in digitized form. (Analysis) Distance between signals from homologous segments and from non-homologous segments are measured and compared (from [2]). (B) Chromosome pairing in wild-type, *spo11D* and *spo11-Y135F*. Percentage of cells exhibiting pairing (d ≤ 0.7 µm) at various times during synchronous meiosis was assessed as described above. Also indicated are the cumulative curves for the initiation and completion of meiotic DNA replication.

two regions would influence one another specifically when they were exactly homologous at the DNA level and not when they differed from one another. In fact, such an effect could be observed in premeiotic G1 cells [10]. It should be possible to extend this approach to provide definitive evidence of homology-dependent recogition in such regions.

Meiotic DSBs occur preferentially in certain domains of the chromosomes [11], and specifically within nucleosome-free regions in those domains [12]. We therefore searched for *trans* effects at the level of meiotic DSB formation, using the same DNAse hypersensitive site analyzed for *trans* effects on chromatin structure. Once again, evidence of *trans* effects could be found [13]. Also, the pattern of effects was somewhat more complex in the case of DSBs which led to the speculation that two types of effects were involved, one analogous (and presumably identical) to that observed at the level of chromatin structure in premeiotic cells and a second that was unique to the DSB process. *trans* effects were also observed in a second DSB/recombination study [14]. The exact basis for these effects is still under investigation.

Cre/loxP site-specific recombination

During site-specific recombination, a recombinase protein mediates reciprocal crossing over between a pair of identical genetically-defined sites that may be located in the same or different DNA molecule(s). One such system, derived from bacteriophage P1, involves the action of Cre recombinase on its corresponding pair of *loxP* sites (Fig. 2A). *In vitro,* the rate of Cre-promoted recombination is determined by the concentration of *loxP* sites. If the same situation pertains *in vivo,* the relative rates of Cre-promoted recombination between two or more different pairs of *loxP* sites will reflect, and thus can be used to assess, the relative proximities of the different pairs. For example, if homologs are paired, the rate of recombination between *loxP* sites located at allelic positions on homologs should be higher than the rate of recombination between pairs of *loxP* sites located on non-homologous chromosomes (Fig. 2B). We have developed a genetic system that assesses the rate of Cre/*loxP* recombination in yeast: recombination between two suitably marked *loxP* sites results in fusion of an active promoter to a promoter-less *ura3* gene; these events are detectable in the appropriate strain background as Ura+ prototrophs (Fig. 2C). Application of this assay to vegetatively growing yeast cells has revealed three levels of chromosome organization which are straightforwardly attributable to homolog pairing, centromere clustering (presumably in the Rabl configuration) and proximity of intrachromosomal regions [15]. This assay also detects homolog juxtaposition during meiosis and we hope to use it in the future to isolate mutants defective in DSB-independent pairing.

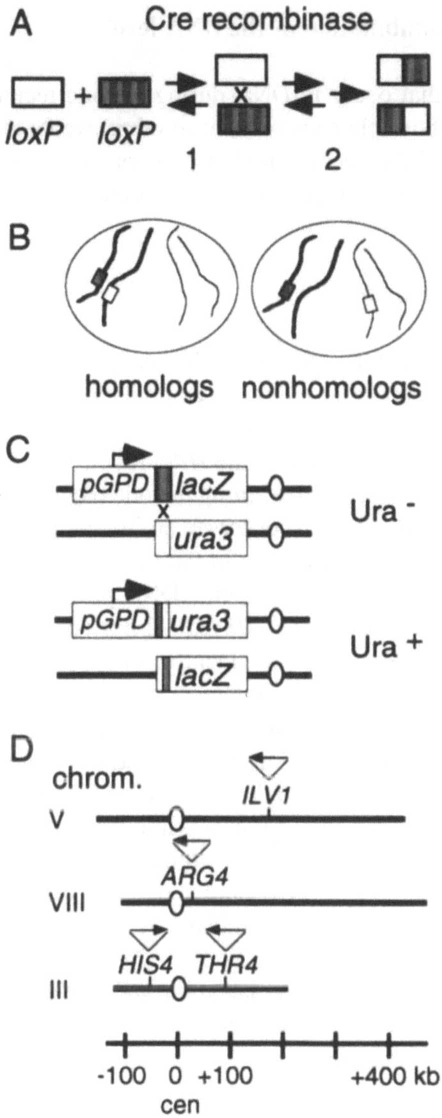

Figure 2. *Cre-loxP* recombination assay. (A) Two steps of the Cre-*loxP* reaction. (B) Recombination rates betwen loci on homologous or non-homologous chromosomes reflect the relative spatial disposition of these sites in the nucleus. (C) *Cre-loxP* recombination creates a *pGPD-ura3* fusion and a Ura+ phenotype. (D) Position and orientation of *loxP* sites integrated into the yeast genome. The orientation of *loxP* results in the reciprocal exchange of chromosome arms following recombination.

Physical assays for recombination at the DNA level

The chemical changes that occur in DNA during meiotic recombination, the time of occurrence of those changes relative to other events of meiosis, and several other features of the recombination process can be investigated by identification and analysis of non-parental DNA species.

DSBs and recombination products

Our laboratory has carried out such studies by analyzing DNA species that arise during meiosis at a particular meiotic recombination hot spot. This work evolved from the initial finding that a molecular insertion of a yeast *LEU2* gene at an "ectopic" position adjacent to the *HIS4* locus, with a bit of bacterial DNA at the fusion junction, created a strong meiotic recombination hot spot. Also, meiosis-specific DSBs could be observed at this site [16]. In this case, a programmed DSB at a particular position in the genome was defined experimentally as an end that is present in extracted DNA prior to cleavage with a restriction enzyme, and a DSB at the *HIS4LEU2* hot spot is observable by Southern blotting analysis as a DNA fragment of the appropriate size that is detected by a probe from the region of interest. Subsequent analysis revealed that meiotic DSBs are generated by Spo11p by a transesterification reaction that leaves protein covalently attached at the 5' strand termini (e.g. [17]). The 5' strands are then specifically resected, leaving behind a 3' single stranded tail which is biochemically suitable to serve as a primer for extension by a DNA polymerase (e.g. [18, 19]). (Fig. 3). Other applications of gel analysis also failed to detect nicking at the *HIS4LEU2* hot spot, consistent with the notion that recombination is initiated at the chemical level almost entirely by DNA double strand breaks (e.g. [13]).

Since a culture of yeast cells can be taken through meiosis as a relatively synchronous population, the time of occurrence of DSBs (and other events of recombination) can be determined relative to the classical cytologically-defined stages of meiosis. Analysis has revealed that DSB formation occurs at leptotene which, in yeast, is characterized by short segments of silver-staining axial elements [20].

Gel analysis can also be used to define mature recombination products. When maternal and paternal homologs differ with respect to particular restriction sites, the occurrence of crossover recombination products can be detected. Recombinational interactions can, however, also generate "non-crossover" products, with the biochemical interactions occurring between two DNA molecules but being resolved to yield a pair of intact chromosomes that have not undergone exchange of flanking chromosome arms. Non-crossover recombination products can also be identified with suitable markers [19]. Temporal

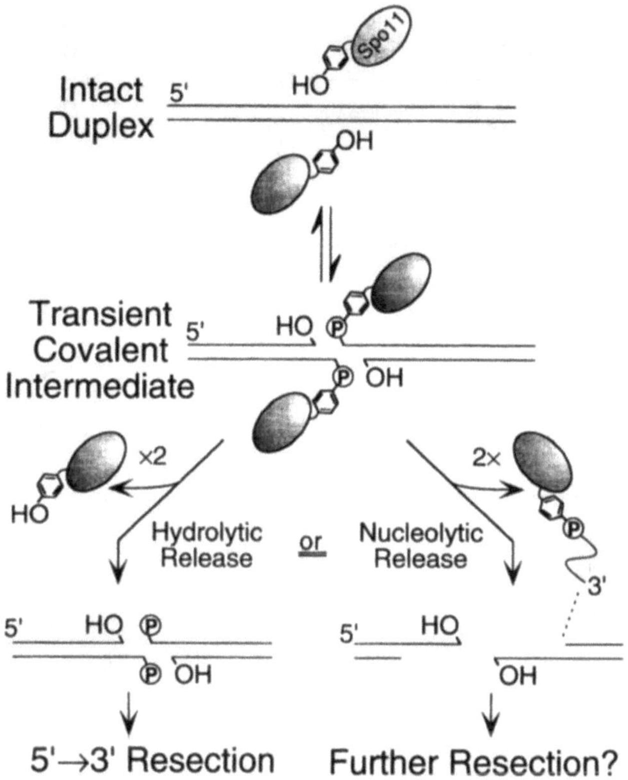

Figure 3. Model for the Mechanism of Meiotic DSB Formation Spo11 (represented by shaded elipses) is proposed to cleave DNA by a revesible transesterase reaction in which a tyrosine side chain on the protein attacks the phosphodiester backbone, forming a covalent phosphodiester linkage between the protein and the 5' terminus of the broken strand and releasing a free 3' OH. Two Spo11 monomers act in concert to generate a pair of closely spaced nicks on opposite strands. The cleavage reaction is driven irreversibly forward by the release of Spo11, either by hydrolysis of the tyrosine phosophodiester, which would release intact Spo11 monomers and a 5' phosphate terminus on the cleaved strand, or by a downstream single-strand nucleolytic cleavage, which would release oligonucleotide-bound Spo11 monomers and a (partially or fully) resected DSB. (from [17], Fig. 6)

analysis has shown that crossover and noncrossover recombination products appear contemporaneously at about the end of pachytene [20]. While it is difficult to be absolutely certain, it is very likely that the recombination process is complete prior to SC disassembly; in any case, mature recombinant products occur with normal timing in cells that arrest at pachytene [22, 13, 23], suggesting that exit from this stage is not required for recombinant maturation. Interestingly, several hours elapse between the time of DSB formation and the time at which recombination products appear, probably because progression of meiotic prophase determines progression of the recombination reaction from one stage to the next [24].

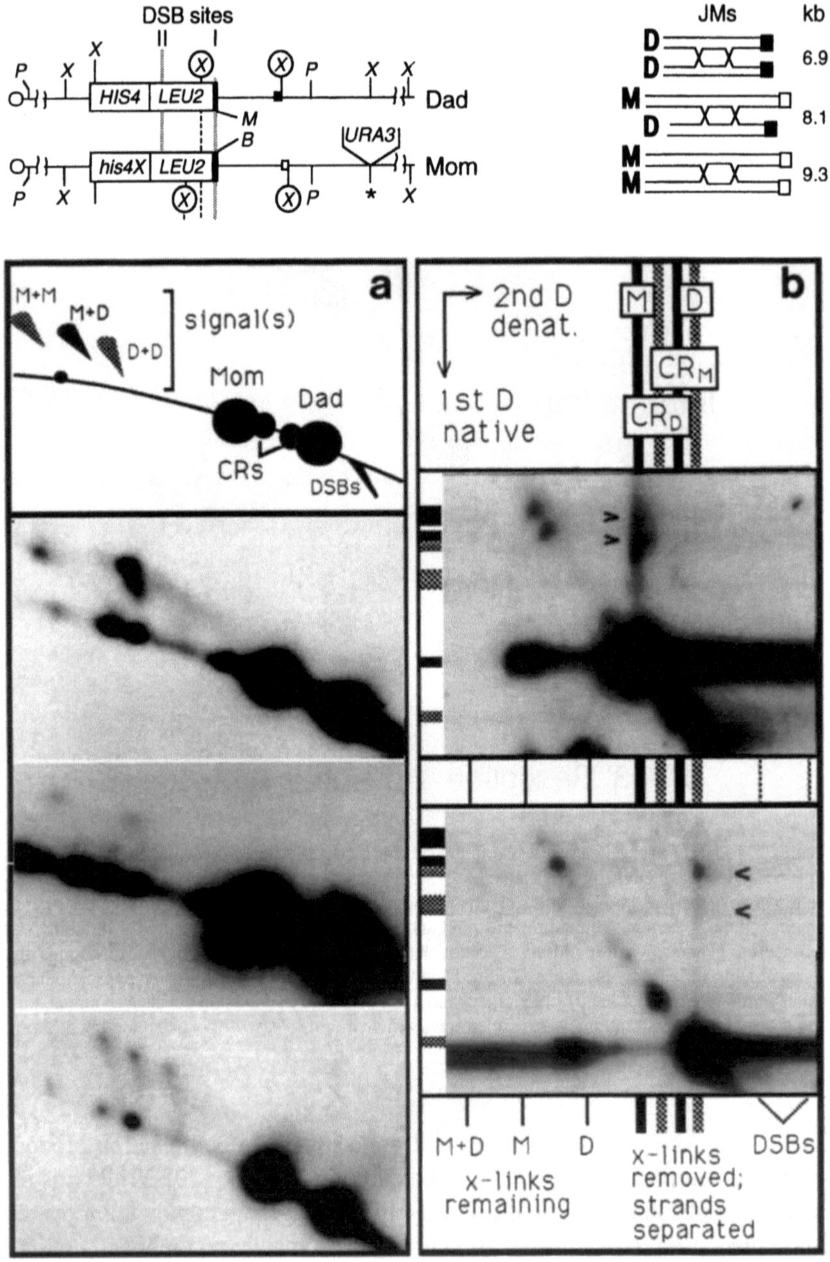

Figure 4. Top: (left): A diploid yeast strain in which homologs differ at genetic and physical markers in the vicinity of a prominent meiotic double strand break/recombination hot spot (*HIS4LEU2*). XhoI restriction sites that flank DSB site I are indicated by circled X's. (right): Double Holliday junctions that have formed between the circled XhoI sites, after excision from the DNA by restriction digestion, are of three types, of differing molecular masses, according to whether the events have occurred

It is also important that no trace of non-crossover recombination products could be observed at earlier stages of meiosis, leptotene or zygotene. This and other findings (e.g. the failure to detect heteroduplex DNA at earlier stages: see [25, 26]) argue against the idea that an early recombinational process that yields only non-crossovers might promote homolog juxtaposition, with crossovers occurring later by a second process [21]. Whether DSB-provoked strand exchange contributes to homolog juxtaposition and whether non-crossover recombination (also known as "gene conversion") has an intrinsic evolutionary significance remains to be determined.

Branched molecules

DSBs and mature recombination products are readily detectable by standard one-dimensional gel electrophoresis. At intermediate stages, however, recombination necessarily involves branched molecules. Byers and colleagues pioneered the use of a specific type of two-dimensional gel electrophoresis for the identification of such molecules [27, 28]. Restriction enzyme-digested DNA is electrophoresed in the first dimension under conditions that separate molecules more or less strictly according to mass; electrophoresis is then carried out in a second dimension under conditions where branched forms lag behind unbranched molecules of the same mass. Byers and colleagues identified two main types of branched structures: those appearing to be double Holliday junctions and those exhibiting a unique "fused junction" form. When combined with Southern blotting with suitable probes, this method reveals branched molecules that arise from DSBs at the *HIS4LEU2* hot spot, and detailed analysis has confirmed that these molecules are recombination intermediates and have the properties expected of double Holliday junctions ([26]; Fig. 4).

---

between sister (M+M and D+D) or non-sister (M+D) chromatids. (Bottom) DNA is extracted from a culture of yeast cells undergoing meiosis, subjected to restriction digestion and analyzed by gel electrophoresis; DNA species from the region of interest are identified specifically by Southern hybridization. (a) Three types of double Holliday junction species. Electrophoresis in the first dimension (left to right) separates molecules according to mass; electrophoresis in the second dimension (top to bottom) separates molecules according to shape, with branched molecules lagging behind the arc of linear forms. The three species shown at top (right) separate from one another in the first dimension and are all retarded, similarly, relative to linear molecules in the second dimension. (from [37]) (b) Strand composition of the three double Holliday junction species. Electrophoresis in the first dimension (top to bottom) separates molecules according to mass; bar key at left shows positions of M+M, M+D and D+D double Holliday junctions and the two parental linear species. DNA is then subjected to denaturing conditions and subjected to electrophoresis in the second dimension (left to right). Positions expected for single DNA strands of various genetic compositions, "parental" (M or D) or "recombinant" (CR_M or CR_D) with respect to the XhoI sites that define their ends. Top and bottom gel panels are probed with sequences present specifically on the "Dad" parent (closed box; top, left) or the "Mom" parent (open box; top, left) respectively. Interhomolog double Holliday junctions, in contrast to several other types of branched species, are expected to contain only the two parental type strands and no recombinant type strands; each intersister double Holliday junction is expected to contain only the single corresponding parental type strand. Results fit these expectations. (from [26]).

These double Holliday junction forms arise during zygotene/early pachytene, between the times of DSBs and mature products. An extension of this analysis has recently revealed a species that likely represents an intermediate between DSBs and double Holliday junctions, "single end invasions", in which one DSB end has interacted with an intact partner duplex.

What's missing?

Two remaining basic questions concerning the nature of meiotic recombination are:

First, do all DSBs progress to mature products via double Holliday junctions? It is generally assumed that double Holliday junctions do give rise to crossover recombination products, though this is not rigorously shown. On the other hand, there are a variety of pathways by which non-crossover products might arise without passing through that type of structure (e.g. [29]).

Second, how are double Holliday junctions resolved? Enzymes that specifically recognize and cleave double Holliday junctions in ways that would give crossover and/or non-crossover products are known in bacteria and in the mitochondria of budding and fission yeasts [30, 31]. As yet, however, no such enzymes have been found in higher organisms that would work on the chromosomes. Alternatively, DNA topoisomerases might be involved in resolving double Holliday junctions [32].

Recombination in context

A definition of the DNA events of meiotic recombination is a necessary prerequisite to analysis of how recombination is integrated with other aspects of meiotic chromosome morphogenesis (for further discussion see [24]). Interesting questions for future study include:

(1) What is the temporal and functional relationship between DSB-independent pairing and the recombination process? Do the two processes overlap at all in time or is there a "pairing stage" followed by a "recombination stage"? Is full homolog colocalization achieved (or achievable) by the non-recombinational pairing process alone or does DNA strand invasion/exchange contribute significantly during normal meiosis? One possibility that can explain diverse findings would be that homolog pairing does most of the work of bringing homologs into joint domains, thereby minimizing the probability of entanglements [2], with recombination occurring regularly in such cases. In regions where the early process fails, however, and corresponding regions of homologs remain far apart and separated by other chromosomes, later events, including chromosome movement, might help resolve the situation, with

recombinational interactions providing the primary juxtaposition of homologs in these "leftover" regions.

(2) Progression of recombination is temporally coordinated with development of chromosome structure during meiotic prophase. How is this coordination achieved? Does the cell cycle provoke changes in chromosome structure which then provoke progression of recombination biochemistry? Does the cell cycle modify the recombination machinery directly? Existing findings already suggest that feedback mechanisms exist to ensure that specific stages of recombination must be completed in order for certain cell cycle transitions to occur. In particular, when recombination is blocked at the DSB stage, or is defective in certain other ways, meiotic yeast cells arrest (or, in some cases, exhibit a significant delay) at pachytene and/or diplotene [19, 33–35].

(3) Recombinational interactions occur in the context of the chromosome axes. When and how does this association occur? Cytological evidence suggests that axis association is established no later than the DSB stage [24, 36].

(4) Recombination occurs preferentially between non-sister chromatids rather than between sisters. This is biologically sensible because crossovers between non-sisters ensure homolog disjunction while crossovers between sisters do not. This bias has been analyzed in wild type and mutant strains with the double Holliday junction assay [37]. With homologs marked by the usual restriction site differences, interhomolog and the two forms of intersister double Holliday junctions are each distinguishable according to their unique molecular masses. The results of this study suggest that there exists an interhomolog-only recombination pathway in which homolog bias is set up prior to (or possibly during) DSB formation; a DSB that forms along this pathway is predestined to interact with a non-sister chromatid. In addition, however, homolog bias must be maintained until the double Holliday junction stage, or perhaps somewhat earlier, at which point partner choice is no longer biochemically reversible. Interestingly, the meiosis-specific recA homolog Dmc1 appears to play a role at this stage. Interhomolog bias clearly involves chromosome structure, however. The abundant axis-associated protein Red1, discovered by Roeder and colleagues [38, 39], is required for at least two aspects: to implement DSB formation along the interhomolog-only pathway and to ensure that DSBs do not interact stably with a partner until Dmc1, with its interhomolog interaction function, has been loaded onto the DSBs. Exactly how each of these genetically defined features is achieved remains to be determined.

(5) Crossover Control. One of the most fascinating aspects of meiosis is the fact that crossovers occur non-randomly both along and among chromosomes [40, 41]. A somewhat oversimplified description [24] considers two main aspects: Every bivalent (pair of homologs) receives an "obligatory" first crossover; then, if additional crossovers occur, they tend to be as far apart as possible. The latter feature corresponds to the classical phenomena of chiasma and crossover interference. Some situations appear to disrupt interference but

not formation of the obligatory crossover (reviewed in [1]). When and how is the recombination process modulated so as to achieve the two crossover control features?

One way of understanding the crossover control process is to say that, at a certain point along the recombination pathway, each recombinational interaction (or potential recombinational interaction) is marked, according to the specific rules of the control process, such that it is destined to become either a crossover or a non-crossover. It has classically been assumed that this decision occurs at the time of (double) Holliday junction resolution ([42]; more recently [43]). Our laboratory has suggested, however, that the decision might occur earlier, at the DSB to double Holliday junction transition [44].

In our model, the fate of a recombinational interaction is determined by the geometry with which a pair of DSB ends invades a partner duplex; this geometry in turn generates two types of double Holliday junctions which, if resolved similarly, will yield crossover and non-crossover products respectively. The two proposed double Holliday junction structures are topological isomers, which are in fact the same two envisioned by Sobell [45]. This model was provoked in part by cytological evidence which suggests that the "crossover control" decision occurs no later than zygotene (reviewed in [1]). Electron microscopy has defined two general classes of recombination nodules, "early/zygotene" and "late/pachytene". The former correlate in timing, number, molecular composition with DSBs and/or double Holliday junctions; the latter correlate in number and position with crossovers. The transition between the two forms can occur as early as mid-zygotene (D. Zickler, personal communication).

Of course, the crossover control decision could be implemented much earlier, even prior to DSB formation, with the outcome (e.g. the geometry of DSB attack) predetermined similarly to interhomolog bias. There is already an indication that some type of control might occur at an early stage because early (pre-SC) associations between homologs in *Allium* occur relatively regularly along the chromosomes, rather than randomly [46]. Whether this cytological non-randomness reflects non-randomness at the DNA level, or only with regard to structures upon which the DNA events are placed, remains an open question, however.

Exactly how might crossover control be implemented? The identification of late pachytene recombination nodules made it clear that crossover control must be implemented no later than pachytene. The first specific suggestion after this point was that of Egel ([47] also [48]) who proposed that SC polymerization might provide the necessary communication along the bivalent; this view has since been supported by Roeder [43]. King and Mortimer [49] subsequently suggested a model in which interference resulted from spreading of another molecule along/atop the SC; also, this model accounted for the obligatory crossover and ensuing interference as a kinetic effect, specifically as a race between nucleation of crossover recombination complexes at recombination sites and elimination of the potential for crossing over at such sites by traversing of the SC. We have recently proposed a different model, motivated by a possible analogy between meiotic chromosomes and a situation often encountered in the physical world in which tension is imposed along the interface between an elastic beam and a thin, brittle film containing flaws ([44, 24, 1], Kleckner, Jones, Henle, Hutchinson and Zickler in preparation). In this model, crossover control is implemented by the imposition and relief of stress along the interface between the structural axes of the homologs and a supra-axial meshwork. Notably, the synaptonemal complex is not required.

Cytological images of late-stage meiotic chromosomes (e.g. at diplotene/diakinesis) give the clear impression that the presence of many and/or closely spaced crossovers (chiasmata) would strongly impede full shortening of the homologs as required for clean separation to opposite poles at telophase I (see [24], for images). Such images presumably contributed to early models for crossover interference which proposed that the final distribution of chiasmata resulted from the effects of late-stage compaction forces. Such models became untenable when it became clear that control was implemented at/before pachytene. Nonetheless, our stress and stress-relief model is an intellectual grandchild of these early ideas because we suppose that crossover control results from axial compaction stress which triggers higher order chromosome coiling/folding along mitotic chromosomes and which, at meiotic prophase, is constrained by the reinforced linear nature of the chromosome axes at that point, thus generating the tension required by our model.

Are meiotic interhomolog interactions derived from mitotic intersister interactions?

We have put forward elsewhere the general idea that many aspects of meiotic interhomolog interactions are evolved from, and mechanistically related to, mitotic intersister interactions [24]. This idea originated during consideration of meiotic crossover control and of the axis-association of meiotic recombinational interactions. The process of meiotic crossover control could be considered to be one of "minimization" – a large number of total recombinational interactions is reduced to a much smaller number of crossovers. One of the simplest predictions of a relationship between interhomolog and intersister interactions is that, in mitotic cells, the force which triggers the onset of higher order chromosome compaction also triggers a minimization of intersister connections. Given the recent discovery of specific molecules involved in mitotic and meiotic sister chromatid cohesion (see [50, 51], for reviews [52]), this and other predictions of our ideas can now be examined experimentally.

Acknowledgements

The author gratefully acknowledges the members of her laboratory for their ideas, experiments, energy and help in putting this article together. Our work on meiosis is funded by grants from the N.I.H. (R01 GM44794; R37 GM25326).

References

1 Zickler D, Kleckner N (1998) The leptotene-zygotene transition of meiosis. *Annu Rev Genet* 32: 619–697
2 Kleckner N, Weiner BM (1993) Potential advantages of unstable interactions for pairing of chromosomes in meiotic, somatic and premeiotic cells. *Cold Spring Harbor Symp Quant Biol* 58: 553–565
3 Scherthan H, Loidl J, Schuster T, Schweizer D (1992) Meiotic chromosome condensation and pairing in *Saccharomyces cerevisiae* studied by chromosome painting. *Chromosoma* 101: 590–595
4 Scherthan H, Eils R, Trelles-Sticken E, Dietzel S, Cremer T, Walt H, Jauch A (1998) Aspects of three-dimensional chromosome reorganization during the onset of human male meiotic prophase. *J Cell Sci* 111: 2337–2351
5 Schwarzacher T, Leitch AR, Bennett MD, Heslop-Harrison JS (1989) *In situ* localization of parental genomes in a wide wheat hybrid. *Ann Bot* 64: 315–324
6 Weiner BM, Kleckner N (1994) Chromosome pairing via multiple interstitial interactions before and during meiosis in yeast. *Cell* 77: 977–991
7 Burgess SM, Kleckner N, Weiner BM (1999) Somatic pairing of homologs in budding yeast: existence and modulation. *Gene Develop* 13: 1627–1641
8 Cha RS, Weiner BM, Keeney S, Dekker J, Kleckner N (1999) Progression of meiotic DNA replication is regulated by interchromosomal interaction proteins, negatively by Spo11p and positively by Rec8p. *Gene Develop* 14: 493–503
9 Stack SM, Brown WV (1969) Somatic pairing, reduction and recombination: an evolutionary hypotehsis of meiosis. *Nature* 222: 1275–1276
10 Keeney S, Kleckner N (1996) Communication between homologous chromosomes: genetic alterations at a nuclease-hypersensitive site can alter mitotic chromatin structure at that site both in *cis* and in *trans*. *Genes Cells* 1: 475–489
11 Baudat F, Nicolas A (1997) Clustering of meiotic double-strand breaks on yeast chromosome III. *Proc Natl Acad Sci USA* 94: 5213–5218
12 Wu TC, Lichten M (1994) Meiosis-induced double-strand break sites determined by yeast chromatin structure. *Science* 263: 515–518
13 Xu L, Kleckner N (1995) Sequence non-specific double strand breaks and interhomolog interactions prior to double strand break formation at a meiotic recombination hot spot in yeast. *EMBO J* 14: 5115–5128
14 Rocco V, Nicolas A (1996) Sensing of DNA non-homology lowers the initiation of meiotic recombination in yeast. *Genes Cells* 1: 645–661
15 Burgess SM, Kleckner N (1999) Collisions between yeast chromosomal loci *in vivo* are governed by three layers of organization. *Gene Develop* 13: 1871–1883
16 Cao L, Alani E, Kleckner N (1990) A pathway for generation and processing of double-strand breaks during meiotic recombination in *S. cerevisiae*. *Cell* 61: 1089–1101
17 Keeney S, Giroux CN, Kleckner N (1997) Meiosis-specific DNA double-strand breaks are catalyzed by Spo11, a member of a widely conserved protein family. *Cell* 88: 375–384
18 Alani E, Padmore R, Kleckner N (1990) Analysis of wild type and rad50 mutants of yeast suggests an intimate relationship between meiotic chromosome synapsis and recombination. *Cell* 61: 419–436
19 Bishop DK, Park D, Xu L, Kleckner N (1992) *DMC1*: a meiosis-specific yeast homolog of bacterial *recA* required for meiotic recombination, synaptonemal complex formation and cell cycle progression. *Cell* 69: 439–456
20 Padmore R, Cao L, Kleckner N (1991) Temporal comparison of recombination and synaptonemal complex formation during meiosis in *S. cerevisiae*. *Cell* 66: 1239–1256
21 Storlazzi A, Xu L, Cao L, Kleckner N (1995) Crossover and noncrossover recombination during

meiosis: timing and pathway relationships. *Proc Natl Acad Sci USA* 92: 8512–8516
22 Xu L, Ajimura M, Padmore R, Klein C, Kleckner N (1995) *NDT80*: a meiosis-specific gene required for exit from pachytene in yeast. *Mol Cell Biol* 15: 6572–6581
23 Rose D, Holm C (1993) Meiosis-specific arrest revealed in DNA topoisomerase II mutants. *Mol Cell Biol* 13: 3445–3455
24 Kleckner N (1996) Meiosis: how could it work? *Proc Natl Acad Sci USA* 93: 8167–8174
25 Goyon C, Lichten M (1993) Timing of molecular events in meiosis in *Saccharomyces cerevisiae*: stable heteroduplex DNA is formed late in meiotic prophase. *Mol Cell Biol* 13 (1): 373–382
26 Schwacha A, Kleckner N (1995) Identification of double Holliday junctions as intermediates in meiotic recombination. *Cell* 83: 783–791
27 Bell LR, Byers B (1983) Homologous association of chromosomal DNA during yeast meiosis. *Cold Spring Harbor Symp Quant Biol* 47: 829–840
28 Byers B, Hollingsworth NM (1994) Meiosis. DNA branching during meiotic recombination. *Curr Biol* 4: 448–451
29 Nassif N, Penney J, Pal S, Engels WR, Gloor GB (1994) Efficient copying of nonhomologous sequences from ectopic sites via P-element-induced gap repair. *Mol Cell Biol* 14: 1613–1625
30 Whitby MC, Dixon J (1998) Substrate specificity of the SpCCE1 holliday junction resolvase of *Schizosaccharomyces pombe*. *J Biol Chem* 273: 35063–35073
31 Kleff S, Kemper B, Sternglanz R (1992) Identification and characterization of yeast mutants and the gene for a cruciform cutting endonuclease. *EMBO J* 11: 699–704
32 Wang JC, Caron PR, Kim RA (1990) The role of DNA topoisomerases in recombination and genome stability: a double-edged sword? *Cell* 62: 403–406
33 Lydall D, Nikolsky Y, Bishop DK, Weinert T (1996) A meiotic recombination checkpoint controlled by mitotic checkpoint genes. *Nature* 383: 840–843
34 Xu L, Weiner BM, Kleckner N (1996) Meiotic cells monitor the status of the interhomolog recombination complex. *Gene Develop* 11: 106–118
35 Sym M, Engebrecht J, Roeder GS (1993) ZIP1 is a synaptonemal complex protein required of meiotic chromosome synapsis. *Cell* 72: 365–378
36 Moens PB, Pearlman RE, Heng HHQ (1998) Chromosome cores and chromatin at meiotic prophase. *Curr Topic Develop Biol* 37: 241–262
37 Schwacha A, Kleckner N (1997) Interhomolog bias during meiotic recombination: meiotic functions promote a highly differentiated interhomolog-only pathway. *Cell* 90: 1123–1135
38 Smith AV, Roeder GS (1997) The yeast Red1 protein localizes to the cores of meiotic chromosomes. *J Cell Biol* 136: 957–967
39 Bailis JM, Roeder GS (1998) Synaptonemal complex morphogenesis and sister-chromatid cohesion require Mek1-dependent phosphorylation of a meiotic chromosomal protein. *Gene Develop* 12: 3551–3563
40 Jones GH (1984) The control of chiasma distribution. *In*: CW Evans, HG Dickinson (eds): *Controlling Elements in Meiosis. Symp Soc Exp Biol* 38: 293–320
41 Jones GH (1987) Chiasmata. *In*: PB Moens (ed.): *Meiosis*. Academic Press, New York, 213–244
42 Holliday R (1964) A mechanism for gene conversion in fungi. *Genet Res* 5: 282–304
43 Roeder GS (1997) Meiotic chromosomes: it takes two to tango. *Gene Develop* 11: 2600–2621
44 Storlazzi A, Xu L, Schwacha A, Kleckner N (1996) Synaptonemal complex component Zip1 plays a role in meiotic recombination independent of SC polymerization along the chromosomes. *Proc Natl Acad Sci USA* 93: 9043–9048
45 Sobell HM (1974) Concerning the stereochemistry of strand equivalence in genetic recombination. *In*: RF Grell (ed.): *Mechanisms in Recombination*. Plenum Press, New York, 433–438
46 Albini SM, Jones GH (1987) Synaptonemal complex spreading in *Allium cepa* and *A. fistulosum*. I. The initiation and sequence of pairing. *Chromosoma* 95: 324–338
47 Egel R (1978) Synaptonemal complex and crossing over: structural support or interference? *Heredity* 41: 233–237
48 Egel R (1995) The synaptonemal complex and the distribution of meiotic recombination events. *Trends Genet* 11: 206–208
49 King JS, Mortimer RK (1990) A polymerization model of chiasma interference and corresponding computer simulation. *Genetics* 126: 1127–1138
50 Biggins S, Murray AW (1998) Sister chromatid cohesion in mitosis. *Curr Opin Cell Biol* 10: 769–775
51 Hirano T (1998) SMC protein complexes and higher-order chromosome dynamics. *Curr Opin Cell*

Biol 10: 317–322
52 Parisi S, McKay MJ, Molnar M, Thompson A, van der Spek PJ, van Drunen-Schoenmaker E Kanaar R, Lehman E Hoeijmakers JHJ, Kohli J (1999) Rec8p, a meiotic recombination and sister chromatid cohesion phosphoprotein of the Rad21p family conserved from fission yeast to humans. *Mol Cell Biol* 19: 3515–3528

Chromosomes Today, Vol. 13
ed. by E. Olmo and C.A. Redi

Meiosis in budding yeast and in multicellular eukaryotes – similarities and differences

Josef Loidl

Institute of Botany, University of Vienna, Rennweg 14, A-1030 Vienna, Austria

For many decades, the investigation of meiotic structures and processes has been almost exclusively a domain of cytology. Meiosis was studied preferably in organisms with large chromosomes and/or easily and abundantly available meiocytes, with favourite systems being monocotyledon pollen mother cells and locust and mammalian spermatocytes. During the last few years, however, we have witnessed the rapidly increasing role of molecular biology, and together with it, the rise of the budding yeast, *Saccharomyces cerevisiae*, as the now best-studied meiotic system. Yeast offers the advantage of elaborate genetics and inducible and highly synchronous meioses, which have allowed a wealth of studies on structural and regulatory genes involved in meiosis. Its initially poor amenability to cytological investigation, owing to the smallness and low degree of condensation of its chromosomes, has been overcome by the development of techniques for electron microscopical analysis of synaptonemal complexes (SCs), and the application of fluorescence *in situ* hybridization (FISH), immunostaining, and green fluorescent protein (GFP)-tagging of nuclear components.

It has to be kept in mind, however, that yeast is a highly specialized unicellular microorganism with the smallest known genome of all eukaryotes. It has about 13 Mbp (megabasepairs) of DNA (1C) and close to 6000 genes. Therefore, the question arises whether the meiotic processes of homologous chromosome recognition, pairing and recombination that have been studied in yeast can be taken as models for the corresponding processes in higher eukaryotes with much larger genomes. Humans, for instance, possess 3400 Mbp of DNA and probably close to 100 000 genes, and the DNA content of some plants and amphibia is well over 20 000 Mbp. In fact, whereas the microscopical structure of the SC is virtually identical, there seem to be considerable differences in the way the chromatin is organized along the SC. An increasing number of genes have been identified in higher eukaryotes, which are homologous to known yeast meiotic genes. On the other hand, cytologically similar components of the SC seem to be composed of completely different proteins. Moreover, recent studies have shown that although the same molecular processes seem to be involved in meiotic pairing and recombination, the order in which these events occur may be different. Here, some recent findings on meiosis in yeast will be reviewed and their significance for our understanding of higher eukaryotic meiosis will be evaluated.

Homology recognition and presynaptic alignment

Between organisms there seem to exist different strategies of how and when homologous contacts, which eventually lead to meiotic chromosome pairing, are made. For the majority of organisms it is believed that homologous encounters happen during meiotic prophase and for many it has been explicitly reported that the first visible contact between homologs occurs at leptotene of meiotic prophase [1–4]. There is, however, a long history of reports from a variety of organisms on pairing of whole chromosomes or subchromosomal regions in non-meiotic nuclei (for review see Ref. [5]).

The most striking case for somatic (vegetative) pairing is in *Drosophila*, where homologous chromosomes start to associate after about 14 mitotic generations in the embryo [6]. In some organisms, association of homologs seems to occur during several premeiotic cell generations or premeiotic interphase. In the fission yeast, *Schizosaccharomyces pombe*, where zygotic meiosis is the rule, the association of homologous chromosome territories immediately following karyogamy, may be a first step towards meiotic pairing. This association is maintained even under conditions where an extended diploid phase follows karyogamy [7]. For *S. pombe* and *Drosophila* males, both of which lack an SC, one could argue that premeiotic pairing was installed as a mechanism to compensate for this loss.

Loidl et al. [8] and Weiner and Kleckner [9] observed the association of homologous chromosome regions in budding yeast nuclei as early as the point of induction of meiosis by transfer to sporulation medium. These cells, however, had undergone growth in a medium that conditions them for entering meiosis. Therefore, this association cannot be considered as truly vegetative. Keeney and Kleckner [10] provided additional circumstantial evidence that homologous chromosomes in vegetatively growing yeast communicate in a way that could be explained by their direct physical interaction. We could not detect unequivocal vegetative pairing in our lab (Q.-W. Jin, unpublished observations). Instead, we have attempted to explain the weak tendency of homologous chromosome regions to colocalize by the sorting of chromosome arms according to their lengths which is dictated by the polarized Rabl-like arrangement of chromosomes in interphase nuclei [11]. Only after irradiation-induced DNA breakage have we observed an increase in vegetative pairing (J. Fuchs, unpublished observations). This can be explained as the association of damaged chromosome regions with their homologs which are used as templates for recombinational repair in unreplicated G_0 cells.

Whatever the nature and biological function of vegetative pairing may be, it is doubtful whether it contributes to the initiation of meiotic pairing, since it has been shown to be abolished during premeiotic S-phase [9]. However, homologous association is re-established early in meiotic prophase and it clearly precedes synapsis (see Refs [9, 12–14]). Also in many other organisms, presynaptic alignment of homologous chromosomes is an early feature of meiosis (see, e.g., Refs [5, 15, 16]). It was postulated to be mediated by the interaction of intact DNA molecules, allowing the rapid and low energy-cost generation of transient contacts. These "kissing interactions" would seem to be an efficient mechanism of primary homology-sensing in the course of screening the whole genome [9, 17–19]. If the number of such weak interactions between homologous chromosomes is sufficiently large, the resulting alignment will be stable enough to allow the next steps in homologous pairing and recombination to occur.

SC structure

The SC occurs in most eukaryotes and is a highly conserved structure (Fig. 1). As can be seen from Figure 2, the width of the SC (as measured between the centers of lateral elements) is almost identical in plants, man, and yeast. Also many details of its ultrastructure are similar. In the few organisms where it is completely missing, like in *Schizosaccharomyces pombe*, *Aspergillus* or in male *Drosophila*, or where it is limited to short chromosome segments, like in the flatworm *Mesostoma* [20], this is clearly due to a secondary evolutionary loss.

An increasing number of homologs of genes that were first recognized as encoding structural or regulatory components involved in yeast meiosis, are also being identified in higher eukaryotes. Among them are RAD51, DMC1 and SPO11 which function in the recombination pathway and whose products colocalize with recombination nodules. The yeast Red1p and the rat Scp2p [21] are components of the lateral elements of the SC and show similarities at the protein level. Zip1p and SCP1p, coiled-coil proteins with similar predicted secondary structures, have been identified as a major constituent of the transversal filaments of the SC in yeast and in rat. However, their amino acid sequences did not show notable homologies [22, 23]. If Zip1p/SCP1p do turn out to be the key elements for the physical link between the lateral elements of the SC, then we would witness the remarkable situation where merely analogous proteins can substitute each other as essential components of an otherwise highly conserved structure.

The temporal order of synapsis and recombination

The classical view of meiosis was that synapsis precedes recombination and it was taken as obvious that the synaptonemal complex provides the framework

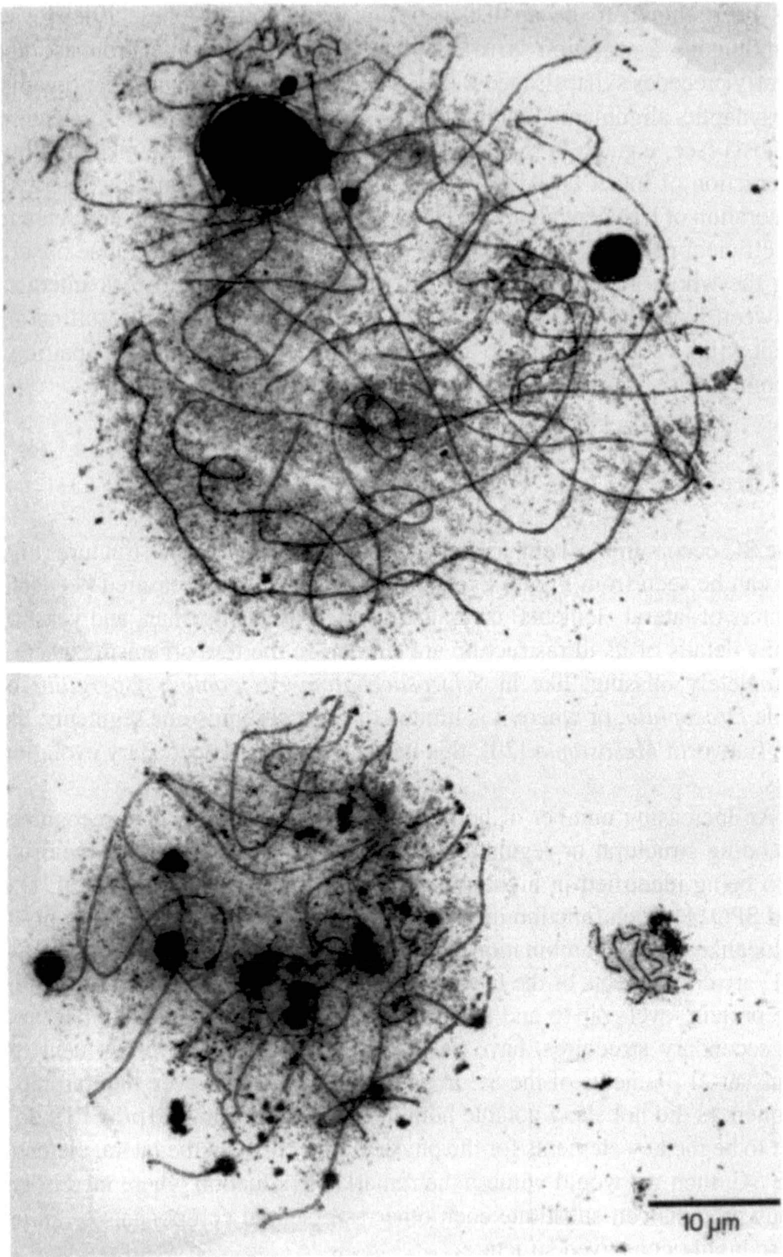

Figure 1. Comparison of SCs (spread, silver-stained, low-magnification electron microscopic images) of (a) the plant *Allium*, (b) human male and (c) *S. cerevisiae*. The measured SC lengths in these cells are of the ratio 30: 8: 1. The width of the SC of yeast looks smaller than that of the other organisms in this particular image because the lateral elements of yeast SCs are fainter and, therefore, the space between them is only poorly contrasted.

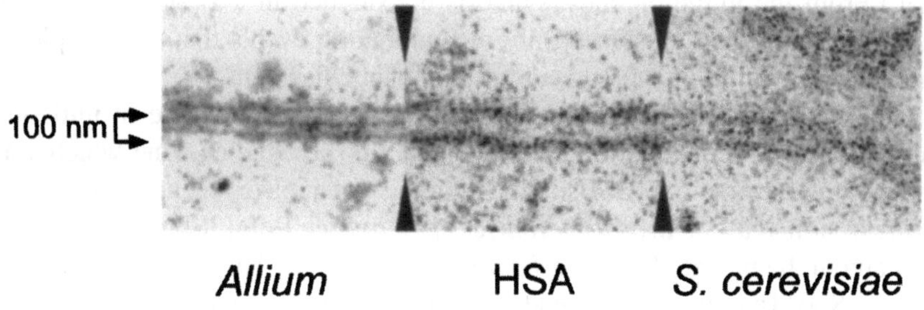

Figure 2. Widths of SCs in *Allium*, human (HSA) and yeast (*S. cerevisiae*). SC widths are highly conserved in a wide variety of species and are around 100 nm.

which holds homologous chromosomes tightly together, thereby allowing the physical exchange of sequences to take place (see Ref. [5]). This view was then challenged by studies in yeast, which showed that DNA double-strand breaks (DSBs), the first step in meiotic recombination, occur well ahead of synapsis and that synapsis is greatly impaired in mutants which are incapable of initiating recombination. On the other hand, mutations which confer a structural defect in the SC had only mild effects on recombination (for review see Ref. [24]). Thus, it seems that in yeast recombination intermediates are a structural or regulatory requirement for SC formation and that synapsis is initiated at sites of incipient recombination. Recently it was shown that SC formation initiates at a few chromosomal sites where Zip2 protein is associated with chromosomes. Zip2p is present only when DSBs have occurred, and the frequency and distribution of Zip2p aggregates suggest that they correspond to DSB sites [14].

DSBs seem to appear only after homologous chromosomal regions have been aligned, because DSBs can in some way sense the presence of a homolog which could serve as a template for their repair. If no homologous chromosome is available (as, e.g., in a haploid), DSB formation is reduced [25, 26]. Those DSBs which do appear in the absence of a homologous chromosome, undergo delayed repair, probably by using the sister chromatid as a template [27, 28].

Thus, from yeast the following chain of meiotic pairing events is emerging: First, homologous chromosomes become aligned by multiple but weak associations (see above). Then DSBs occur along the chromosomes. They widen into gaps which are flanked by 3' single-stranded tails. Strand exchange proteins bind to the single-stranded DNA regions and mediate recombinational repair of the gaps (see, e.g., Ref. [29]) by using the homologous chromosome as a template. In the course of recombinational repair, crossovers are frequently formed. At the same time, SC is initiated at the sites of DSBs. (The DSB alone is sufficient to nucleate homologous SC formation, for in the mutant rad50S, which accumulates unmended DSBs, synapsis is nearly normal.) From the initiation sites, synapsis is extended in a zipper-like fashion along the entire chromosome. Only recombination events that are completed in the context of an intact SC lead to functional chiasmata.

Thus in yeast, meiotically induced DSBs play a central role in initiating most if not all recombination. Homologs of Spo11p, the enzyme which was shown to induce meiotic DSBs in the budding yeast, have recently been found in the fission yeast *Schizosaccharomyces pombe* (see Ref. [30]), in *Drosophila* [31] and in the nematode *Caenorhabditis elegans* [32]. The *Drosophila* and *C. elegans* Spo11p homologues are also required for recombination. If in a *C. elegans* spo11 null mutant breaks are induced by radiation, then recombination is partially restored [32], which suggests that the normal role of Spo11p in *C. elegans* is to generate DSBs. Further evidence for a common role of meiotic DSBs in yeast and higher eukaryotes comes from human-derived yeast artificial chromosomes (YACs). It has been shown that DSBs appear in YACs, and their extent correlates well with the degree of meiotic recombination measured at these loci in humans [33].

On the other hand, higher eukaryotes may differ from yeast with respect to the function of recombination in initiating synapsis. It was recently shown in *Drosophila* that meiotic synapsis occurs in the absence of any detectable form of recombination [34]. Likewise, in *C. elegans* the homologue of Spo11p is not required for homologous synapsis [32]. Although homologous synapsis may not require DSB formation and initiation of recombination in higher eukaryotes, early recombination nodules (which contain components of the recombination machinery [35–37]) are formed at the sites where the axial elements of homologs converge and synapsis starts [38, 39]. This makes it difficult to decide whether synapsis and the initiation of recombination occur more or less simultaneously or whether synapsis precedes the initiation of recombination by a small but significant period of time, which would agree with the classical model of recombination taking place between synapsed chromosomes.

It is conceivable that in higher eukaryotes with a much larger DNA content and a higher proportion of repeated DNA sequences, additional measures are necessary to avoid unequal recombination of dispersed homologous sequences. Hence synapsis could serve to preselect homologous chromosomes (or large pieces thereof) as legitimate partners for exchange. In the mouse, 1–2 megabase λ-phage transgenes were inserted at ectopic chromosomal loci. At pachytene these inserts did not associate and no evidence for chiasmata between these regions were found at metaphase. It was concluded that 1–2 megabase of DNA is insufficient to support ectopic recombination in mice and that homology is assessed within the broader context of the chromosome to initiate synapsis at meiotic prophase [40]. In yeast, on the other hand, short artificially inserted DNA tracts of several hundred bp can autonomously check homology and recombine irrespective of their chromosomal position [41]. However, a preference does exist for sequences in closer vicinity on the same chromosome [42]. Thus, a major difference between yeast and higher eukaryotic meioses seems to be that in yeast short homologous stretches can exchange, in principle, with homologous partners anywhere in the genome, whereas in higher eukaryotes homologous sequences undergo recombination only if they are present within the context of synapsed homologous chromosomes.

In yeast, the inherent susceptibility to ectopic recombination of unsynapsed sequences is partially suppressed by several mechanisms. First, telomeric regions (which are particularly rich in repetitive sequence elements) are sequestered to peripheral regions of the nucleus, such that nuclear architecture may prevent excessive harmful ectopic recombination [43]. Second, homologous chromosomes are aligned prior to the stage when recombination occurs, and perhaps even premeiotically, which does not seem to be the case for all eukaryotes (see above). Third, mismatch repair reduces ectopic recombination between homologous sequences [44, 45]. Fourth, dispersed repeats may undergo silencing by "heterochromatinization" that both represses transcription and recombination [46–48]. Together these conditions may make ectopic recombination negligible in normal yeast meiosis. However, the absence of a homologous chromosome (as for instance in haploid meiosis) stimulates ectopic recombination [49].

The centromere – telomere clustering switch

In mitotically active tissues of many higher eukaryotes, interphase chromosomes assume the so-called Rabl-orientation, with their centromeres assembled near one pole of the nucleus and the telomeres extending towards the other pole. This polarized arrangement of chromosomes is commonly interpreted as a consequence of the pole-directed movement of centromeres and the trailing of chromosome arms at preceding anaphase (for review see Ref. [11]). In *S. cerevisiae*, centromere clustering reminiscent of Rabl-orientation was found in vegetative cells [11]. It is still unknown whether this Rabl-like arrangement is a remnant of anaphase chromosome polarization or whether it can be formed *de novo* during interphase.

A different type of chromosome arrangement, the bouquet, where all telomeres are assembled in a peripheral region of the nucleus, occurs at meiotic prophase. The bouquet has been described for a wide variety of higher organisms (for review see Ref. [50]). Its presence in yeast has been observed sporadically and was recently established by FISH with a pan-telomeric probe [51]. Both in yeast and in higher organisms it occurs around zygotene and chromosome ends assemble in the vicinity of the microtubule organizing centers (the spindle pole body in fungi).

Rabl-orientation of chromosomes exists also in wheat [16]. During floral development, the centromeres remain clustered whereas the association of telomeres is less obvious. Very early in meiotic prophase, however, telomeres start to cluster whereas centromeres lose their association [15, 52]. A very similar phenomenon was observed in *Schizosaccharomyces pombe*. Here, centromeres are clustered near the spindle pole body in mitotic nuclei. During karyogamy, which is at the beginning of the meiotic pathway, the telomeres also associate with the spindle pole body. Later, the centromeres detach, and only the telomeres remain clustered [53]. Recently, the observation of the clustering of centromeres in vegetative nuclei in budding yeast, the resolution of the centromere cluster in meiosis [11], and the clustering of telomeres at the nuclear membrane [51] has shown that this centromere telomere clustering switch may be a common event (Fig. 3).

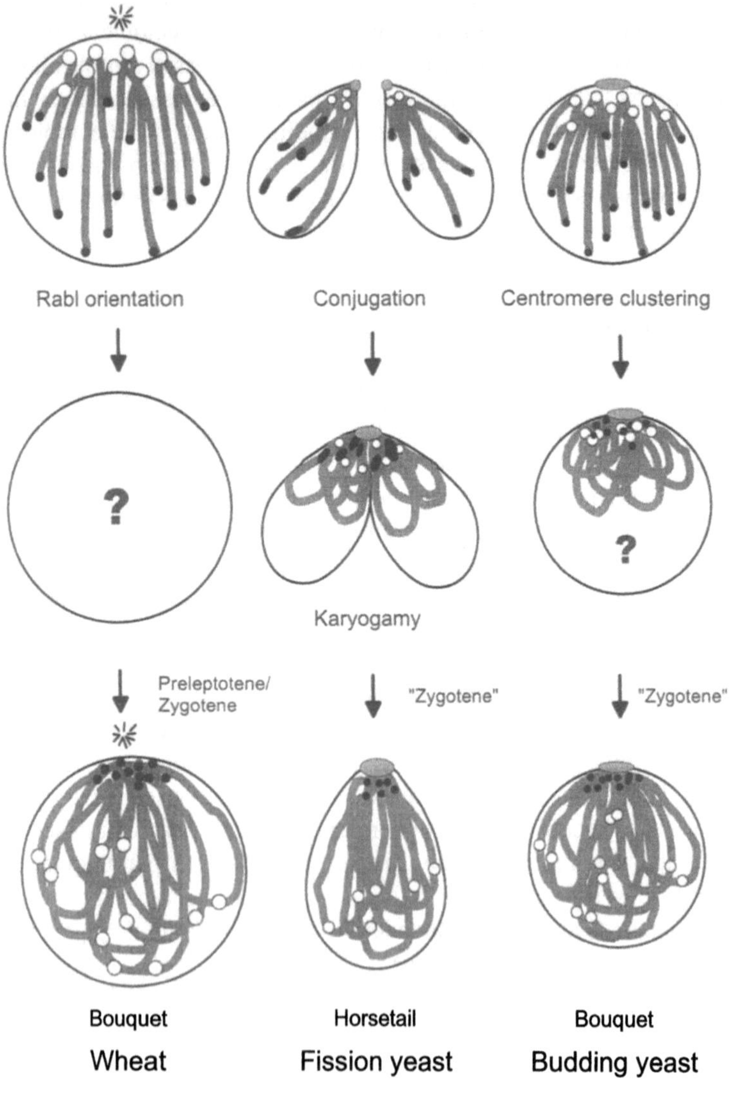

A connection between meiotic chromosome condensation and recombination frequency

It appears to be a general phenomenon that recombination frequency per physical map length decreases with increasing genome size. This is very obvious from the comparison of data from members of the fungi, plant and animal kingdoms with very large differences in their DNA contents (Tab. 1, Fig. 4). This negative correlation seems to hold also within closely related groups of

Table 1. Synopsis of data from the literature on 1C DNA contents (in µm), SC lengths (in µm) and mean recombination rates (Centimorgans per kilobase – cM/kb) in a variety of organisms. (Letters in parentheses are used in Fig. 4 as symbols for the organisms.) The ratios of DNA length per SC length are also shown. Data for which no source is given are referenced in [19]. DNA amounts from the literature were converted to µm by using the functions 1 pg $\cong 0.98 \times 10^9$ bp $\cong 3.3 \times 10^5$ µm

	1C DNA (µm)	SC (µm)	DNA/SC	cM/kb
Budding yeast (Y)	5×10^3	28	180	0.26
Neurospora (N)	1.4×10^4	46	300	0.05
C. elegans (C)	2.7×10^4 [67]	41	660	0.004
Drosophila female (D)	7×10^4	110	640	0.0017
Arabidopsis (A)	2×10^5	132[b]	1510	0.003
Tomato (T)	3.6×10^5 [59]	207 [69]	1740	0.001[e]
Zebrafish (Z)	5.3×10^5 [a]	280[c]	2000	0.0017[f]
Pig (P)	9.3×10^5 [68]	167 [70, 71]	5567	0.007[g]
Human male (H)	1×10^6	245[d]	4330	0.009 [72]
Mouse (M)	1×10^6	156	6410	0.006
Zea mays (ZM)	2.5×10^6	353	7082	0.001

Key to references: [a]mean of photometric measurements by Hinegardner and Rosen [73] and J. Greilhuber (personal communication), [b]mean of values given in Refs [19, 74], [c]Loidl, "unpublished observations", [d]mean of measurements in Refs [75, 76], [e]calculated on the basis of a linkage map length of 1063 cM [77], [f]calculated on the basis of a linkage map length of 2720 cM [78], [g]calculated on the basis of a linkage map length of 2000 cM [79].

Figure 3. Centromere-telomere clustering switch in wheat (after Ref. [52]), the fission yeast, *Schizosaccharomyces pombe* (after Ref. [53]), and the budding yeast, *Saccharomyces cerevisiae* [51]. The strongly simplified cartoon highlights the transition from a Rabl-like chromosome orientation with centromeres assembled near the microtubule-organizing centers to a bouquet arrangement early in meiosis. In the case of wheat it is not clear whether Rabl-orientation alone contributes to the centromere clustering or if during zygotene the association of pericentromeric regions, known in many plants as the synizetic knot, also plays a role [65, 66]. In *S. pombe* there exists a stage when both centromeres and telomeres are clustered near the spindle pole. In *S. cerevisiae* a few such nuclei were observed. It is not known whether they represent a short but consistent intermediate state or whether resolution and formation of centromere and telomere clusters occur in an uncoordinated way with only some nuclei accidentally showing both. Also, in wheat the intermediate state is not known. Centromeres are shown as open circles, telomeres as black dots. The centrosome of the wheat is represented by the aster-shaped structure next to the nucleus, the yeast spindle pole bodies are shown as shaded areas on top of the nuclei. Drawing is not to scale.

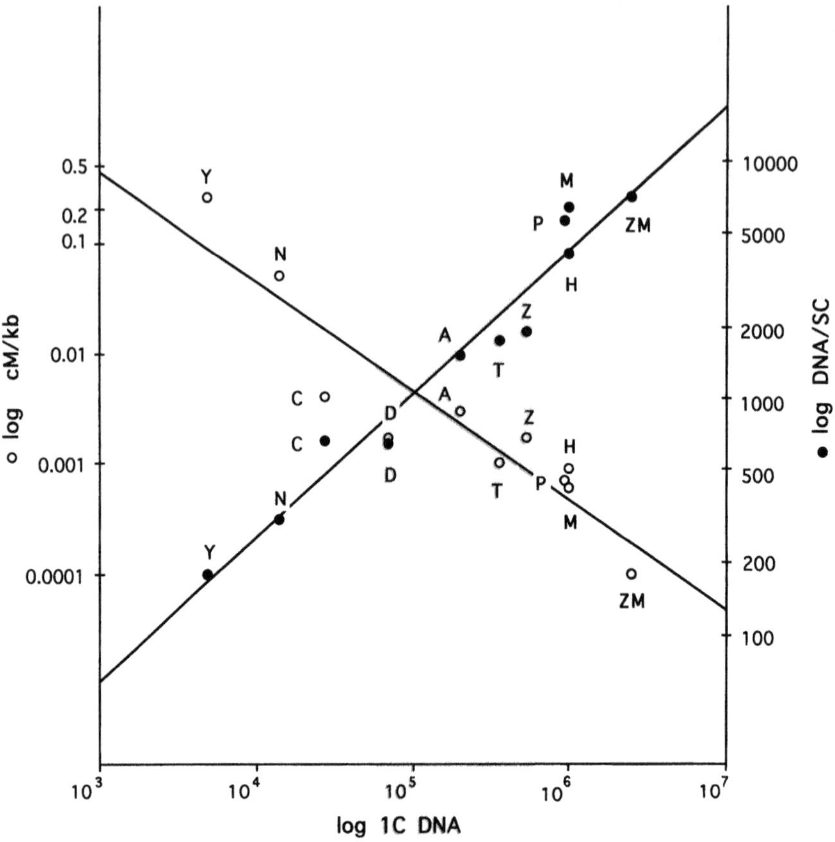

Figure 4. Log/log plot of 1C DNA content versus recombination rate (from Tab. 1) and DNA/SC ratio in a variety of organisms. The letters, as symbols for organisms, correspond to Table 1. There is a negative correlation between DNA content and recombination frequency, a positive correlation between DNA content and packaging density of the DNA along the SC (meiotic chromosome cores) and a negative correlation between packaging density and recombination frequency.

organisms, since Rees and Durrant [54] found a decrease in the chiasma frequency per picogram of DNA with an increase in nuclear DNA amount within the three plant genera *Lathyrus*, *Lolium* and *Petunia*.

Several factors have been proposed to govern the frequency and distribution of meiotic recombination events on a local and a global scale. Local chromatin conformation may determine the susceptibility of DNA to meiotic DSBs as a prelude to recombination [55, 56]. Centromeres, telomeres, heterochromatin bands and other structural features of chromosomes control the intrachromosomal distribution of crossovers/chiasmata (see Ref. [57]). Recombination also depends on chromosome size with smaller members of a chromosome set showing a disproportionally high recombination frequency (e.g., [58]).

Another global control of recombination frequency seems to be exerted by chromatin packaging. As can be seen from Table 1, SC length does not change proportionally with genome size. Therefore, the amount of DNA that is accomodated per unit SC increases with DNA content (Fig. 4). Whereas this relationship becomes apparent when a range of organisms with widely varying DNA contents is compared, the situation is not so obvious within lower systematic entities. For instance, Anderson et al. [59] reported a strong correlation between total SC length and genome size for a variety of higher plants. On the other hand, in human females the SC is twice as long as in males [60], whereas genome size is virtually the same.

Figure 4 illustrates that meiotic recombination frequency is inversely proportional to pachytene chromatin compaction (DNA/SC length ratio). In man, for instance, packaging of DNA along the SC is ~24 times as dense as in yeast whereas the average recombination per unit DNA is ~37 times lower. This rule (see also Refs [19, 61, 62]) is corroborated by the observation that chiasma frequency is lower in spermatocytes in species where spermatocyte SCs are shorter than oocyte SCs (see [61] for references). It can be explained by a model of DNA packaging by the formation of loops at the meiotic chromosome core. When the packaging is loose (many small loops for a given amount of DNA), a higher proportion of DNA may be available for recombination than with tight packaging. Accordingly, human DNA in a yeast artificial chromosome (YAC), which adopts the loose packaging in meiosis that is characteristic of yeast [62], recombines at the same high rates as yeast DNA [63].

Conclusions

The budding yeast, *S. cerevisiae*, has become an important tool for the investigation of meiosis in spite of its rather exceptional position among eukaryotes as a unicellular organism with an extremely small genome. Attempts to generalize its meiotic behaviour have to take into account these properties. There are notable differences between yeast and multicellular organisms with larger genomes in the degree of chromosome compaction and recombination frequency per DNA amount. These are not fundamental and rather seem to be adaptations to ensure appropriate numbers of chiasmata in the presence of highly varying genome sizes. On the other hand, the molecular processes and cytological structures are highly conserved. DNA double-strand breaks, for instance, seem to initiate recombination not only in yeast but also in metazoans and, in addition, the ultrastructure of the SC is virtually identical. Also, at higher cytological levels there exist parallels with respect to the formation of a chromosomal bouquet at meiotic prophase, and presynaptic alignment as a reflection of homology recognition at a premeiotic or very early meiotic stage is found in yeast and in many higher eukaryotes. The only major difference found so far is that in yeast DSBs are required for the initiation of homologous synapsis, whereas in *Drosophila* and *C. elegans* this is not the case. However, as McKim and Hayashi-Hagihara [31] pointed out, this may only reflect a higher capability of these metazoans to convert DSB-independent presynaptic alignment into synapsis. With about 500 genes known to be induced during sporulation [64] the task of getting a clear picture of yeast meiosis seems huge, but achievable. With an increasing number of homologous meiotic genes coming in from eukaryotic sequencing projects the elucidation of the basic mechanisms of meiosis comes within reach.

Acknowledgements

This work was supported by grant S8202-GEN from the Austrian Science Fund (FWF).

References

1 Armstrong SJ, Kirkham AJ, Hultén MA (1994) XY chromosome behaviour in the germ-line of the human male: a FISH analysis of spatial orientation, chromatin condensation and pairing. *Chromosome Res* 2: 445–452
2 Dawe RK, Sedat JW, Agard DA, Cande WZ (1994) Meiotic chromosome pairing in maize is associated with a novel chromatin organization. *Cell* 76: 901–912
3 Scherthan H, Weich S, Schwegler H, Härle M, Heyting C, Cremer T (1996) Centromere and telomere movements during early meiotic prophase of mouse and man are associated with the onset of chromosome pairing. *J Cell Biol* 134: 1109–1125
4 Bass HW, Marshall WF, Sedat JW, Agard DA, Cande WZ (1997) Telomeres cluster *de novo* before the initiation of synapsis: a three-dimensional spatial analysis of telomere positions before and during meiotic prophase. *J Cell Biol* 137: 5–18
5 Loidl J (1990) The initiation of meiotic chromosome pairing: the cytological view. *Genome* 33: 759–778
6 Hiraoka Y, Dernburg AF, Parmelee SJ, Rykowski MC, Agard DA, Sedat JW (1993) The onset of homologous chromosome pairing during *Drosophila melanogaster* embryogenesis. *J Cell Biol* 120: 591–600
7 Scherthan H, Bähler J, Kohli J (1994) Dynamics of chromosome organization and pairing during meiotic prophase in fission yeast. *J Cell Biol* 127: 273–285
8 Loidl J, Klein F, Scherthan H (1994) Homologous pairing is reduced but not abolished in asynaptic mutants of yeast. *J Cell Biol* 125: 1191–1200
9 Weiner BM, Kleckner N (1994) Chromosome pairing via multiple interstitial interactions before and during meiosis in yeast. *Cell* 77: 977–991
10 Keeney S, Kleckner N (1996) Communication between homologous chromosomes: genetic alterations at a nuclease-hypersensitive site can alter mitotic chromatin structure at that site both in cis and in trans. *Genes Cells* 1: 475–489
11 Jin Q-w Trelles-Sticken E, Scherthan H, Loidl J (1998) Yeast nuclei display prominent centromere clustering that is reduced in nondividing cells and in meiotic prophase. *J Cell Biol* 141: 21–29
12 Scherthan H, Loidl J, Schuster T, Schweizer D (1992) Meiotic chromosome condensation and pairing in *Saccharomyces cerevisiae* studied by chromosome painting. *Chromosoma* 101: 590–595
13 Roeder GS (1997) Meiotic chromosomes: it takes two to tango. *Gene Develop* 11: 2600–2621
14 Chua PR, Roeder GS (1998) Zip2, a meiosis-specific protein required for the initiation of chromosome synapsis. *Cell* 93: 349–359
15 Aragón-Alcaide L, Reader S, Beven A, Shaw P, Miller T, Moore G (1997) Association of homologous chromosomes during floral development. *Curr Biol* 7: 905–908
16 Schwarzacher T (1997) Three stages of meiotic homologous chromosome pairing in wheat: cog-

nition, alignment and synapsis. *Sex Plant Reprod* 10: 324–331
17 Camerini-Otero RD, Hsieh P (1993) Parallel DNA triplexes, homologous recombination, and other homology-dependent DNA interactions. *Cell* 73: 217–223
18 Kleckner N, Weiner BM (1993) Potential advantages of unstable interactions for pairing of chromosomes in meiotic, somatic, and premeiotic cells. *Cold Spring Harbor Symp Quant Biol* 58: 553–565
19 Loidl J (1994) Cytological aspects of meiotic recombination. *Experientia* 50: 285–294
20 Oakley HA, Jones GH (1982) Meiosis in *Mesostoma ehrenbergii ehrenbergii* (Turbellaria, Rhabdocoela). I. Chromosome pairing, synaptonemal complexes and chiasma localisation in spermatogenesis. *Chromosoma* 85: 311–322
21 Offenberg HH, Schalk JAC, Meuwissen RLJ, van Aalderen M, Kester HA, Dietrich AJJ, Heyting C (1998) SCP2: a major protein component of the axial elements of synaptonemal complexes of the rat. *Nucl Acid Res* 26: 2572–2579
22 Tung K-S, Roeder GS (1998) Meiotic chromosome morphology and behavior in zip1 mutants of *Saccharomyces cerevisiae*. *Genetics* 149: 817–832
23 Schmekel K, Meuwissen RLJ, Dietrich AJJ, Vink ACG, van Marle J, van Veen H, Heyting C (1996) Organization of SCP1 protein molecules within synaptonemal complexes of the rat. *Exp Cell Res* 226: 20–30
24 Hawley RS, Arbel T (1993) Yeast genetics and the fall of the classical view of meiosis. *Cell* 72: 301–303
25 Bullard SA, Kim S, Galbraith AM, Malone RE (1996) Double strand breaks at the HIS2 recombination hot spot in *Saccharomyces cerevisiae*. *Proc Natl Acad Sci USA* 93: 13054–13059
26 Rocco V, Nicolas A (1996) Sensing of DNA non-homology lowers the initiation of meiotic recombination in yeast. *Genes Cells* 1: 645–661
27 de Massy B, Baudat F, Nicolas A (1994) Initiation of recombination in *Saccharomyces cerevisiae* haploid meiosis. *Proc Natl Acad Sci USA* 91: 11929–11933
28 Gilbertson LA, Stahl FW (1994) Initiation of meiotic recombination is independent of interhomologue interactions. *Proc Natl Acad Sci USA* 91: 11934–11937
29 Gasior SL, Wong AK, Kora Y, Shinohara A, Bishop DK (1998) Rad52 associates with RPA and functions with Rad55 and Rad57 to assemble meiotic recombination complexes. *Gene Develop* 12: 2208–2221
30 Smith KN, Nicolas A (1998) Recombination at work for meiosis. *Curr Opin Genet Develop* 8: 200–211
31 McKim KS, Hayashi-Hagihara A (1998) mei-W68 in *Drosophila melanogaster* encodes a Spo11 homolog: evidence that the mechanism for initiating meiotic recombination is conserved. *Gene Develop* 12: 2932–2942
32 Dernburg AF, McDonald K, Moulder G, Barstead R, Villeneuve A (1998) Meiotic recombination in *C. elegans* initiates by a conserved mechanism and is dispensable for homologous chromosome synapsis. *Cell* 94: 387–398
33 Klein S, Zenvirth D, Sherman A, Ried K, Rappold G, Simchen G (1996) Double-strand breaks on YACs during yeast meiosis may reflect meiotic recombination in the human genome. *Nat Genet* 13: 481–484
34 McKim KS, Green-Marroquin BL, Sekelsky JJ, Chin G, Steinberg C, Khodosh R, Hawley RS (1998) Meiotic synapsis in the absence of recombination. *Science* 279: 876–878
35 Anderson LK, Offenberg HH, Verkuijlen WMHC, Heyting C (1997) RecA-like proteins are components of early meiotic nodules in lily. *Proc Natl Acad Sci USA* 94: 6868–6873
36 Moens PB, Chen DJ, Shen ZY, Kolas N, Tarsounas M, Heng HHQ, Spyropoulos B (1997) Rad51 immunocytology in rat and mouse spermatocytes and oocytes. *Chromosoma* 106: 207–215
37 Plug AW, Peters AHFM, Keegan KS, Hoekstra MF, De Boer P, Ashley T (1998) Changes in protein composition of meiotic nodules during mammalian meiosis. *J Cell Sci* 111: 413–423
38 Albini SM, Jones GH (1987) Synaptonemal complex spreading in *Allium cepa* and *A. fistulosum*. I. The initiation and sequence of pairing. *Chromosoma* 95: 324–338
39 Anderson LK, Stack SM (1988) Nodules associated with axial cores and synaptonemal complexes during zygotene in *Psilotum nudum*. *Chromosoma* 97: 96–100
40 Moens PB, Heddle JAM, Spyropoulos B, Heng HHQ (1997) Identical megabase transgenes on mouse chromosomes 3 and 4 do not promote ectopic pairing or synapsis at meiosis. *Genome* 40: 770–773
41 Haber JE, Leung WY, Borts RH, Lichten M (1991) The frequency of meiotic recombination in

yeast is independent of the number and position of homologous donor sequences: implications for chromosome pairing. *Proc Natl Acad Sci USA* 88: 1120–1124
42 Goldman ASH, Lichten M (1996) The efficiency of meiotic recombination between dispersed sequences in *Saccharomyces cerevisiae* depends upon their chromosomal location. *Genetics* 144: 43–55
43 Louis EJ, Gorham HC, Timbrell AC (1998) Telomere recombination in meiosis: sequestering and non-random choice of partners. *J Exp Bot* 49 (Supplement): 82
44 Selva EM, New L, Crouse GF, Lahue RS (1995) Mismatch correction acts as a barrier to homeologous recombination in *Saccharomyces cerevisiae. Genetics* 139: 1175–1188
45 Datta A, Adjiri A, New L, Crouse GF, Jinks-Robertson S (1996) Mitotic crossovers between diverged sequences are regulated by mismatch repair proteins in *Saccharomyces cerevisiae. Mol Cell Biol* 16: 1085–1093
46 Nevo-Caspi Y, Kupiec M (1994) Transcriptional induction of Ty recombination in yeast. *Proc Natl Acad Sci USA* 91: 12711–12715
47 Fritze CE, Verschueren K, Strich R, Esposito RE (1997) Direct evidence for SIR2 modulation of chromatin structure in yeast rDNA. *EMBO J* 16: 6495–6509
48 Nicolas A (1998) Relationship between transcription and initiation of meiotic recombination: toward chromatin accessibility. *Proc Natl Acad Sci USA* 95: 87–89
49 Loidl J, Nairz K (1997) Karyotype variability in yeast caused by nonallelic recombination in haploid meiosis. *Genetics* 146: 79–88
50 Dernburg AF, Sedat JW, Cande WZ, Bass HW (1995) Cytology of telomeres. *In*: EH Blackburn., CW Greider (eds): *Telomeres*. Cold Spring Harbor Monograph Series, CSH Laboratory Press, Cold Spring Harbor, 295–338
51 Trelles-Sticken E, Loidl J, Scherthan H (1999) Bouquet formation in budding yeast: initiation of recombination is not required for meiotic telomere clustering. *J Cell Sci* 112: 651–658
52 Moore G (1998) To pair or not to pair: chromosome pairing and evolution. *Curr Opin Plant Biol* 1: 116–122
53 Chikashige Y, Ding D-Q, Imai Y, Yamamoto M, Haraguchi T, Hiraoka Y (1997) Meiotic nuclear reorganization: switching the position of centromeres and telomeres in the fission yeast *Schizosaccharomyces pombe. EMBO J* 16: 193–202
54 Rees H, Durrant A (1986) Recombination and genome size. *Theor Appl Genet* 73: 72–76
55 de Massy B, Nicolas A (1993) The control in cis of the position and the amount of the ARG4 meiotic double-strand break of *Saccharomyces cerevisiae. EMBO J* 12: 1459–1466
56 Wu T-C, Lichten M (1994) Meiosis-induced double-strand break sites determined by yeast chromatin structure. *Science* 263: 515–517
57 Jones GH (1987) Chiasmata. *In*: PB Moens (ed.): *Meiosis*. Academic Press, Orlando, 213–244
58 Kaback DB, Guacci V, Barber D, Mahon JW (1992) Chromosome size-dependent control of meiotic recombination. *Science* 256: 228–232
59 Anderson LK, Stack SM, Fox MH, Zhang C (1985) The relationship between genome size and synaptonemal complex length in higher plants. *Exp Cell Res* 156: 367–378
60 Wallace BMN, Hultén MA (1985) Meiotic chromosome pairing in the normal human female. *Ann Hum Genet* 49: 215–226
61 Peterson DG, Stack SM, Healy JL, Donohoe BS, Anderson LK (1994) The relationship between synaptonemal complex length and genome size in four vertebrate classes (Osteicthyes, Reptilia, Aves, Mammalia). *Chromosome Res* 2: 153–162
62 Loidl J, Scherthan H, den Dunnen JT, Klein F (1995) Morphology of a human-derived YAC in yeast meiosis. *Chromosoma* 104: 183–188
63 Sears DD, Hegemann JH, Hieter P (1992) Meiotic recombination and segregation of human-derived artificial chromosomes in *Saccharomyces cerevisiae. Proc Natl Acad Sci USA* 89: 5296–5300
64 Chu S, DeRisi J, Eisen M, Mulholland J, Botstein D, Brown PO, Herskowits I (1998) The transcriptional program of sporulation in budding yeast. *Science* 282: 699–705
65 Cawood AH, Jones JK (1980) Chromosome behaviour during meiotic prophase in the Solanaceae. *Chromosoma* 80: 57–68
66 de Jong JH, Stam P (1985) The association of centromeres of nonhomologous chromosomes at meiotic prophase in *Beta vulgaris* L. *Can J Genet Cytol* 27: 165–171
67 Kleinig H, Sitte P (1986) Zellbiologie. 2. Aufl. G Fischer Verlag, Stuttgart, New York
68 Schmitz A, Chaput B, Fouchet P, Guilly MN, Frelat G, Vaiman M (1992) Swine chromosomal

DNA quantification by bivariate flow karyotyping and karyotype interpretation. *Cytometry* 13: 703–710

69 Sherman JD, Stack SM (1992) Two-dimensional spreads of synaptonemal complexes from solanaceous plants. V. Tomato (*Lycopersicon esculentum*) karyotype and idiogram. *Genome* 35: 354–359

70 Schwarzacher G (1984) Untersuchungen zur Organisation der Synaptonemalen Komplexe in Oberflächen-gespreiteten Meiocyten einiger Angiospermen- und Säuger-Arten. PhD Thesis, University of Vienna

71 Villagómez DAF (1993) Zygotene-pachytene substaging and synaptonemal complex karyotyping of boar spermatocytes. *Hereditas* 118: 87–99

72 Morton NE (1991) Parameters of the human genome. *Proc Natl Acad Sci USA* 88: 7474–7476

73 Hinegardner R, Rosen DE (1972) Cellular DNA content and the evolution of teleostean fishes. *Amer Naturalist* 106: 621–644

74 Albini SM (1994) A karyotype of the *Arabidopsis thaliana* genome derived from synaptonemal complex analysis at prophase I of meiosis. *Plant J* 5: 665–672

75 Holm PB, Rasmussen SW (1977) Human meiosis. I. The human pachytene karyotype analyzed by three-dimensional reconstruction of the synaptonemal complex. *Carlsberg Res Commun* 42: 283–323

76 Solari AJ (1980) Synaptonemal complexes and associated structures in microspread human spermatocytes. *Chromosoma* 81: 315–337

77 Tanksley SD, Mutschler MA (1990) Linkage map of the tomato (*Lycopersicon esculentum*) (2N = 24). *In*: SJ O'Brien (ed.): *Genetic maps: locus maps of complex genomes. Book 6, plants.* Cold Spring Harbor Laboratory Press, Cold Spring Harbor, 6.3–6.15

78 Postlethwait JH, Johnson SL, Midson CN, Talbot WS, Gates M, Ballinger EW, Africa D, Andrews R, Carl T, Eisen JS et al (1994) A genetic linkage map for the zebrafish. *Science* 264: 699–703

79 Ellegren H, Chowdhary BP, Johannson M, Marklund L, Fredholm M, Gustavsson I, Andersson L (1994) A primary linkage map of the porcine genome reveals a low rate of genetic recombination. *Genetics* 137: 1089–1100

70 Kleckner N [illegible]

[illegible]

Chromosomes Today, Vol. 13
ed. by E. Olmo and C.A. Redi

Meiotic pairing sites and genes involved in segregation of the X and Y chromosomes of *Drosophila melanogaster*

Bruce D. McKee, Chia-sin Hong, and Siuk Yoo

University of Tennessee, Department of Biochemistry and Cellular and Molecular Biology, Knoxville, TN 37996-0840, USA

Introduction

A poorly understood aspect of chromosome structure and function is the role of specific chromosomal sites in meiotic pairing and segregation. Homologous chromosomes begin aligning early in meiotic prophase and by mid-prophase are fully synapsed, a state in which a ladder-like nucleoprotein structure known as synaptonemal complex connects the aligned chromosomal axes [1]. Alignment is necessary both for recombination and for accurate segregation at anaphase. However, it is not clear whether homology recognition and pairing takes place generally or at specific sites. In a variety of organisms, homologous chromosomes align early in meiotic prophase, prior to synapsis, sometimes making multiple string-like associations (reviewed in [2]). Fluorescent *in situ* hybridization (FISH) studies have shown that the earliest associations between homologous chromosomes of yeast also occur at multiple sites [3]. However, it has not been possible to determine in any organism whether the initial sites of association are specific or randomly chosen.

Another unanswered question about chromosome alignment is the nature of the initial connections. A variety of models have been proposed that involve protein-protein interactions and various types of nucleic acid interactions [2, 4]. Clearly, homologous nucleic acid interactions do occur in the course of a "normal" meiosis since gene conversion and crossing-over must result from intermediates in which the parental duplexes are joined. Moreover, recent data from yeast indicate that these interhomolog DNA interactions begin quite early in meiotic prophase [5], but the question of whether the initial interhomolog interactions are of this type remains unresolved. One of the difficulties in solving this problem is that the processes of pairing and recombination are intertwined temporally, so that the specific requirements of the latter are difficult to disentangle from those of the former.

Although meiotic pairing is accompanied by recombination in most cases, there are exceptions, the most familiar being *Drosophila* males in which pairing and segregation occur without formation of a synaptonemal complex and without significant levels of either interhomolog crossing-over [6, 7] or gene conversion [8]. This exceptional behavior provides a valuable system for analysis, since the interhomolog interactions in this case apparently serve only the purpose of pairing and segregation.

We have utilized *Drosophila* male meiosis to begin addressing the questions raised above, namely the specificity of meiotic pairing sites, and the nature of pairing interactions. In the following paragraphs, we will first briefly review work on the localization and identification of sex chromosomal meiotic pairing sites, second, review the present state of knowledge concerning the types of genes involved in male meiosis, and third, present some new data relevant to the hypothesis that pairing involves DNA interactions and will therefore require the participation of a variety of DNA metabolic enzymes.

Analysis of sex chromosome pairing sites in *Drosophila* males

Mapping pairing sites

The sex chromosomes of *Drosophila* are heteromorphic and pairing interactions are confined to a relatively discrete region that encompasses portions of the centric heterochromatin of the X (Xh) and the short arm of the Y (YS) near the centromere (Fig. 1A) [9]. The nucleolus organizers (NOs) are located in these regions, each containing 200–250 copies of an 8 kb transcription unit responsible for the 18S, 28S and 5.8S rRNAs interspersed with 3–4 kb "intergenic spacers" (IGS) (Fig. 1B, C) [10]. Flanking sequences include large blocks of satellite repeats and numerous transposons [11].

Mapping studies have implicated the rDNA as the major pairing site. All deficiencies that remove the entire rDNA array induce X-Y non-disjunction, while all deficiencies that retain part of the rDNA block segregate regularly from the Y (reviewed in [12]). Analysis of the pairing behavior of a large number of mini-X chromosomes that consist mostly of heterochromatin also points to the rDNA as the major pairing site, since these chromosomes retain pairing capacity only if they retain rDNA [13]. The proximal region of Xh, the region flanking the centromere, appears to be without pairing capacity, as shown by the random segregation of X chromosomes that lack the rDNA and distal heterochromatin but retain the proximal heterochromatin [9, 14, 15]. Since the proximal Xh region consists largely of various satellite families, it has been concluded that satellite DNAs are probably not involved in meiotic pairing, at least in males [16]. The heterochromatic region distal to the rDNA does have some pairing capacity as indicated by the fact that deficiencies for the rDNA array which retain a substantial portion of distal Xh exhibit elevated X-Y non-disjunction, but not random disjunction. As nothing is known about the nature of the pairing capacity in this distal region of Xh, it will not be discussed further.

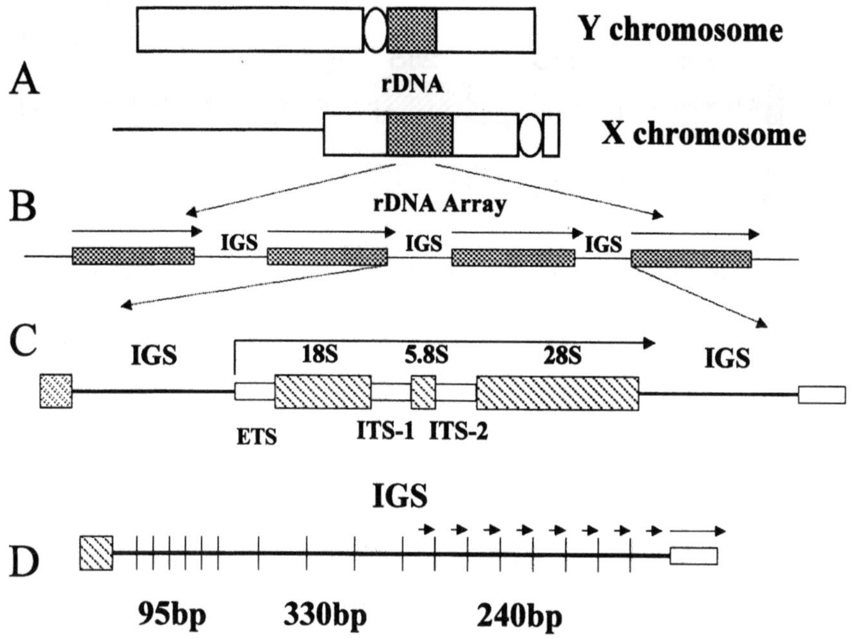

Figure 1. A. Sex chromosomes of *Drosophila melanogaster*. Rectangles = heterochromatic regions; line = euchromatin of X; ovals = centromeres. B. The rDNA arrays of the X and Y chromosomes. Solid arrows represent transcription units. IGS = intergenic spacer. C. An rDNA repeat with one transcription unit flanked by intergenic spacers. Hatched rectangles = structural gene regions of rDNA; unfilled rectangles = transcribed spacers; ETS = external transcribed spacer; ITS-1 and ITS-2 = internal transcribed spacers 1 and 2. D. Structure of the intergenic spacer. Small arrows represent spacer promoters; longer arrow represents the start of the rDNA transcription unit.

The heterochromatic region of the X chromosome also participates in pairing and segregation of non-exchange X chromosomes in *Drosophila* female meiosis. Surprisingly, however, the X heterochromatic pairing sites are not the same in males and females, despite the fact that in both cases heterochromatic pairing is used to segregate achiasmatic pairs. In females, homology for heterochromatin proximal to the rDNA, including the major satellite blocks, is adequate to ensure regular disjunction, whereas rDNA homology is not [17, 15]. The basis for this difference in specificity is unknown.

Molecular analysis of pairing sites

We have used transgenic methods to test the hypothesis that the rDNA mediates X-Y pairing. A single rDNA repeat, which encompassed a complete transcription unit plus flanking spacer sequences (Fig. 2), was cloned into a standard P element vector [18] and introduced into the germ line by P-mediated

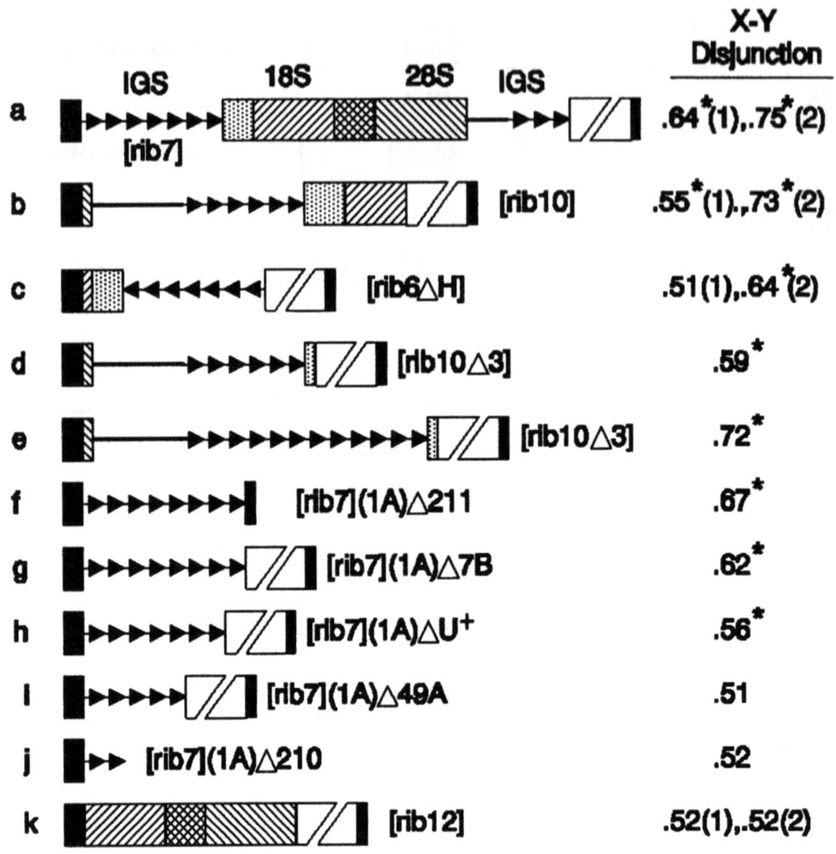

Figure 2. rDNA transgenes and X-Y Disjunction. Filled rectangles = P element sequences; open rectangles = *rosy+* eye-color marker; hatched and shaded rectangles = parts of the rDNA transcription unit; arrowheads = 240 bp repeats; solid line = upstream IGS region. In X-Y disjunction column, decimal numbers represent fraction of disjunctional (X or Y) sperm in testes of males carrying Df(1)X-1 with indicated transgene insertion and B^SYy^+; number in parentheses = copy number of transgene; * = statistically significantly different from control with no transgene (.51). Data and calculations are from [19] (line a), [20] (lines b–e); [21] (lines f–j); [22] (line k).

microinjection methods. Several X chromosomal insertions were recovered and recombined onto an Xh deficiency (Df(1)X-1) that lacks most of Xh and that disjoins randomly from the Y. Single copies of this rDNA transgene stimulated X-Y pairing and disjunction, so that around 65% of secondary spermatocytes received an X or a Y and only 35% were non-disjunctional. X chromosomes with two copies of this vector disjoined even more frequently, with about 75% of secondary spermatocytes being normal and only 25% non-disjunctional. Stimulation of X-Y disjunction was accompanied by enhancement of X-Y pairing in primary spermatocytes (based on cytological analysis), and pairing could be shown to occur at the site of the rDNA insertion. Thus the ectopic rDNA insertions act as mobile pairing sites. Insertions at a variety of sites on the X chromosome were able to stimulate pairing, and double insertions did not need to be in tandem to provide additional stimulation [19].

Further analysis showed that intact rDNA repeats are not needed for stimulation of X-Y pairing [20–22]. rDNA fragments missing part, or all, of the transcription unit could stimulate pairing as long as the intergenic spacer was present (Fig. 2, lines b–h). Conversely, fragments internal to the transcription unit were incapable by themselves of stimulating pairing (Fig. 2, line k), even when present in multiple copies on the X. The intergenic spacer is internally repetitious, containing repeats of 95, 330 and 240 bp (Fig. 1D) [23]. The 240 bp repeats are located immediately upstream of each transcription unit and are present in 6–10 copies per spacer. These repeats each contain a functional copy of the RNA polymerase I promoter [23, 24]. Several transgenes that contained only 240 bp repeats were found to stimulate pairing of the heterochromatically deficient X. A minimum of six copies of the repeat is required for this stimulation; above six copies the frequency of X-Y disjunction is proportional to 240 bp repeat copy number. Thus the major X-Y pairing site apparently resides within a relatively short defined sequence in the rDNA spacer. This sequence is present on both the X and Y chromosomes of *D. melanogaster* at a copy number typically around 2000, but as just described, X chromosomes with as few as six copies exhibit detectable pairing ability. Interestingly, the Y chromosome of the sibling species *D. simulans* lacks rDNA repeats (the only rDNA array is on the X) but contains a large (3 Mb) tandem block of 240 bp repeats [25]. The location of this block, the tip of the long arm of the Y, corresponds to the cytologically identified Y chromosomal pairing site [26]. This observation thus supports the idea that 240 bp intergenic spacer (IGS) repeats are the X-Y pairing sites in *Drosophila* males and that the rDNA *per se* is not required for pairing. Analysis of the mechanism by which the 240 bp repeats mediate pairing is ongoing in our laboratory.

Trans-acting mutations and X-Y pairing

Mutations that cause X-Y non-disjunction

A handful of mutations have been described that cause elevated X-Y non-disjunction, but no genes clearly required for X-Y pairing have been identified. *mei-081* [27] and *mei-I1* [28] were reported to disrupt pairing of all chromosome pairs, but were lost before their products could be identified. Also lost were the *mei-I3* and *mei-G17* mutations that caused non-disjunction of the X and autosomes [28, 29]. A large group of X-linked mutations that cause elevated X-Y non-disjunction were recovered following ethyl methanesulphonate (EMS) treatment, but all proved unstable and reverted [30]. A dominant mutation, *Double-or-nothing (Dub),* causes meiosis I non-disjunction in both sexes;

however, the loss-of-function phenotype for this gene is not known, and it has not been cloned [31].

Other mutations cause X-Y non-disjunction for reasons unrelated to pairing. For example, deletions of the Y-linked *Su(Ste)* locus derepress the X-linked *Stellate* repeats, leading to elevated non-disjunction of all chromosomes due to formation of anaphase chromatin bridges and chromosome fragmentation, but no pairing defects [32–34]. Mutations in the *ord* and *mei-S332* genes, which encode proteins required for sister chromatid cohesion in both male and female meiosis cause elevated non-disjunction of all chromosomes due to premature sister chromatid separation, but homologous pairing appears to be normal [35]. Mutations in *polo*, which encodes a kinase required both in mitosis and meiosis, undergo non-disjunction at both meiotic divisions due to abnormal centrosome behavior [36]. Mutations in *UbcD1* and *peo* cause chromosome fragmentation and loss due to failure to resolve non-homologous telomere attachments [37, 38]. These and other mutations are of great value in understanding various aspects of meiosis, but are not likely to provide insight into the problem of chromosome pairing.

Genes involved in DNA metabolism

The types of genes involved in X-Y pairing will depend upon the mechanism of pairing. If pairing involves DNA level associations, then it is likely that genes encoding a variety of enzymes of DNA metabolism will be required for X-Y segregation. For example, one model for achiasmatic (non-recombinational) pairing invokes formation of DNA hemicatenanes by the combined action of a DNA pairing/strand transfer protein like Rad51 or Dmc1 and topoisomerase I [21]. Another model postulates interlocks created and, perhaps resolved, by a topoisomerase II-like activity such as Spo11 [39].

The possibility of a requirement for DNA metabolism genes in X-Y pairing has not been extensively explored. Some genes involved in aspects of DNA metabolism have been shown to be required for female meiosis but not male meiosis. *mei41*, *mei9* and *mei-W68* encode *Drosophila* homologs of ATM (a protein involved in both DNA damage repair and a damage checkpoint), Rad1p (an endonuclease required for excision repair), and Spo11 (the endonuclease thought to induce the double-strand breaks that initiate meiotic recombination), respectively [40–42]. Mutations in these genes severely reduce meiotic recombination in female meiosis but have no known phenotypes in male meiosis. However, there are numerous genes that encode enzymes of DNA metabolism, such as *mus* genes and *spindle*-group genes, which have not been examined with respect to male meiotic phenotypes.

mus genes: *mus* mutations confer enhanced sensititvity to mutagens, including radiation and various chemical agents [43, 44]. Most cause defects in one or more pathways of DNA repair, including nucleotide excision repair (NER), post-replication repair, cross-link repair and double-strand break repair [45–47]. Some *mus* mutations have been shown to affect female meiosis, mostly causing increases in rates of X chromosome non-disjunction and/or loss [43, 47, 48]. Several *mus* genes have been cloned in the last few years, and all have turned out to encode enzymes involved in some aspect of DNA metabolism. *mus205* encodes the *Drosophila* homolog of Rev3, an error-prone polymerase implicated in post-replication repair [49, 46]. *mus209* encodes proliferating cells nuclear antigen (PCNA), an essential component of the eukaryotic DNA replication complex which has also proven to function in DNA mismatch repair and, curiously, in modification of position effect variegation (mosaic expression of genes relocated to the vicinity of heterochromatin) [50]. *mus308* encodes a novel protein which shows homology both to eukaryotic DNA helicases and to the prokaryotic DNA polymerase I family; it is uniquely involved in repair of DNA cross-links [51, 52]. *mus309* encodes the *Drosophila* homolog of Ku70, a DNA-end-binding protein needed for various pathways of double-strand break repair; the *Drosophila* protein also binds to the inverted repeats of P elements [53]. Thus *mus* genes encode enzymes involved in a variety of aspects of DNA metabolism. If meiotic pairing in *Drosophila* males involves DNA interactions, important clues to the mechanism of pairing may emerge from identification of DNA metabolic genes required for male meiosis.

Table 1B shows the results of tests of the role of *mus205, 206, 207* [44] and *mus301, 302, 304, 305, 306, 307* and *308* [43] in X-Y segregation in male meiosis; tests of other *mus* genes will be reported elsewhere. In the tests reported in Table 1, males carrying a normal X chromosome and the structurally normal, but marked, Y chromosome B^SYy^+ [54] were mated singly to females carrying normal X chromosomes, and the progeny scored to determine the frequency of non-disjunctional (XY or O) gametes in the tested males. In otherwise normal genotypes such as the two controls presented in Table 1A the non-disjunction frequency is typically less than 0.5%. Males in part B were either homozygous for a *mus* mutation (*mus205*^{A1}, *206*^{A1}, *207*^{A1}, *306*^{D1} and *307*^{D1}) or heterozygous for two mutations in the same gene (*mus 301, 302, 305, 305* and *308*). These trans-heterozygotes were generated from both reciprocal crosses to control for possible X chromosome effects or maternal effects. The tested alleles for these genes are listed in the legend to Table 1.

The results indicate little if any role of *mus* genes in X-Y segregation. Males homozygous for *mus205*^{A1}, *mus206*^{A1}, *mus207*^{A1}, *mus306*^{D1} and *mus307*^{D1} exhibited non-disjunction frequencies in or very near the normal range (0.1–0.5%). Thus none of the *mus* alleles tested as homozygotes causes significant levels of X-Y non-disjunction. For *mus302, mus304* and *mus305*, none of the trans-heterozygous combinations exhibited elevation of the X-Y non-disjunction frequency. For *mus301* and *mus308*, most of the trans-heterozygous combinations were also without effect. However, trans-heterozygotes containing *mus301*^{D2} along with any of the other *mus301* alleles (but not when

Table 1. X-Y non-disjunction frequencies in males homozygous or trans-heterozygous for *mus* and *spn*-group alleles

		Sperm Genotype				
		X	Y	XY	O	%NDJ[a]
A. **Controls**[b]	OR/B^SYy^+	5,228	4,489	7	22	0.30
	CS/B^SYy^+	4,297	3,716	2	5	0.087
B. **mus genes**	**Alleles**					
mus205	*A1/A1*	753	517	0	3	0.23
mus206	*A1/A1*	567	435	1	6	0.69
mus207	*A1/A1*	1,008	860	0	1	0.054
mus301	*D2*/other[c]	4,280	3,232	3	65	0.90
	D2/+	291	253	0	1	0.18
	other[c]	17,161	12,554	8	52	0.20
mus302	All[d]	22,052	19,049	7	29	0.088
mus304	*D1/D3*	4,699	3,718	1	11	0.14
mus305	All[e]	4,420	3,676	2	5	0.086
mus306	*D1/D1*	652	560	1	6	0.57
mus307	*D1/D1*	318	296	2	0	0.32
mus308	*D1*/other[f]	1,593	920	1	33	1.33
	other/*D1*[f]	3,210	1,947	1	18	0.37
	+/D1	835	542	2	4	0.43
	Other[f]	3,635	2,260	3	9	0.20
C. **spn genes**						
spnA	*1*	228	150	0	1	0.26
spnB	*1*	465	408	0	1	0.11
spnD	2	372	319	0	0	0.00
hls	*616*	1191	746	20	65	4.2
okr	*WS/Cy*	1416	1189	0	3	0.11
	WS/WS	99	77	0	0	0
	WS/RU	1797	1645	2	4	0.17
	WS/AA	699	544	1	1	0.16
	WS/Df[g]	985	898	0	5	0.27

mus alleles were obtained from the Bloomington and Bowling Green stock centers. These included *mus205*^{A1}, *mus206*^{A1}, *mus207*^{A1}, *mus301*$^{D1, D2, D3, D4 \text{ and } D5}$, *mus302*$^{D1, D2, D3, D4, D5 \text{ and } D6}$, *mus304*$^{D1 \text{ and } D3}$, *mus305* $^{D1, D2 \text{ and } D3}$, *mus306*^{D1}, *mus307*^{D1}, and *mus308*$^{D1, D2, D4 \text{ and } D5}$. spn-group alleles spnA^1, spnB^1, spnD^2, and hls^{E616} were obtained from the Bloomington stock center. okr^{WS} was donated by the Umea stock center, and okr $^{RU \text{ and } AA}$ along with Df(2R)JS17, were obtained from Dr. T. Schupbach.
X-Y disjunction was measured in males carrying structurally normal unmarked X chromosomes and the marked Y chromosome *B^SYy^+* [54]. Males carrying the various mutations to be tested were crossed to y w females (one male to two females); the crosses were incubated at 23 °C, the parents removed by day ten and the progeny counted until day 20. Non-disjunctional progeny are recognized as y w males and B females. Reported results are combined counts from all pairwise combinations of indicated alleles. [a]%NDJ = (XY+O/Total)x100. [b]OR is an X chromosome derived from the Oregon R stock; CS is an X chromosome derived from the Canton S stock. [c]Other alleles of *mus301* are *D1, D3, D4* and *D5*. Results of reciprocal crosses are combined. [d]The tested alleles of *mus302* are *D1, D2, D3, D4, D5* and *D6*. [e]The tested alleles of *mus305* are *D1, D2* and *D3*. [f]Other alleles of *mus308* are *D2, D4* and *D5*. Maternally derived allele is listed first. [g]Df = Df(2R)JS17, an *okr* deficiency.

heterozygous with *mus301*$^{+}$) exhibited a slightly elevated recovery of nullo-XY sperm but not XY sperm, indicative of a recessive effect on sex chromosome loss. Elevations were observed in both reciprocal crosses (results are grouped in Tab. 1), which rules out the possibility that they are due to the X chromosome in the *mus301*^{D2} stock. Males trans-heterozygous for *mus301*^{D2} also exhibited markedly reduced fertility. A similar increase in sex chromosome loss but with no accompanying fertility deficit was seen in trans-heterozygotes containing *mus308*^{D1} in combination with other *mus308* alleles. However, in this case the effect was seen only when the *mus308*^{D1} allele was transmitted from the female. This raises the possibility that the stimulation is actually due to the X chromosome in the *mus308*^{D1} stock, a possibility that is under investigation. Thus the data indicate that *mus301*^{D2}, and possibly *mus308*^{D1}, when heterozygous with other alleles of the same gene, cause a slight increase in the frequency of sex chromosome loss. It is not clear how significant these effects are. The rather weak phenotypes, combined with their allele-specificity, preclude any direct interpretation.

spindle (spn)-group genes: spn mutations were identified as female steriles that cause defects in dorsal structures of eggs and an abnormal spindle-like egg shape [55–57]. All members of this group are required for establishment of dorsal-ventral polarity during oogenesis and for proper localization of key molecular determinants of polarity. This group is of interest in the present context because of recent evidence that three members (*spnB, spnD* and *okr*) are also required for normal levels of recombination and chromosome segregation in female meiosis, and two (*spnB* and *okr*) encode enzymes related to known members of the double-strand break repair pathway [58]. *okr* encodes the *Drosophila* homolog of Rad54, a member of the Snf/Swi family of DNA helicases required for double-strand break repair [59], and *spnB* encodes a homolog of *Dmc1*, a meiosis-specific member of the RecA-like family [58, 60]. Yeast homologs of both genes are required for meiosis [61, 62].

Table 1C shows the results of tests of the involvement of *spnA, spnB, spnD, spnE* and *okr* in X-Y segregation. The *spnA*^{1}, *spnB*^{1}, *spnD*^{2}, *hls*^{E616}, and *okr*^{WS} alleles were tested in homozygous males carrying B^SYy^+. As shown in Table 1C, none of the alleles significantly elevated X-Y non-disjunction except *hls*^{E616}, an allele of the *homeless (hls)* gene (also known as *spindle-E (spnE)*). Males carrying this allele exhibited 4.2% non-disjunction, an order of magnitude above the background frequency. Unlike the cases of *mus301*^{D2} and *mus308*^{D1}, this increase can be attributed to non-disjunction rather than (or in addition to) chromosome loss since both the XY and nullo-XY classes increased proportionally. The *okr*^{WS} chromosome proved to carry a semi-penetrant male sterile associated with defects in meiotic chromatin condensation (unpublished observations). This phenotype is probably not due to the *okr*^{WS} allele, however, as trans-heterozygous combinations of *okr*^{WS} with other *okr* alleles (*okr*^{RU}, *okr*^{AA} and Df(2R)JS17) were fully fertile. Neither *okr*^{WS}

homozygotes nor any of the tested trans-heterozygous combinations of *okr* alleles caused any elevation of X-Y non-disjunction.

The most interesting result in Table 1 is certainly the stimulation of X-Y non-disjunction by the hls^{E616} mutation. The level of non-disjunction in our test, 4.2%, although only moderate, represents an order of magnitude increase over control levels, and is certainly high enough to merit further study. Other alleles of this gene exhibit a similar phenotype (unpublished observations), so this effect is not allele-specific. This result was something of a surprise as *hls* has not previously been reported to have any involvement either in spermatogenesis or in meiosis. The gene was identified in two separate screens for female sterile mutations [55, 57]. Alleles of *hls* cause a variety of defects in oogenesis, including ventralization of the chorion, mispositioning of the oocyte, failure to localize and/or activate key RNAs and proteins involved in establishment of dorsal/ventral and anterior/posterior polarity in the oocyte, and failure to lay eggs [57, 63]. Complete loss-of-function alleles are lethal, indicating a role in development [57]. An effect on female meiosis has not been reported, but has not been ruled out either. The discovery of a male meiotic function adds to the functional complexity of this already pleiotropic gene. The *hls* gene has been cloned; it encodes a protein homologous to the DE-H-box family of RNA helicases, some of which are involved in mRNA splicing [57]. The *Drosophila* Mle protein, which binds to the X chromosome and helps mediate dosage compensation, is also a member of this family [64]. However, the level of homology between the Hls protein and any members of this family is not strong enough to infer a probable biochemical function for Hls [57]. We are carrying out further studies of the male meiotic function of Hls identified here; these studies will be reported elsewhere.

Concluding remarks: what types of interactions underlie X-Y pairing in achiasmatic meiosis?

The identification of a discrete sequence that functions as an X-Y pairing site provides an opportunity to explore the mechanisms that underlie meiotic pairing of these chromosomes. The fact that identical 240 bp repeats are present on the Y as well as the X, and at approximately the same copy number, argues that pairing is based on DNA homology. However, the question of whether the DNA sequences associate directly or indirectly with each other is unresolved. DNA pairing reactions typically lead to recombination, and there is no evidence in *Drosophila* male meiosis for either crossing-over or gene conversion. DNA pairing models that do not involve recombination can be envisioned (e.g., [22]) but lack empirical support.

The results in Table 1 provide little encouragement for models of X-Y pairing that involve DNA level interactions. None of the *mus* genes tested proved to have a substantial effect on X-Y disjunction; included are genes thought to encode an error-prone polymerase (*mus205*) [46] and an enzyme with both DNA helicase and DNA polymerase activity (*mus308*) [52]. Most or all of the other *mus* genes are likely to also encode enzymes involved in DNA metabolism, based on their defects in various DNA repair pathways [45, 46]. Some, such as *mus301, mus304* and *mus305* [43, 48] have also been shown to be needed for normal levels of meiotic recombination in female meiosis. One mutation in *mus301*, and possibly one in *mus308*, did stimulate X-Y non-disjunction slightly, in combination with other alleles. This result may suggest a role for these genes in male meiosis, but further analysis will depend upon finding alleles or assay conditions in which the phenotype is stronger. Also ineffective were mutations in *okr* and *spnB*, which encode DNA metabolic enzymes involved in female meiosis, and *spnD*, a female meiotic gene of unknown function [58]. The lack of a male meiotic phenotype of the *spnB* mutation is inconsistent with hypotheses that invoke DNA pairing by members of the RecA family; *spnB* encodes a *Drosophila* homolog of Dmc1, a meiosis-specific member of this family. This gene is essential for normal meiosis in *Drosophila* females, but evidently not in *Drosophila* males. However, it remains possible that other members of the RecA family function in male meiotic pairing. We have reported the cloning of a *Drosophila* member of the Rad51 subfamily of RecA-like proteins [65]; however, as yet no mutations in DRad51 have been identified. Our finding of elevated non-disjunction in *hls* mutant males does not support the DNA-interaction hypothesis because *hls* encodes a product more likely to be concerned with RNA than DNA metabolism.

These results are far from conclusive, as numerous other DNA metabolic genes remain to be tested, and since functional redundancies could mask roles of tested genes. However, these negative results suggest that other types of interactions should be considered. An alternative is that pairing interactions are mediated by chromosomal proteins that are distributed differentially on different chromosomes, so as to provide homolog-specific patterns of association. In light of the presence of RNA polymerase I promoters in the 240 bp pairing sites, it may be worth considering scenarios related to transcription. For example, perhaps pairing could be mediated directly by complexes of transcription factors assembled on the spacer promoters. The well-established tendency of transcription factors to form dimers and higher-order complexes provides a possible mechanism for bridging homologous chromosomes. Other possible scenarios could involve roles of spacer transcripts or transcript binding proteins.

The underlying basis for chromosomal pairing in *Drosophila* remains elusive. It will be of value to test additional genes involved in DNA metabolism, but it also seems warranted to expand the scope of mutational analyses to focus on genes involved in other aspects of cellular metabolism, such as transcriptional regulation and RNA metabolism. In addition, continued analysis of the role of 240 bp IGS repeats may provide clues to the mechanism of X-Y pairing.

Acknowledgments

We wish to thank Dr. T. Schupbach and the stock centers at Bloomington IN, Bowling Green OH and Umea, Sweden for providing stocks of *Drosophila*. This work was supported by grant # R01 GM40489 from the U.S. Public Health Service.

References

1 von Wettstein D, Rasmussen SW, Holm PB (1984) The synaptonemal complex in genetic segregation. *Annu Rev Genet* 18: 331–414
2 Loidl J (1990) The initiation of meiotic chromosome pairing: the cytological view. *Genome* 33: 759–778
3 Weiner BM, Kleckner N (1994) Chromosome pairing via multiple interstitial inter- actions before and after meiosis in yeast. *Cell* 77: 977–991
4 Kleckner N, Padmore R, Bishop DK (1991) Meiotic chromosome metabolism: one view. *Cold Spring Harbor Symp Quant Biol* 56: 729–743
5 Padmore R, Cao L, Kleckner N (1991) Temporal comparison of recombination and synaptonemal complex formation during meiosis in *S. cerevisiae*. *Cell* 66: 1239–1256
6 Morgan TH (1912) Complete linkage in the second chromosome of *Drosophila melanogaster. Science* 36: 719–720
7 Meyer GF (1960) The fine structure of spermatocyte nuclei of *Drosophila melanogaster*. *In*: AL Houwink, BJ Spit (eds): *Proceedings of the European Regional Conference on Electron Microscopy*. Die Nederlandse Verening voor Electronmicroscopic Delft, Delft, 951–954
8 Chovnick A, Ballantyne GH, Baillie DL, Holm DG (1971) Gene conversion in higher organisms: half-tetrad analysis of recombination within the rosy cistron of *Drosophila melanogaster*. *Genetics* 66: 315–329
9 Cooper KW (1964) Meiotic conjunctive elements not involving chiasmata. *Proc Natl Acad Sci USA* 52: 1248–1255
10 Long EO, Dawid IB (1980) Repeated genes in eukaryotes. *Annu Rev Biochem* 49: 727–764
11 Lohe AR, Hilliker AJ, Roberts PA (1993) Mapping simple repeated DNA sequences in heterochromatin of *Drosophila melanogaster*. *Genetics* 134: 1149–1174
12 McKee BD (1996) The license to pair: identification of meiotic pairing sites in *Drosophila*. *Chromosoma* 105: 135–141
13 Park H-S, Yamamoto M-T (1995) The centric region of the X chromosome rDNA functions in male meiotic pairing in *Drosophila melanogaster*. *Chromosoma* 103: 700–707
14 McKee B, Lindsley DL (1987) Inseparability of X-heterochromatic functions responsible for X:Y pairing, meiotic drive, and male fertility in *Drosophila melanogaster* males. *Genetics* 116: 399–407
15 Karpen GH, Le M-H, Le H (1996) Centric heterochromatin and the efficiency of achiasmate disjunction in *Drosophila* female meiosis. *Science* 273: 118–122
16 McKee BD (1998) Pairing sites and the role of chromosome pairing in meiosis and spermatogenesis in male *Drosophila*. *Curr Topic Develop Biol* 37: 77–115
17 Hawley RS, Irick H, Zitron AE, Haddox DA, Lohe A, New C, Whitley MD, Arbel T, Jang J, McKim K et al (1993) There are two mechanisms of achiasmate segregation in *Drosophila* females, one of which requires heterochromatic homology. *Develop Genet* 13: 440–467
18 Rubin GM, Spradling AC (1983) Vectors for P element-mediated gene transfer in *Drosophila*. *Nucl Acid Res* 11: 6341–6351
19 McKee BD, Karpen GH (1990) *Drosophila* ribosomal RNA genes function as an X-Y meiotic pairing site during male meiosis. *Cell* 61: 61–72
20 Merrill CJ, Chakravarti D, Habera L, Das S, Eisenhour L, McKee BD (1992) Promoter-containing

ribosomal DNA fragments function as X-Y meiotic pairing sites in *D. melanogaster* males. *Develop Genet* 13: 468–484
21 McKee BD, Habera L, Vrana JA (1992) Evidence that intergenic spacer repeats of *Drosophila melanogaster* rRNA genes function as X-Y pairing sites in male meiosis, and a general model of achiasmatic pairing. *Genetics* 132: 529–544
22 Ren X-J, Eisenhour L, Hong C-S, Lee Y, McKee BD (1997) Roles of rDNA spacer and transcription unit sequences in X-Y meiotic pairing in *Drosophila melanogaster. Chromosoma* 106: 29–36
23 Simeone A, La Volpe A, Boncinelli E (1985) Nucleotide sequence of a complete ribosomal spacer of *D. melanogaster. Nucl Acid Res* 13: 1089–1101
24 Lohe AR, Roberts PA (1990) An unusual Y chromosome of *Drosophila simulans* carrying amplified rDNA spacer without rRNA genes. *Genetics* 125: 399–406
25 Coen ES, Dover GA (1982) Multiple polI initiation sequences in rDNA spacers of *Drosophila melanogaster. Nucl Acid Res* 10: 7017–7026
26 Ault JG, Rieder CL (1994) Meiosis in *Drosophila* males. I. The question of separate conjunctive mechanisms for the XY and autosomal bivalents. *Chromosoma* 103: 352–356
27 Sandler L, Lindsley DL, Nicoletti B, Trippa G (1968) Mutants affecting meiosis in natural populations of *Drosophila melanogaster. Genetics* 60: 525–558
28 Ivy J (1981) *Mutations that disrupt meiosis in males of Drosophila melanogaster*. Ph.D. thesis, University of California at San Diego, CA
29 Gethmann R (1974) Meiosis in male *Drosophila melanogaster. Genetics* 78: 1127–1142
30 Baker BS, Carpenter ATC (1972) Genetic analysis of sex chromosomal meiotic mutants in *Drosophila melanogaster. Genetics* 71: 255–286
31 Moore DP, Miyazaki WY, Tomkiel JE, Orr-Weaver TL (1994) *Double or nothing*: a *Drosophila* mutation affecting meiotic chromosome segregation in both females and males. *Genetics* 136: 953–964
32 Livak KJ (1984) Organization and mapping of a sequence on the *Drosophila melanogaster* X and Y chromosomes that is transcribed during spermatogenesis. *Genetics* 107: 611–634
33 Hardy RW, Lindsley DL, Livak KJ, Lewis B, Silversten AL, Joslyn GL, Edwards J, Bonaccorsi S (1984) Cytogenetic analysis of a segment of the Y chromosome of *Drosophila melanogaster. Genetics* 107: 591–610
34 Palumbo G, Bonnacorsi S, Robbins L, Pimpinelli S (1994) Genetic analysis of *Stellate* elements of *Drosophila melanogaster. Genetics* 138: 1181–1197
35 Orr-Weaver TL (1995) Meiosis in *Drosophila*: seeing is believing. *Proc Natl Acad Sci USA* 92: 10443–10449
36 Sunkel CE, Glover DM (1988) *polo*, a mitotic mutant of *Drosophila* displaying abnormal spindle poles. *J Cell Sci* 89: 25–38
37 Cenci G, Rawson RB, Belloni G, Castrillon DH, Tudor M, Petrucci R, Goldberg ML, Wasserman SA, Gatti M (1997) *UbcD1*, a *Drosophila* ubiquitin-conjugating enzyme required for proper telomere behavior. *Gene Develop* 11: 863–875
38 Cenci G, Belloni G, Santolamazza C, Dimitri P, Gatti M (1998) Genetic, cytological and molecular analysis of *pendolino*, a gene that controls telomere behavior in *Drosophila melanogaster. A Conf Dros Res* 39: 777B
39 Keeney S, Giroux CN, Kleckner N (1997) Meiosis-specific DNA double-strand breaks are catalyzed by Spo11, a member of a widely conserved protein family. *Cell* 88: 375–384
40 Hari KL, Santerre A, Sekelsky JJ, McKim KS, Boyd JB, Hawley RS (1995) The *mei-41* gene of *D. melanogaster* is a structural and functional homolog of the human *ataxia telangiectasia* gene. *Cell* 82: 815–821
41 Sekelsky JJ, McKim KS, Chin GM, Hawley RS (1995) The *Drosophila* meiotic recombination gene *mei-9* encodes a homologue of the yeast excision repair protein Rad1. *Genetics* 141: 619–627
42 McKim KS, Hayashi-Hagihara A (1998) *mei-W68* in *Drosophila melanogaster* encodes a Spo11 homolog: evidence that the mechanism for initiating meiotic recombination is conserved. *Gene Develop* 12: 2932–2942
43 Boyd JB, Golino MD, Shaw KES, Osgood CJ, Green MM (1981) Third chromosome mutagen-sensitive mutants of *Drosophila melanogaster. Genetics* 97: 607–623
44 Henderson DS, Bailey DA, Sinclair DA, Grigliatti TA (1987) Isolation and characterization of second chromosome mutagen-sensitive mutations in *Drosophila melanogaster. Mutat Res* 177: 83–93
45 Dusenberry RL, Smith PD (1996) Cellular responses to DNA damage in *Drosophila melanogaster. Mutat Res* 364: 133–145

46 de Buendia PG (1998) Search for DNA repair pathways in *Drosophila melanogaster. Mutat Res* 407: 67–84
47 Sekelsky JJ, Burtis KC, Hawley RS (1998) The pleiotropy of DNA repair genes in *Drosophila melanogaster. Genetics* 148: 1587–1598
48 Sekelsky J, Brodsky M, Rubin G, Hawley S (1998) *mus304* encodes a RecQ helicase-related protein required for a cell cycle checkpoint and normal meiotic chromosome behavior. *A Conf Dros Res* 39: 43
49 Eeken JC, deJong A, Romeijn R, Pastink A (1997) Molecular and genetic characterization of the MMS-sensitive locus *mus205*, the *Drosophila* homolog of the *S. cerevisiae REV3* gene, encoding a non-essential DNA polymerase. *A Conf Dros Res* 38: 271C
50 Henderson DS, Banga SS, Grigliatti TA, Boyd JB (1994) Mutagen sensitivity and suppression of position-effect variegation result from mutations in *mus209*, the *Drosophila* gene encoding PCNA. *EMBO J* 13: 1450–1459
51 Boyd JB, Sakaguchi K, Harris PV (1990) *mus*308 mutants of *Drosophila* exhibit hypersensitivity to DNA cross-linking agents and are defective in a deoxyribonuclease. *Genetics* 125: 813–819
52 Harris PV, Mazina OM, Leonhardt EA, Case RB, Boyd JB, Burtis KC (1996) Molecular cloning of *Drosophila mus308*, a gene involved in DNA cross-link repair with homology to prokaryotic DNA polymerase I genes. *Mol Cell Biol* 16: 5764–5771
53 Beall EL, Rio DC (1996) *Drosophila* IRBP/Ku70 corresponds to the mutagen-sensitive *mus309* gene and is involved in P-element excision *in vivo. Gene Develop* 10: 921–933
54 Lindsley DL, Zimm G (1992) *The genome of* Drosophila melanogaster. Academic Press, New York, London
55 Tearle R, Nusslein-Volhard C (1987) Tubingen mutants stocklist. *Drosophila Inform Service* 66: 209–226
56 Schupbach T, Wieschaus E (1991) Female sterile mutations on the second chromosome of *Drosophila melanogaster*. II. Mutations blocking oogenesis or altering egg morphology. *Genetics* 129: 1119–1136
57 Gillespie DE, Berg CA (1995) *homeless* is required for RNA localization in *Drosophila* oogenesis and encodes a new member of the DE-H family of RNA-dependent ATPases. *Gene Develop* 9: 2495–2508
58 Ghabrial A, Ray RP, Schupbach T (1998) okra and spindle-B encode components of the RAD52 DNA repair pathway and affect meiosis and patterning in *Drosophila* oogenesis. *Gene Develop* 12: 2711–2723
59 Kooistra R, Vreeken K, Zonneveld JBM, de Jong A, Eeken JCJ, Osgood CJ, Buerstedde JM, Lohman PHM, Pastink A (1997) The *Drosophila* RAD54 homolog, *DmRAD54*, is involved in the repair of radiation damage and recombination. *Mol Cell Biol* 17: 6097–6104
60 Cummings WJ, Zolan ME (1998) Functions of DNA repair genes during meiosis. *Curr Topic Develop Biol* 37: 117–141
61 Bishop DK, Park D, Xu L, Kleckner N (1992) DMC1: A meiosis-specific yeast homolog of *E. coli* RecA required for recombination, synaptonemal complex formation and cell cycle progression. *Cell* 79: 1081–1092
62 Klein HL (1997) *RDH54*, a RAD54 homologue in *Saccharomyces cerevisiae*, is required for mitotic diploid-specific recombination and repair and for meiosis. *Genetics* 147: 1533–1543
63 Gonzalez-Reyes A, Elliott H, St Johnston D (1997) Oocyte determination and the origin of polarity in *Drosophila*: the role of the *spindle* genes. *Development* 124: 4927–4937
64 Kuroda MI, Kernan MJ, Kreber R, Ganetzky B, Baker BS (1991) The maleless protein associates with the X chromosome to regulate dosage compensation in *Drosophila. Cell* 66: 935–947
65 McKee BD, Ren X-J, Hong C-S (1996) A *recA*-like gene in *Drosophila melanogaster* that is expressed at high levels in female but not male meiotic tissues. *Chromosoma* 104: 479–488

Part four
Chromosome evolution

Tomás Naranjo

Lurdes F. de Almeida-Toledo

Teresa Capriglione

Horst Hameister

Chromosomes Today, Vol. 13
ed. by E. Olmo and C.A. Redi

Genome evolution in Triticeae

Belén Maestra and Tomás Naranjo

Departamento de Genética, Facultad de Biología, Universidad Complutense, 28040 Madrid, Spain

Introduction

The Triticeae, a tribe of the family *Poaceae (Gramineae)*, contains about five hundred annual and perennial species, among which the three major cereals, wheat, barley and rye, and a number of forage and pasture grasses are included. Wild species of the tribe are valuable sources of germplasm that can be used for gene introgression due to the many possibilities for interspecific and intergeneric hybridisation. Diploid species diverged from a common ancestor with a basic number of x = 7 chromosomes. Many species of the tribe are polyploids. The polyploid species consist of autopolyploids that evolved through genome duplication or allopolyploids that arose after hybridisation between different species. The existence of autopolyploid species of various ploidy level and the complexity of allopolyploids with several genomes involved, in addition to the possible existence of recurrent polyploidisation, make it difficult to understand the phylogenetic structure of the whole tribe. Gene trees based on three nuclear loci and the chloroplast genome led to a reconstruction of the phylogeny of genera with diploid species [1]. The common ancestor of the Triticeae gave rise to the lineage leading to present-day *Critesion, Hordeum,* and *Psathyrostachys*. The genus *Secale* diverged next. Subsequent divergence resulted in two branches, one leading to *Pseudoroegneria, Agropyron,* and *Australopyrum,* and the other leading to *Aegilops, Crithopsis, Lophopyrum, Taeniatherum,* and *Thinopyrum* (the "ACTL clade"). Diploid species of the genus *Triticum* are assumed to have generated from introgression that occurred between a member of the ACTL clade and an early-diverging member of the tribe, perhaps one closely related to *Secale*.

Chromosomes from different genomes, that preserve a residual homology because of their common origin, are designated homoeologous. The identification of genetic relationships between such chromosomes may provide valuable information on the chromosome evolution of the tribe as well as for the introgression of useful alien genes in cultivated species. Common bread wheat, *Triticum aestivum* (2*n* = 6x = 42, genome AABBDD), is an allohexaploid that arose from two hybridisation events involving three different diploid progenitors. Hybridisation between *T. urartu* (AA) [2] and *Aegilops speltoides* (SS) gave rise to the tetraploid wheat *T. turgidum* ($2n = 4x = 28$, AABB). The second hybridisation involved durum wheat, a cultivated form of *T. turgidum*, and the wild *Ae. tauschii* (DD) [3, 4]. The economic importance of cultivated wheats makes their relationships to other species of particular interest. Different approaches have been used to identify the genetic affinity between chromosomes from different genomes. Substitution-compensation tests, which were developed by Sears [5] to establish the seven homoeologous groups of wheat, have been used to assign chromosomes from other species to the seven homoeologous groups. Analysis of homoeologous pairing at metaphase I in interspecific hybrids allows the identification of the arm homoeology between chromosomes from different genomes [6]. Genetic maps constructed using molecular markers provide information on the degree of conservation of genetic colinearity between genomes [7]. The chromosomal location of DNA sequences using *in situ* hybridisation and the C-banding karyotype have also been used as alternative approaches [8–10].

The cytogenetic features of bread wheat allow the use of meiotic pairing in the identification of genetic relationships between wheat and related species. Firstly, chromosomes of wheat show a diagnostic C-banding pattern [11, 12]. In interspecific hybrids, the arms of wheat and related species chromosomes associated at metaphase I can be identified by means of C-banding (Fig. 1). Secondly, chromosome pairing can be manipulated. Control of chromosome pairing in bread wheat is under the action of several promoter and suppressor genes [13]. The *Ph1* gene, located in the long arm of chromosome 5B, is the major pairing suppressor and is responsible for the regular bivalent formation at metaphase I. Another pairing suppressor gene, *Ph2*, is located in the short arm of chromosome 3D. *Ph2* is, however, less effective than *Ph1*. Induction of homoeologous pairing in hybrids between wheat and related species may be achieved through nullisomy for chromosomes 5B or 3D or using the deletion mutants *ph1b* or *ph2b* [14–16]. Interspecific hybrids carrying both genes (*Ph1* and *Ph2*) hybrids lacking *Ph2*, and hybrids lacking *Ph1* show low, intermediate, and high levels of homoeologous pairing at metaphase I, respectively.

The chromosome structure of the genomes of wheat and related species may be comparatively studied by analysing meiotic pairing at metaphase I in interspecific hybrids. Pairing between homoeologous chromosome arms is expected to occur when such chromosome arms maintain the same structure. The occurrence of pairing between arms from different homoeologous groups would indicate the existence of structural differences between genomes that arose as a result of translocations. Since chromosome pairing is considered a good measure of the degree of affinity between genomes, the analysis of homoeologous pairing also provides information on the degree of relatedness between bread wheat and its relatives. In addition, the frequency of homoeologous pairing may be used as an estimate of the frequency of recombination between wheat chromosomes and alien chromosomes, which may have implications for breeding [17, 18].

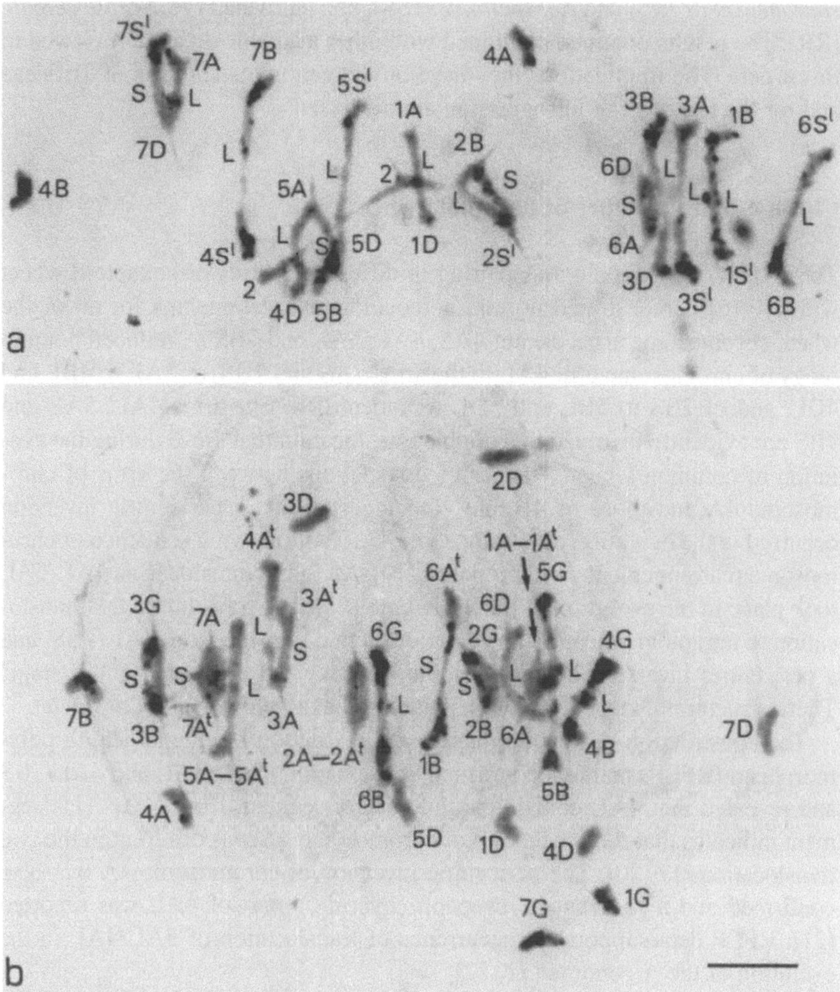

Figure 1. C-banded metaphase I cells from interspecific hybrids of wheat. Individual chromosomes and the arms associated are identified. (a) Cell from a *ph1b T. aestivum* × *Ae. longissima* hybrid. The 4S^{l}L arm, which carries a translocated segment from 7S^{l}L, is associated with 7BL. Chromosomes 2A and 2D (2) could not be distinguished from one another. (b) Cell from a standard *T. aestivum* × *T. timopheevii* hybrid. The trivalent configuration 3A^t-3A-4A^t and the association of 6A^{t}S-1BS are the result of translocations 3A^{t}L/4A^{t}L and 6A^{t}S/1GS, respectively, of *T. timopheevii*. Individual chromosomes of pairs 1A-1A^t, 2A-2A^t, and 5A-5A^t could not be distinguished from one another. Bar represents 10 μm.

Meiotic pairing analysis was first used in the identification of the arm homoeology of bread wheat chromosomes. Later, the analysis was extended to tetraploid wheats, *T. turgidum* (AABB) and *T. timopheevii* (A^{t}A^{t}GG), and to different diploid species, three wild species, *Aegilops longissima* (S^{l}S^l), *Ae.*

sharonensis ($S^{sh}S^{sh}$) and *Ae. speltoides* (SS), and cultivated rye, *Secale cereale* (RR). The results obtained combined with other available data are reviewed in this article. The implications they have on the genomic evolution of Triticeae and for the use of gene introgression are discussed.

Chromosome structure of hexaploid wheat

The results of homoeologous pairing in different hybrids of hexaploid wheat with rye [6] demonstrated normal homoeologous relationships for all of the wheat chromosome arms except 4AS, 4AL, 5AL and 7BS. A reduced homoeology of chromosome arm 4AL to the arms 7AS and 7DS, of 5AL to 4BL and 4DL, and of 7BS to 5BL and 5DL was identified. The arms 4AL, 5AL and 7BS are evidently involved in a double translocation that arose during the evolution of common wheat. The scarce associations between the arms of chromosome 4A and those of 4B and 4D suggested that a pericentric inversion occurred in the evolution of chromosome 4A. The following sequence of chromosome rearrangements was proposed [19]. An initial translocation 4AL/5AL took place in the evolution of the A genome. *T. urartu* transmitted this translocation to tetraploid wheat, *T. turgidum*. A second translocation, 4AL/7BS, and a pericentric inversion of chromosome 4A occurred at the tetraploid stage. These structural changes were transmitted to hexaploid wheat, *T. aestivum*.

The construction of genetic maps using restriction fragments length polymorphism (RFLP) markers confirmed translocations 5AL/4AL and 4AL/7BS and revealed that 4AL contains an intercalary segment from 5AL. This segment indicates that the breakpoint of chromosome 4A was different in the two translocations [7, 20]. The pericentric inversion of chromosome 4A was also confirmed and a paracentric inversion covering a part of 4AL was reported [21]. RFLP data support the occurrence of translocation of 5AL/4AL in the evolution of the A genome [21, 22].

Relationships of *S. cereale*, *Ae. longissima*, *Ae. sharonensis* and *Ae. speltoides* to wheat

Chromosome structure of the R, S^l, S^{sh} and S genomes

Chromosomes of rye were assigned to the seven homoeologous groups on the basis of the ability to compensate when substituted for chromosomes of common wheat [23, 24]. Arm homoeology of rye chromosomes to wheat was established from the results of homoeologous pairing at metaphase I in *ph1b* mutant wheat × rye hybrids combined with data on the chromosomal location of homoeoloci series in wheat and rye [25, 18]. Only the arms 1RS, 1RL, 2RL, 3RS, 4RS and 5RS showed normal homoeologous relationships. The remaining arms of rye appeared to be involved in chromosome rearrangements that occurred during the evolution of the genus *Secale*. A reciprocal translocation between 3RL and 6RL, three translocations involving the arms 4RL, 5RL, 6RS and 7RS, and at least two more rearrangements involving 2RS, 6RL and 7RL were inferred. An RFLP-based genetic map of rye confirmed these structural modifications and a sequence of rearrangements consisting of six translocations and a pericentric inversion was proposed [26].

The homoeologous relationships of *Ae. longissima* chromosomes to bread wheat were established from the genetical and physiological compensation in wheat-alien substitution lines as well as from the chromosomal location of marker genes [27, 28]. Normal homoeologous relationships were identified for chromosomes $1S^l$, $2S^l$, $3S^l$, $5S^l$, and $6S^l$. Chromosomes $4S^l$ and $7S^l$ showed homoeology to both group 4 and 7 chromosomes of wheat, which suggested the presence of a translocation in *Ae. longissima* relative to wheat. Substitution-compensation tests identified the homoeology of chromosome $4S^{sh}$ of *Ae. sharonensis* to wheat chromosomes 4B and 4D [29]. The remaining six chromosomes of *Ae. sharonensis* and the chromosomes of *Ae. speltoides* were assigned to the seven homoeologous groups on the basis of similarities in C-banding pattern and morphology with the chromosomes of *Ae. longissima* [10].

Homoeologous pairing at metaphase I was studied in standard, *ph2b,* and *ph1b* hybrids of hexaploid wheat with *Ae. longissima, Ae. sharonensis*, and *Ae. speltoides* [30–33]. The results confirmed the homoeology of chromosomes $1S^l$, $2S^l$, $3S^l$, $5S^l$, and $6S^l$ of *Ae. longissima* to wheat group 1, 2, 3, 5, and 6 chromosomes and that the arms $4S^lL$ and $7S^lL$ underwent an unequal translocation in the evolution of the S^l genome. A long segment from $7S^lL$ was translocated to $4S^lL$ while the $7S^lL$ arm received a very short, or no, segment from $4S^lL$. The arms of the seven chromosomes of the S^{sh} and S genomes showed normal homoeologous relationships to wheat, which implies that no apparent chromosome rearrangement occurred in the evolution of *Ae. sharonensis* and *Ae. speltoides*.

The origin of the B genome of wheat

The mean frequency of the different types of homoeologous associations in wheat-*Aegilops* and wheat-rye hybrids carrying the ph1b mutation are shown in Table 1. Among the wheat chromosomes, those of the A and D genomes showed preferential pairing. This pattern of pairing persisted in the four types of hybrids although the extra genome changed. Alien (*Aegilops* or rye) chromosomes showed preferential pairing with chromosomes of the B genome. This result is consistent with previous reports suggesting that the S genome species are more closely related to the B genome than to the A or D genomes [34]. *Aegilops* chromosomes paired much more frequently with wheat chromosomes than rye chromosomes, which agrees with the proposal that rye is a

Table 1. Average frequency (%) of the different arm association types in interspecific hybrids of *ph1b* mutant wheat with *Ae. longissima* (S^lS^l), *Ae. sharonensis* ($S^{sh}S^{sh}$), *Ae. speltoides* (SS), and *S. cereale* (RR)

Association type	Hybrids				
	$ABDS^l$	$ABDS^{sh}$	ABDS	ABDR	
A-B-D	1.9	0.7	0.4	4.2	
A-B	7.6	6.4	6.8	5.8	
B-D	13.1	10.7	9.8	10.7	
A-D	48.5	53.4	55.2	53.4	
W-W-S/R	2.7	1.2	1.4	0.0	(0.0)*
A-S/R	7.0	5.9	7.7	1.8	(2.7)*
D-S/R	10.8	8.8	10.7	1.7	(2.7)*
B-S/R	33.3	38.8	48.9	3.5	(5.4)*

W = A-, B- or D- genomes.
* Mean pairing frequencies of the rye arms without structural modifications.

more distant relative to wheat than *Aegilops* species [1]. All of these data support the assumption that the level of pairing between different genomes reflects their degree of closeness. Under this assumption we have analysed the origin of the B genome of wheat.

Different studies suggested that species belonging to the *Sitopsis* section of genus *Aegilops, Ae. bicornis, Ae. longissima, Ae. sharonensis, Ae. speltoides,* and *Ae. searsii,* contributed to the B genome [34]. Translocation $4S^l/7S^l$ of *Ae. longissima* excludes this species to be the B genome progenitor. Molecular studies carried out in the last decade on the interspecific variation or the chromosomal location of different DNA sequences suggest that the B genome was most likely derived from *Ae. speltoides* [8, 35–39]. The results of homoeologous pairing between group 5 chromosomes support such an origin.

The mean frequencies of wheat-*Aegilops* associations for the short and the long arms of the seven homoeologous groups in the hybrids of *ph1b* mutant wheat with *Ae. longissima, Ae. sharonensis* and *Ae. speltoides* are represented in Figure 2. With the exception of group 5 chromosomes, B-S pairing was much more frequent than A-S pairing or D-S pairing in all of the hybrids. Translocation $4S^l/7S^l$ of *Ae. longissima* accounts for the reduction in pairing between group 4 chromosomes in the hybrids of this species. In group 5 chromosomes, B-S pairing and D-S pairing showed similar frequencies in the hybrids of *Ae. longissima* and *Ae. sharonensis* but, in the hybrids of *Ae. speltoides,* B-S pairing was more frequent than D-S pairing. That is to say, while the group 5 chromosomes of *Ae. longissima* and *Ae. sharonensis* show a similar degree of relatedness to 5B and 5D, chromosome 5 of *Ae. speltoides* is more closely related to 5B than to 5D. Chromosome 5B has played an important role in the evolution of polyploid wheats since this chromosome carries

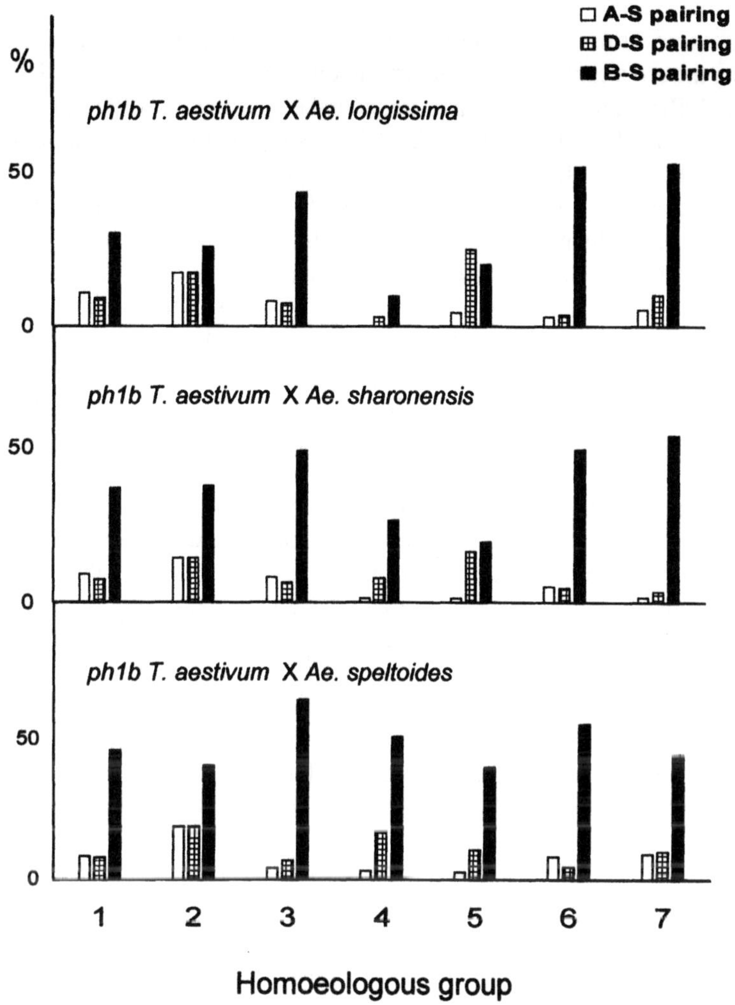

Figure 2. Mean frequency (%) of the different wheat-*Aegilops* arm associations in the seven homoeologous groups in hybrids of *ph1b* mutant wheat with three *Aegilops* species.

the *Ph1* gene that ensures the disomic inheritance of these species. The closer relationship between 5B and 5S supports that *Ae. speltoides* contributed to the B genome.

Implications for breeding

The presence of telomeric C-heterochromatin blocks in the chromosomes of rye was found to hinder wheat-rye pairing in *ph1b* wheat × rye hybrids.

Wheat-rye pairing also showed a considerable variation between groups and chromosome arms in the absence of telomeric heterochromatin [18, 25]. Among the arms carrying no structural modification, the short arms 1RS, 3RS, 4RS and 5RS yielded an average frequency of pairing of 3.5% and the long arms 1RL and 2RL a frequency of 25.3%. Among the arms involved in translocations, 2RS and 6RS did not pair with wheat chromosomes while the frequencies of pairing of 7RS, 3RL, 4RL, 6RL and 7RL ranged from 0.9% to 3.2%. These results provide evidence that structural modifications reduce the frequency of homoeologous pairing. The exception was 5RL. This arm paired with wheat chromosomes in 25.4% of cells. Its translocated segment derived from 4RL should be quite long. The behaviour of chromosome 4 from *Ae. longissima, Ae. sharonensis* and *Ae. speltoides* indicates a negative effect of structural modifications on homoeologous pairing [31–33]. In *ph1b* hybrids, associations between wheat group 4 chromosomes and the arms $4S^{sh}L$ or 4SL reached a frequency of 44.3% and 85%, respectively. These frequencies are much higher than that of 2% of the translocated chromosome arm $4S^{l}L$ of *Ae. longissima.* On the other hand, chromosomes of rye and *Aegilops* paired more often with the wheat chromosomes of the B genomes than with those of the A or D genomes (Tab. 1). All of these data have two main implications for the use of controlled introgression of genes from alien species by induction of homoeologous pairing. First, structural modification of the alien chromosomes from which genetic material should be transferred to wheat may strongly reduce the possibility of recovering the desired recombinants. Second, the probability of achieving the desired transfer of genes depends on which genomes of wheat the alien genes are being introduced into.

Diphyletic origin of tetraploid wheats

The tetraploid wheats are divided into two groups. Domesticated and wild forms of *T. turgidum*, ssp. *durum* and ssp. *dicoccoides* (genome AABB) comprise the emmer group. Similarly, domesticated and wild forms of *T. timopheevii*, ssp. *timopheevii* and ssp. *araraticum* (genome A^tA^tGG) comprise the timopheevi group. From the variation of different repeated DNA sequences in polyploid wheats and related species, *T. urartu* was identified as the donor of both the A- and A^t-genomes [2]. *Ae. speltoides*, the B genome progenitor, is also the most closely related extant diploid species to the G genome of *T. timopheevii* as based on molecular studies using nuclear DNA [8, 36, 37–40], chloroplast DNA [41, 42] and mitochondrial DNA [43].

Chromosome pairing analysis in hybrids of *T. timopheevii* with *T. turgidum* or *T. aestivum* revealed translocation differences between the emmer and *timopheevii* wheats that involved both the A and B or G genomes [44–46]. Such differences do not include translocation 5AL/4AL. The ribosomal RNA gene loci location [8] and the regular bivalent pairing of chromosomes $5A\text{-}5A^t$ in interspecific hybrids [47] support the theory that both tetraploids inherited this translocation from *T. urartu*. Structural differentiation occurred at the tetraploid stage. Translocation 4AL/7BS and the pericentric inversion of chromosome 4A, which occurred in the emmer lineage, do not exist in *T. timopheevii* [8, 9, 46, 47]. There is, however, no information on the presence or absence in *T. timopheevii* of a paracentric inversion like that of 4AL of *T. turgidum* and *T. aestivum*. Other chromosome rearrangements appeared in the *timopheevii* lineage. An analysis by sequential N-banding and genomic *in situ* hybridisation of three accessions of *T. timopheevii* ssp. *timopheevii* and three accessions of *T. timopheevii* ssp. *araraticum* led to the conclusion that the arms $6A^tS$, 1GS and 4GS are involved in two intergenomic translocations, $6A^tS$/1GS and 1GS/4GS, which were considered to be fixed in this species [9]. Two lines of *T. timopheevii* ssp. *timopheevii* differed from *T. turgidum* or *T. aestivum* in four interchanges based on C-banding analysis of meiotic pairing in interspecific hybrids [47]. In addition to translocations $6A^tS$/1GS and 1GS/4GS, translocations 4GS/$4A^tL$ and $4A^tL$/$3A^tL$ were identified. These four translocations most likely arose in that order. The fact that chromosomes 4GS, $4A^tL$ and $3A^tL$ of these two lines showed the same morphology and C-banding pattern as chromosomes of many accessions of *T. timopheevii* ssp. *araraticum* with the standard karyotype [48] suggests that translocations $4A^tL$/4GS and $4A^tL$/$3A^tL$ are also species-specific.

Two different hypotheses have been proposed on the origin of emmer and *timopheevii* wheats. One assumes that they had a monophyletic origin. Both lineages diverged at the tetraploid stage after a single hybridisation event [44, 49, 50]. According to the second hypothesis, tetraploid wheats had a diphyletic origin involving two independent hybridisation events. The different chromosome rearrangements which occurred at the tetraploid level in both lineages are consistent with a diphyletic origin. The restriction endonuclease analyses of chloroplast, mitochondrial and nuclear DNA [41–43, 51] also support this hypothesis and suggest that *T. turgidum* originated much earlier than *T. timopheevii*.

Genomic evolution in diploid and polyploid species

A summary of the chromosome rearrangements reported in species of the tribe Triticeae is shown in Table 2. All of these changes are relative to the D genome of wheat, whose structure is used as the standard reference in the comparisons between the genomes of bread wheat and its relatives [7, 25]. Species listed in Table 2 belong to the genera *Hordeum* and *Secale* and the group *Triticum/Aegilops*, which are representative members of separated lineages in the phylogeny of Triticeae and originated in that order [1]. Genetic synteny was highly conserved during the radiation of diploid species of the tribe since very distant relatives, as are barley and some *Aegilops* species, differ only in

Table 2. Number of evolutionary chromosome rearrangements detected in Triticeae species relative to the D genome of hexaploid wheat

Species	No. of rearrangements	Reference
Hordeum vulgare (HH)	2 (I)	[52]
Secale cereale (RR)	7 (T+I)	[25, 26]
Aegilops umbellulata (UU)	11 (T+I)	[53]
Aegilops longissima (S^lS^l)	1 (T)	[28, 30]
Aegilops sharonensis ($S^{sh}S^{sh}$)	0	[32]
Aegilops speltoides (SS)	0	[33]
Aegilops tauschii (DD)	0	[54]
Triticum monococcum (AA)	1 (T)	[52]
Triticum urartu (AA)	1 (T)	[22]
Triticum turgidum (AABB)	4 (T+I)	[6, 21]
Triticum timopheevii (A^tA^tGG)	5 (T)*	[9, 46, 47]
Triticum aestivum (AABBDD)	4 (T+I)	[6, 19, 21]

T = translocation; I = inversion.
* Two translocations should be confirmed in wild populations of *T. timopheevii*.

two small paracentric inversions [32, 33, 52, 54]. Nevertheless, some diploid species such as rye or *Ae. umbellulata* have accumulated a remarkable number of chromosome rearrangements, mainly translocations [25, 26, 53]. Variation in the number of rearrangements clearly indicates that structural chromosome differentiation was not parallel to genomic diversification.

The number of evolutionary chromosome rearrangements produced in both tetraploid wheats contrasts with the highly conserved structure of their parental species, *T. urartu* and *Ae. speltoides*. These structural changes probably arose at a very early stage because the transient structures were not found. Most likely, these rearrangements are only part of an overall genomic reorganisation produced immediately after the polyploid formation. The formation of new polyploids is often accompanied by extensive genomic modifications within a short period of time [55]. Accordingly, rapid elimination of low-copy DNA sequences was demonstrated in wheat by Feldman et al. [56] who suggested that polyploidisation *per se* may act as a releasing factor of genomic rearrangements, which may accelerate the diploidisation of allopolyploids. On the other hand, in instances of recurrent polyploidy, the magnitude of genomic reorganisation may also impede the genetic flow between polyploids and lead to the formation of new species. Multiple origins of polyploids have been reported in different plant taxa [55]. Polyploids with the same genome constitution, but with independent origins, may produce sterile hybrids if they underwent a rapid structural differentiation. Due to the reproductive isolation, such sister polyploids give rise to different species. Tetraploid species such as *T. turgidum* and *T. timopheevii* may be a representative example of this type of evolution.

Acknowledgements

We thank Dr. I. King for the critical reading of the manuscript. This work has been supported by grant No. PB96-0632 of the Dirección General de Enseñanza Superior (DGES) of Spain.

References

1 Kellogg EA, Appels R, Mason-Gamer RJ (1996) When genes tell different stories: the diploid genera of Triticeae (Gramineae). *Syst Bot* 21: 321–347
2 Dvořák J, di Terlizzi P, Zhang HB, Resta P (1993) The evolution of polyploid wheats: identification of the A genome donor species. *Genome* 36: 21–31
3 Kihara H (1944) Discovery of the DD analyser, one of the ancestors of *Triticum vulgare* (In Japanese). *Agric Hortic* 19: 13–14
4 McFadden ES, Sears ER (1946) The origin of *Triticum spelta* and its free-threshing hexaploid relatives. *J Hered* 37: 81–89
5 Sears ER (1966) Nullisomic-tetrasomic combinations in hexaploid wheat. *In*: R Riley, KR Lewis (eds): *Chromosome manipulation and plant genetics. Heredity* (Suppl) 20: 29–45
6 Naranjo T, Roca A, Goicoechea PG, Giráldez R (1987) Arm homoeology of wheat and rye chromosomes. *Genome* 29: 873–882
7 Liu CJ, Atkinson MD, Chinoy CN, Devos KM, Gale MD (1992) Nonhomoeologous translocations between group 4, 5, and 7 chromosomes within wheat and rye. *Theor Appl Genet* 83: 305–312
8 Jiang J, Gill BS (1994) New 18S·26S ribosomal RNA gene loci: chromosomal landmarks for the evolution of polyploid wheats. *Chromosoma* 103: 179–185
9 Jiang J, Gill BS (1994) Different species-specific chromosome translocations in *Triticum timopheevii* and *T. turgidum* support the diphyletic origin of polyploid wheats. *Chromosome Res* 2: 59–64
10 Friebe B, Gill BS (1996) Chromosome banding and genome analysis in diploid and cultivated polyploid wheats. *In*: PP Jauhar (ed.): *Methods of genome analysis in plants*. CRC Press, Boca Raton, New York, London, Tokyo, 39–60
11 Lukaszewski AJ, Gustafson JP (1983) Translocations and modifications of chromosomes in Triticale × wheat hybrids. *Theor Appl Genet* 64: 239–248
12 Gill BS, Friebe B, Endo TR (1991) Standard karyotype and nomenclature system for description of chromosome bands and structural aberrations in wheat (*Triticum aestivum*). *Genome* 34: 830–839
13 Sears ER (1976) Genetic control of chromosome pairing in wheat. *Ann Rev Genet* 10: 31–51
14 Sears ER (1977) An induced mutant with homoeologous pairing in wheat. *Can J Genet Cytol* 19: 585–593
15 Sears ER (1982) A wheat mutation conditioning an intermediate level of homoeologous chromosome pairing. *Can J Genet Cytol* 24: 715–719
16 Sears ER (1984) Mutations in wheat that raise the level of meiotic chromosome pairing. *In*: JP Gustafson (ed.): *Gene manipulation in plant improvement*. The 16th Stadler Genetics Symposium. Plenum Press, New York, 295–300
17 Naranjo T, Fernández-Rueda P, Goicoechea PG, Roca A, Giráldez R (1989) Homoeologous pairing and recombination between the long arm of group 1 chromosomes in wheat × rye hybrids. *Genome* 32: 293–301
18 Naranjo T, Fernández–Rueda P (1996) Pairing and recombination between individual chromosomes of wheat and rye in hybrids carrying the *ph1b* mutation. *Theor Appl Genet* 93: 242–248
19 Naranjo T (1990) Chromosome structure of durum wheat. *Theor Appl Genet* 79: 397–400
20 Anderson JA, Ogihara Y, Sorrels ME, Tanksley SD (1992) Development of a chromosomal arm map for wheat based on RFLP markers. *Theor Appl Genet* 83: 1035–1043
21 Devos KM, Dubcovsky J, Dvořák J, Chinoy CN, Gale MD (1995) Structural evolution of wheat chromosomes 4A, 5A and 7B and its impact on recombination. *Theor Appl Genet* 91: 282–288
22 King IP, Purdie KA, Liu CJ, Reader SM, Pittaway TS, Orford SE, Miller TE (1994) Detection of interchromosomal translocations within the Triticeae by RFLP analysis. *Genome* 37: 882–887
23 Zeller FJ, Hsam SLK (1983) Broadening the genetic variability of cultivated wheat by utilizing rye

chromatin. *In*: S Sakamoto (ed.): *Proc 6th Int Wheat Genet Symp*, Kyoto, 161–173
24 Miller TE (1984) The homoeologous relationships between the chromosomes of rye and wheat. Current status. *Can J Genet Cytol* 26: 578–589
25 Naranjo T, Fernández-Rueda P (1991) Homoeology of rye chromosome arms to wheat. *Theor Appl Genet* 82: 577–586
26 Devos KM, Atkinson MD, Chinoy CN, Francis HA, Harcourt RL, Koebner RMD, Liu CJ, Masojć P, Xie DX, Gale MD (1993) Chromosomal rearrangements in the rye genome relative to that of wheat. *Theor Appl Genet* 85: 673–680
27 Hart GE, Tuleen N (1983) Characterizing and selecting alien genetic material in derivatives of wheat-alien species hybrids by analyses of isozyme variation. *In*: S Sakamoto (ed.): *Proc 6th Int Wheat Genet Symp*, Kyoto, 377–385
28 Friebe B, Tuleen N, Jiang J, Gill BS (1993) Standard karyotype of *Triticum longissimum* and its cytogenetic relationship with *T. aestivum*. *Genome* 36: 731–742
29 Miller TE, Hutchinson J, Chapman V (1982) Investigation of a preferentially transmitted *Aegilops sharonensis* chromosome in wheat. *Theor Appl Genet* 61: 27–33
30 Naranjo T (1995) Chromosome structure of *Triticum longissimum* relative to wheat. *Theor Appl Genet* 91: 105–109
31 Naranjo T, Maestra B (1995) The effect of *ph* mutations on homoeologous pairing in hybrids of wheat with *Triticum longissimum*. *Theor Appl Genet* 91: 1265–1270
32 Maestra B, Naranjo T (1997) Homoeologous relationships of *Triticum sharonense* chromosomes to *T. aestivum*. *Theor Appl Genet* 94: 657–663
33 Maestra B, Naranjo T (1998) Homoeologous relationships of *Aegilops speltoides* chromosomes to bread wheat. *Theor Appl Genet* 97: 181–186
34 Kerby K, Kuspira J (1987) The phylogeny of the polyploid wheats *Triticum aestivum* (bread wheat) and *Triticum turgidum* (macaroni wheat). *Genome* 29: 722–737
35 Gill BS, Appels R (1987) Relationships between NOR-loci from different Triticeae species. *Plant Syst Evol* 160: 77–89
36 Dvořák J, Zhang HB (1990) Variation in repeated nucleotide sequences sheds light on the phylogeny of the wheat B and G genomes. *Proc Natl Acad Sci USA* 87: 9640–9644
37 Badaeva ED, Friebe B, Gill BS (1996) Genome differentiation in *Aegilops*. 1. Distribution of highly repetitive DNA sequences on chromosomes of diploid species. *Genome* 39: 293–306
38 Badaeva ED, Friebe B, Gill BS (1996) Genome differentiation in *Aegilops*. 2. Physical mapping of 5S and 18S·26S ribosomal RNA gene families in diploid species. *Genome* 39: 1150–1158
39 Daud HM, Gustafson JP (1996) Molecular evidence for *Triticum speltoides* as a B-genome progenitor of wheat (*Triticum aestivum*). *Genome* 39: 543–548
40 Sasanuma T, Miyashita NT, Tsunewaki K (1996) Wheat phylogeny determined by RFLP analysis of nuclear DNA. 3. Intra- and interspecific variations of five *Aegilops* Sitopsis species. *Theor Appl Genet* 92: 928–934
41 Ogihara Y, Tsunewaki K (1988) Diversity and evolution of chloroplast DNA in *Triticum* and *Aegilops* as revealed by restriction fragment analysis. *Theor Appl Genet* 76: 321–332
42 Miyashita NT, Mori N, Tsunewaki K (1994) Molecular variation in chloroplast DNA regions in ancestral species of wheat. *Genetics* 137: 883–889
43 Terachi T, Ogihara Y, Tsunewaki K (1990) The molecular basis of genetic diversity among cytoplasms of *Triticum* and *Aegilops*. 7. Restriction endonuclease analysis of mitochondrial DNAs from polyploid wheats and their ancestral species. *Theor Appl Genet* 80: 366–373
44 Feldman M (1966) Identification of unpaired chromosomes in F1 hybrids involving *Triticum aestivum* and *T. timopheevii*. *Can J Genet Cytol* 8: 144–151
45 Hutchinson J, Miller TE, Jahier J, Shepherd KW (1982) Comparison of the chromosomes of *Triticum timopheevii* with related wheats using the techniques of C-banding and *in situ* hybridization. *Theor Appl Genet* 64: 31–40
46 Gill BS, Chen PD (1987) Role of cytoplasm-specific introgression in the evolution of the polyploid wheats. *Proc Natl Acad Sci USA* 84: 6800–6804
47 Maestra B, Naranjo T (1999) Structural chromosome differentiation between *Triticum timopheevii* and *T. turgidum* and *T. aestivum*. *Theor Appl Genet* 98: 744–750
48 Badaeva ED, Badaev NS, Gill BS, Filatenko AA (1994) Intraspecific karyotype divergence in *Triticum araraticum* (*Poaceae*). *Plant Syst Evol* 192: 117–145
49 Wagenaar EB (1961) Studies on the genome constitution of *Triticum timopheevi* Zhuk. I. Evidence for genetic control of meiotic irregularities in tetraploid hybrids. *Can J Genet Cytol* 3: 47–60

50 Tanaka M, Kawahara T, Sano J (1978) The evolution of wild tetraploid wheats. *In*: S Ramanujan (ed.): *Proc. 5th Int. Wheat Genet. Symp.*, Indian Soc Genet Plant Breeding, India, 73–80
51 Mori N, Liu YG, Tsunewaki K (1995) Wheat phylogeny determined by RFLP analysis of nuclear DNA. 2. Wild tetraploid wheats. *Theor Appl Genet* 90: 129–134
52 Dubcovsky J, Luo MC, Zhong GY, Bransteitter R, Desai A, Kilian A, Kleinhofs A, Dvořák J (1996) Genetic map of diploid wheat, *Triticum monococcum* L. and its comparison with maps of *Hordeum vulgare* L. *Genetics* 143: 983–999
53 Zhang H, Jia J, Gale MD, Devos KM (1998) Relationships between the chromosomes of *Aegilops umbellulata* and wheat. *Theor Appl Genet* 96: 69–75
54 Riley R, Chapman V (1960) The D genome of hexaploid wheat. *Wheat Inf Serv* 11: 18–19
55 Leitch IJ, Bennett MD (1997) Polyploidy in angiosperms. *Trends Plant Sci* 2: 470–476
56 Feldman M, Liu B, Segal G, Abbo S, Levy AA, Vega JM (1997) Rapid elimination of low-copy DNA sequences in polyploid wheat: a possible mechanism for differentiation of homoeologous chromosomes. *Genetics* 147: 1381–1387

Chromosomes Today, Vol. 13
ed. by E. Olmo and C.A. Redi

Karyotypic evolution in Neotropical freshwater fish

Lurdes F. de Almeida-Toledo[1], Fausto Foresti[2] and Silvio de A. Toledo-Filho[1]

[1] *Departamento de Biologia, Instituto de Biociências, Universidade de São Paulo, Caixa Postal 11.461 - 05422-970 São Paulo, SP, Brasil*
[2] *Departamento de Morfologia, Instituto de Biociências, UNESP, Campus de Botucatu, 18618-000, Botucatu, SP, Brasil*

The largest freshwater fish fauna of the world is found in the Neotropical region and includes 2500 to 2700 known species [1], and a total number of species that may actually reach 5000 [2]. Until the late seventies not much was known about the cytogenetic characteristics of these fish and the cytogenetic data available about the Neotropical fish fauna was limited to knowledge of the chromosome number of about 252 species, most of them obtained from specimens of unknown geographical origin, as listed by Almeida-Toledo [3]. Cytogenetic studies focusing on the Neotropical fish species have greatly developed within the last twenty years, mainly in Brazil. The number of species already karyotyped has reached 706, belonging to 207 genera and 38 families [4]. These data include not only chromosome number and formulae, but also data about the presence of differentiated sex chromosomes, supernumerary B-chromosomes, and the location of nucleolus organizer regions (NORs). For a growing number of species, restriction banding analysis, *in situ* localization of rDNA sequences in the chromosomes [5, 6], of satellite DNA probes (obtained from heterochromatic regions either from pericentromeric sequences [7] or from B-chromosomes [8]) or of telomeric sequences and, more recently, immunocytogenetic analyses [9] are being carried out.

The amount of data now available about the extraordinary chromosome variability found in the Neotropical fish fauna allows the identification of the main chromosome evolutionary trends and of some peculiar cytogenetic features of this fauna.

Chromosome number and formula

Chromosome numbers in Neotropical freshwater species already analysed vary from $2n = 20$ found in *Pterolebias longipinnis* [10] to $2n = 134$ reported for *Corydoras aeneus* [11].

Considering the total amount of data on chromosome number and formula already available, it is possible to recognize cytogenetically *conservative groups* characterized by very constant chromosome numbers and formulae and cytogenetically *variable groups* which present wide variability in chromosome number and formulae. Hypotheses concerning the relationship between conservative or variable karyotypes and population structure have already been proposed based on theoretical models [12, 13] which correlate population structure with fixation of chromosome rearrangements. The cytogenetic characteristics of neotropical freshwater fish seem to fit these models.

Conservative groups

A very striking example of conservatism in terms of chromosome number and morphology is found in the families Anostomidae, Curimatidae, Hemiodontidae, Parodontidae, Prochilodontidae and Chilodontidae, order Characiformes: from the 54 species already cytogenetically analysed, 50 species present $2n = 54$ bi-armed chromosomes [14] and karyotypes that are almost indistinguishable after conventional staining. Since Anostomidae, Chilodontidae, Curimatidae and Prochilodontidae have been considered sister groups on the basis of morphological characteristics [15], the gross cytogenetic characteristics presented by these species may also reflect a chromosomal relationship among these groups. These groups, on the other hand, are characterized by highly vagile species and a large population size, fitting the proposition that chromosome rearrangements have difficulty in being fixed in large populations. The characteristic, almost homogeneous, karyotype of all bi-armed chromosomes present in all these species could be a product of the accumulation of chromosome pericentric inversions in a process of karyotypic orthoselection [16]; this similarity in karyotypes among species also seems to be in agreement with the phylogenetic proximity suggested by the morphological similarity of these taxa.

Non-conservative groups

Species from fish groups characterized by small population sizes, such as *Eigenmannia, Hoplias* and *Microlepidogaster*, on the other hand, have accumulated a large number of structural changes during their evolutionary diversification, such that even closely related species have species-specific karyotypes, and chromosomes of variable morphology. These species-specific karyotypes may be very useful in cytotaxonomic approaches.

The population structure should not, however, be considered as the only determinant of karyotype structure. Karyotypic homogeneity has also been detected in fish from small populations [14] and this could be explained by the process of canalization of the chromosome morphology as mentioned for high vagility groups.

Natural hybrids

A rare occurrence of natural hybridization was detected involving two sibling species from genus *Eigenmannia*: among individuals with $2n = 36$ and $2n = 38$ chromosomes a $2n = 37$ specimen was detected and the morphological analysis of the chromosomes allowed the identification of the haploid sets from the two parental species.

Supernumerary B chromosomes

Variability in chromosome number may also be due to the presence of B supernumerary chromosomes. The occurrence of B chromosomes in fish was ignored until the early eighties [17], but now a considerable number of occurrences of B-supernumerary chromosomes have already been detected, mainly in Neotropical fish species. These B-chromosomes are either macrochromosomes, that is, their size reaches the size of the largest chromosomes of the A complement, or they are small, even punctual chromosomes. They are usually found along the entire distribution area of the species, as is the case of *Prochilodus lineatus* [18] and *Callichthys callichthys* [19], both presenting micro B chromosomes. On the other hand, in *Astyanax scabripinnis*, while macro B-chromosomes were detected in populations from the Paraná river system in Southeast Brazil [20], micro B-chromosomes restricted to males were found in populations from the so called East Basins [21]. This occurrence of different types of B-chromosomes in populations of the same nominal species was interpreted as a possible incipient genetic differentiation among geographical populations of *A. scabripinnis.*

Poliploidy and genome size variation

The variation in nuclear DNA content in Neotropical fish goes from 1.04 ± 0.09 pg DNA/nucleus in diploid cells found in *Corydoras* cf. *simulatus* [22] to the extremely high value of 3,540% of the mammalian cell value (about 248.0 pg DNA/nucleus) reported for *Lepidosiren paradoxa* [23]. Considering the data already available on the nuclear DNA content of Characiformes and Siluriformes, which constitute about 90% of the Neotropical freshwater fish fauna [24], the nuclear DNA content in these groups present modal numbers of 3.0 pg DNA/nucleus and 3.4 pg DNA/nucleus, respectively [25]. Extreme variabilities are present in certain species, as is the case for *Lepidosiren paradoxa* or in groups such as the genus *Corydoras.* The genus *Corydoras* belongs to the family Callichthyidae and species from five genera of this family, excluding the genus *Corydoras*, present a nuclear DNA content variability ranging from 1.18 ± 0.07 found in *Dianema urostriata* to 2.33 ± 0.19 found in

Brochis splendens. In the genus *Corydoras*, however, from which 27 species have been analysed in terms of nuclear DNA content, a variation from 1.04 ± 0.09 in *Corydoras* cf. *simulatus* to 8.75 ± 1.50 in *Corydoras metae* was detected [25]. In this context, *Corydoras metae* with $2n = 92$ chromosomes and a nuclear DNA content of 8.75 ± 0.0 pg/nucleus is possibly a tetraploid form in relation to species like *C. schwartzi* or *C. trilineatus*, that present with $2n = 46$ and nuclear DNA content of 4.38 ± 0.49 pg/nucleus [22]. Species of the genus *Corydoras* were arranged in five groups on the basis of nuclear DNA content, taking into account the evidence of many events of nuclear DNA content modification that have occurred in the evolutionary history of this group. However, the relationship among groups cannot be explained by simple events of doubling of the chromosome complement. Rather, more complex events were probably involved such as DNA gain or loss not involving the entire genome [22].

Sporadic events of polyploidy in natural populations resulting in natural triploidy have also been detected in a considerable number of specimens from an unrelated species, for example, in the genus *Eigenmannia*, or in some cases in related species as in the genus *Astyanax* [26]. These sporadic triploids are probably derived from the fertilization of unreduced egg cells, possibly originating from thermal shocks that may have occurred during fertilization, thus disturbing the second meiotic division.

Nucleolus organizer region (NOR) polymorphism in Neotropical freshwater fish

One of the most striking characteristics of the nucleolus organizer regions (NORs) of Neotropical freshwater fish species is their polymorphism. This fact was stressed by Foresti et al. for Gymnotiformes [27], and it has been observed that in almost every Neotropical species already analysed, NOR size polymorphism is present. This is usually due to the duplication of the ribosomal DNA sequences, probably caused by unequal crossing over during meiosis. Morphological alterations of NOR chromosomes involving pericentric and paracentric inversions were detected in many species of Gymnotiformes: in *Gymnotus carapo* three morphologically different NOR chromosomes were found, originating from two paracentric inversions involving the NOR region. The combination of the three chromosomes resulted in five NOR phenotypes that presented both inter- and intra-populational variability. All the six populations analysed were in Hardy-Weinberg equilibrium in relation to this characteristic [28]. In *Eigenmannia*, pericentric inversions involving the NOR chromosomes have been found [27], but variant NORs were found only in the heterozygous state.

Another important feature of NORs concerns their localization either in only one pair of chromosomes, giving the species two NORs, or in more than one chromosome pair giving as many as 22 NORs in *Mylossoma paraguayensis,* which has the maximum number of NORs yet detected in a fish species, (Almeida-Toledo et al., unpublished observations). In species with multiple NORs, the maximum number of NORs is rarely present or active and this character is usually very polymorphic: in *Astyanax scabripinnis,* from the nine putative NOR-bearing pairs morphologically detected, 1 to 15 NORs were found to be active, and usually in only one of the homologues [21]. The presence of simple or multiple NORs sometimes characterizes entire groups of fish families and may constitute useful evidence of a possible relationship as occurs in the families Anostomidae, Prochilodontidae and Curimatidae. All these families are phylogenetically related with almost all the species of this group presenting $2n = 54$ bi-armed chromosomes and only one pair of NORs (Oliveira, unpublished observations). On the other hand, the family Characidae, which is a variable group in terms of karyotypic constitution, displays high diversity in NOR location among its subfamilies but conservation within the subfamilies: while Bryconinae and Salminae present only one pair of NORs [29], Serrasalminae and Tetragonopterinae are characterized by multiple NORs, usually in a species-specific position.

Heterochromatin bands in chromosomes

When entire groups of fish species are compared it is possible to observe different evolutionary tendencies in relation to the constitutive heterochromatin behavior: subfamilies Serrasalminae and Bryconinae belonging to the order Characiformes, for instance, are characterized by the occurrence of species with specific heterochromatin patterns and although these species present conservative chromosome number and formulae, heterochromatin reorganization has occurred during their evolutionary history. Genus *Brycon* presents a non-conservative heterochromatin pattern [29], in spite of the conservation of chromosome number and NOR position detected in all the species already analysed. Subfamily Serrasalminae also presents species-specific heterochromatin markers which were found to characterize closely related species as, for instance, in the case of *Piaractus mesopotamicus* and *Colossoma macropomum.* The chromosomes of these two species are indistinguishable by conventional Giemsa staining or by NOR patterns, but five differential C-banded chromosome markers could be identified which allowed the cytogenetic identification of the two species and of their hybrids [30].

Fluorochrome staining

Euchromatic regions of fish chromosomes do not display chromosome bands after fluorochrome staining. Thus, euchromatic fluorescent bands such as, for instance, Quinacrine bands (Q-bands) as observed in higher vertebrates were not found in fish. G+C specific fluorochromes, like Mitramycin and Chromomycin A3, stain nucleolus organizer regions in fish irregardless of their activity during the previous interphase. This characteristic was observed in many Neotropical fish species, such as in species of the genus *Astyanax* [31], corroborating results already reported in many fish species. Besides the NORs, many regions unrelated to NORs are stained by chromomycin A3 in Neotropical fish species [32]. These regions are C-band positive regions and have a strong affinity with G+C specific fluorochromes which stain them brighter than the NORs. Some of these heterochromatin blocks were found to be associated with sex chromosomes.

A+T specific fluorochromes, such as quinacrine and DAPI (4'-6-diamidino-2-phenylindole) stain specific C-band positive regions in fish. In *Eigenmannia* sp.2 the quinacrine staining produces a very homogeneous bright color in the chromosomes with no euchromatin banding pattern, except for two small bright regions located near the NORs. A bright region associated with the Y chromosome of this species was detected after DAPI staining.

Restriction endonuclease (RE) banding

In Neotropical freshwater fish, RE banding with a number of different restriction enzymes was performed, usually resulting in a modified C-band pattern and, more rarely, in a G-band pattern.

Neotropical species belonging to the genus *Astyanax*, *A. bimaculatus*, *A. fasciatus*, *A. schubarti* and *A. scabripinnis,* displayed a modified C-band pattern after digestion with *Alu* I, *Hae* III, *Msp* I and *Pvu* II that is specific for each species and each type of enzyme [33]. *Alu* I and *Hae* III treatment of *Eigenmannia virescens* chromosomes has resulted in the differential digestion of a heterochromatic X-linked region: whilst *Alu* I digested almost all the DNA on the chromosomes except the X-linked region, the enzyme *Hae* III digested only the X-linked region so that this region was completely extracted (Almeida-Toledo, unpublished observations).

Longitudinal euchromatin bands

Unlike chromosomes of higher vertebrates, fish chromosomes do not display euchromatic bands when submitted to the usual G, R or Q band techniques, using fixed chromosome preparations and proteolytic enzyme treatment, incubation in salt solutions, or quinacrine staining. This lack of euchromatic bands has been explained by the presence of differences in the DNA organization in these groups [34] or by the difficulty in obtaining good chromosome preparations from fish; mainly due to the small size and large number of chromosomes frequently present in fish species [35]. A successful alternative technique for obtaining euchromatic bands in fish is incorporation or replication banding using 5-bromodeoxyuridine (BrdU).

Incorporation bands obtained in two related Neotropical species of electric eels, *Eigenmannia* sp.1 and *Eigenmannia* sp.2, permitted the identification of chromosome homologies between these two species. In *Eigenmannia* sp.2, which presents a X_1X_2Y sex chromosome system, it was possible to identify, by the banding pattern, the translocated chromosomes that gave origin to the neo-Y in males [36]. In another species of this same genus, *Eigenmannia virescens,* which presents an XX:XY sex chromosome system, it was also possible to characterize the banding pattern of the differentiated sex chromosomes and compare them to a corresponding undifferentiated sex chromosome pair present in another geographical race of the same species [37]. Using BrdU bands, it was also possible to test the homology of the euchromatin banding pattern in the family Characidae. A very interesting characteristic of the Characidae, observed by Scheel [38] and confirmed by further studies, is the presence of one large bi-armed marker chromosome that was detected in almost all the species already analysed in this group. The R-banding technique was applied to several species of this family (Fig. 1b) and revealed the conservation of a characteristic banding pattern in the marker chromosomes as shown in Figure 1a for *Astyanax bimaculatus, A. fasciatus, A. schubarti, Brycon* cf. *cephalus* and *Probolodus heterostomus.* It was suggested that this similarity could be indicative of a common origin for these marker chromosomes.

Sex chromosomes in Neotropical fish species

Fish are characterized by an extraordinary variability of sex determination systems. Eight types of chromosome systems of sex determination have already been described [39], involving both morphologically undifferentiated and differentiated homologues in XX:XY and ZZ:ZW simple or multiple systems with either male or female heterogamety. In fact, morphologically differentiated sex chromosomes have been detected in only 5.7% of the total number of Neotropical freshwater species already analysed cytogenetically [4], about 35 species. However, the sex chromosome systems described in these species include at least five different types. Differences in the sex homologues from simple systems are usually due to the addition/deletion of heterochromatin blocks or to the occurrence of a pericentric inversion affecting one of the homologues. Multiple systems have been found that involve chromosome rearrangements between the original sex chromosome and one autosome, resulting either from a Robertsonian translocation involving heterochromatin loss [36], or from a tandem translocation [40], with no observable loss of heterochromatin.

Differentiated sex chromosomes probably evolved independently in many Neotropical fish lineages along their evolutionary histories. Some of these

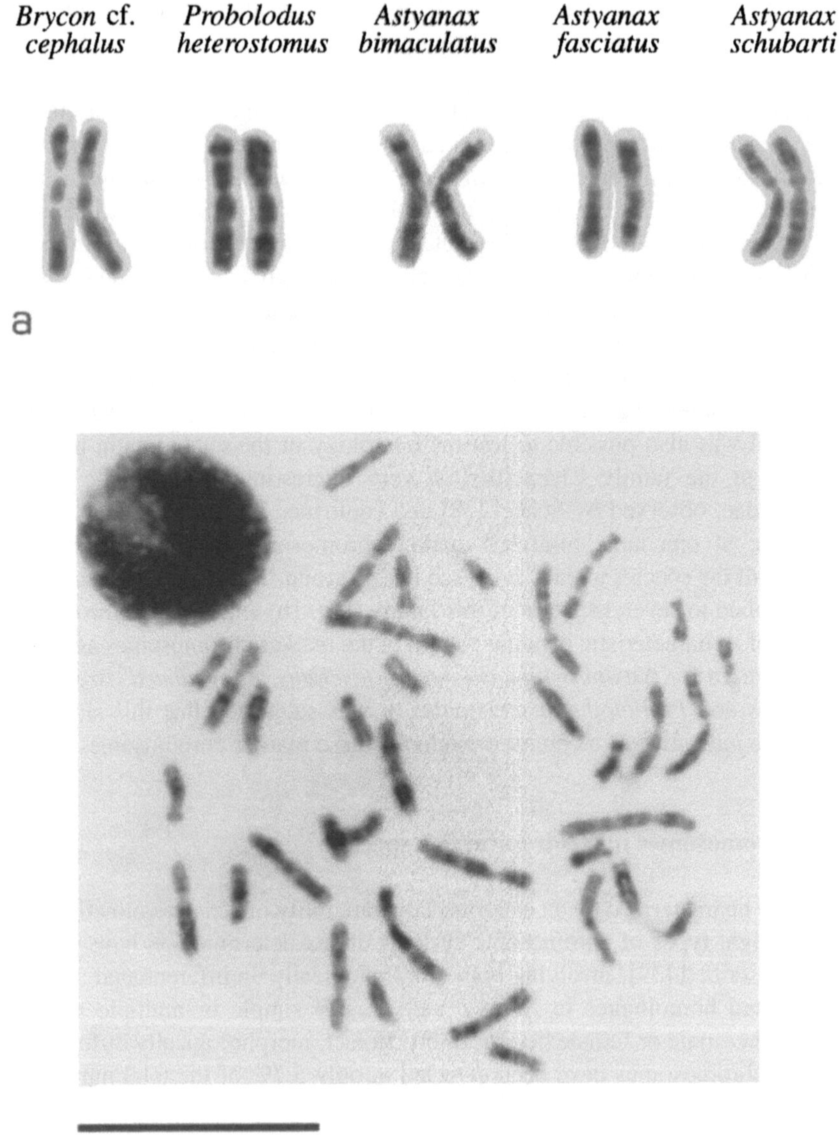

Figure 1. R-banding after 5-BrdU incorporation in Characidae. a) R-banding pattern of the marker chromosome of five Characidae species; b) R-banded metaphase of *Astyanax schubarti*. Bar represents 10 μm.

occurrences are sporadic and probably recent, as is the case for *Semaprochilodus taeniurus* (Prochilodontidae), in which a ZZ:ZW system was reported. The well-differentiated W chromosome contains a large conspicuous heterochromatic block [41]. A further eleven species of Prochilodontidae were analysed and no evidence of differentiated sex chromosomes was found. In fact, the occurrence of differentiated sex chromosomes in fish is, in general, sporadic, that is, it happens in one species of a group with no observable recurrence in the same genus or family.

However, in opposition to this more general rule, two well-documented occurrences were observed involving the presence of sex chromosomes of the same nature in related species from a genus, indicating a monophyletic relationship. In the family Anostomidae, genus *Leporinus*, a monophyletic group of six species, among the 20 species already analysed, share a characteristic ZZ:ZW chromosome system in which the Z and W chromosomes differ by the presence of a large sized heterochromatin block present in the W [42–44]. A similar occurrence was reported in the genus *Triportheus* from which five species with a similar ZZ:ZW system were reported [45, 46].

In some cases, a tendency is detected for differentiated sex chromosomes to occur in related species, but different sex chromosome systems are present indicating probably that these groups are in an incipient stage of sex chromosome differentiation. This is the case in some species of Erythrinidae and Parodontidae, order Characiformes, and of genus *Eigenmannia*, order Gymnotiformes, in which differentiated sex chromosomes occur in closely related species, and either multiple or simple systems with female or male heterogamety may be found, sometimes in geographical races of the same species. In the species complex *Hoplias malabaricus* (family Erythrinidae), XX:XY, $X_1X_1X_2X_2$:X_1X_2Y or XX:XY_1Y_2 systems were detected [14]. In this case there was a tendency towards the occurrence of rearrangements involving the initially undifferentiated X and Y chromosomes. However, either the X is the sex homologue involved in the translocation, resulting in a X_1X_2Y system, or the Y homologue is involved, resulting in a XX:XY_1Y_2 system. The same pattern is observed in Parondontidae: female heterogamety was reported in *Appareiodon affinis,* which has a ZZ:ZW_1W_2 multiple system [48]. In this same family the occurrence of a simple system ZZ:ZW was reported in *Parodon hilarii* [49].

Species of *Eigenmannia,* belonging to the order Gymnotiformes, were found to present different sex chromosome systems. In *E. virescens* from the Upper Paraná river basin, specimens from the Mogi-Guaçu river have undifferentiated sex homologues, while specimens from the Tietê river have an XX:XY system in which sex chromosome differentiation has been attained by addition/deletion of a CMA3 positive heterochromatin block [37]. The same nominal species, *E. virescens*, presents female heterogamety in the São Francisco basin and again male heterogamety in specimens from the Amazon basin. In this last case, the Y chromosome is submetacentric and differs morphologically from the acrocentric Y chromosome characteristic of the Upper Paraná system, this difference being probably due to a pericentric inversion. Another species of Gymnotiformes, *Eigenmannia* sp.2, is characterized by a multiple system of sex chromosomes $X_1X_1X_2X_2$:X_1X_2Y in which the metacentric neo-Y originated by a Robertsonian translocation that probably occurred in an ancestral form, involving an acrocentric Y and an acrocentric autosome [47]. A second occurrence in Gymnotiformes, and the fourth in Neotropical freshwater fish, of a $X_1X_1X_2X_2$:X_1X_2Y sex chromosome system involving Robertsonian translocation was detected in *Brachyhypopomus pinnicaudatus* (unpublished observations).

The analysis of the heterochromatic regions of sex chromosomes using base-specific fluorochromes was carried out in some of these species. In *Eigenmannia* sp.2, a difference in the amount of heterochromatin associated with the sex chromosomes was detected: a CMA3 positive, G+C rich, heterochromatic portion of one of the chromosomes involved in the Robertsonian fusion that gave origin to the bi-armed Y was lost so that female X_1 and X_2 chromosomes have more G+C rich heterochromatin than the corresponding neo-Y in the male. On the other hand, both in *Eigenmannia* sp.2 and *B. pinnicaudatus*, the neo-Y chromosome presented an A+T rich heterochromatin pericentromeric region detected after DAPI staining (Fig. 2 a, b), indicating that these regions, probably present in the original translocated acrocentrics, were not lost. The association of A+T rich heterochromatin with the sex chromosomes was already reported in another Neotropical fish species from the genus *Poecilia* by Haaf and Schmid [50].

The probable meaning of the loss of G+C heterochromatin in *Eigenmannia* sp.2 and the conservation of A+T heterochromatin associated with the sex chromosomes in these two species remains to be explained. However, the finding that, in addition to the occurrence of different systems of sex chromosomes, differences also occur in the molecular nature of the heterochromatin associated with the sex-related regions, reinforces the idea that sex chromosomes may have evolved independently and in different ways in the various lineages of fish, resulting in an extremely rich variability of sex chromosome systems.

The growing use of molecular approaches associated with the cytogenetic analysis of the Neotropical freshwater fish, mainly involving replication bands, fluorochrome staining, restriction enzyme banding, *in situ* hybridization and antibody labeling has permitted not only a better understanding of the evolutionary trends in fishes of the Neotropical region, but also allowed a closer view of the molecular organization of fish chromosomes.

Acknowledgements

We thank Dr. Ann J. Stocker for the critical reading of the manuscript and FAPESP, FINEP and CNPq for providing the financial funds for our research.

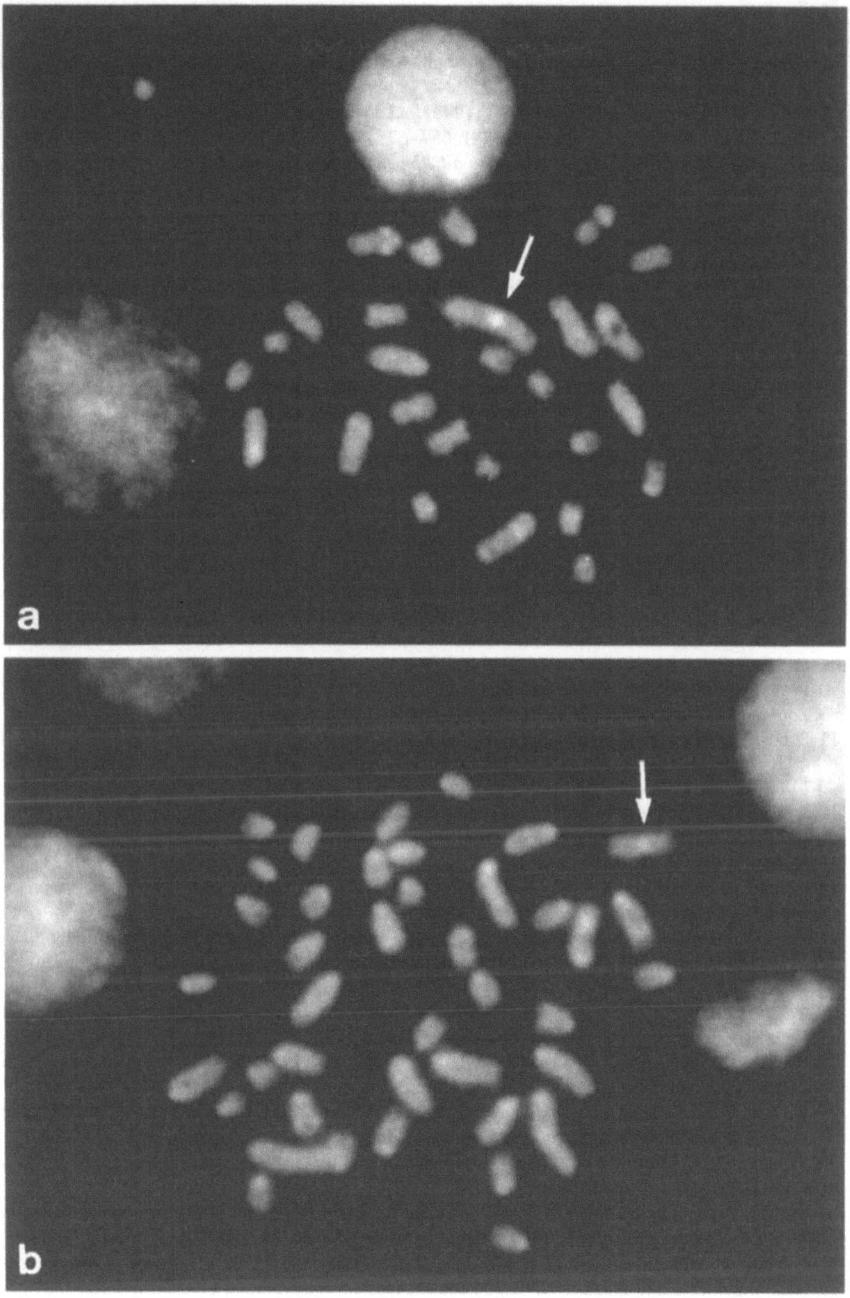

Figure 2. DAPI staining of metaphases of species presenting a $X_1X_1X_2X_2$:X_1X_2Y multiple sex chromosome system. a) *Eigenmannia* sp.2; b) *Brachyhypopomus pinnicaudatus*. The arrows indicate the pericentromeric brightly stained region of the neo-Y chromosomes.

References

1 Géry J (1968) The fresh-water fishes of South America. *In*: EJ Fiitkav (ed.): *Biogeography and ecology in South America II*, vol. 19, Junk NV Publishers, The Hague, Netherlands, 828–847
2 Böhlke JE, Weitzman SH, Menezes NA (1978) Estado atual da sistemática dos peixes de água doce da América do Sul. *Acta Amazonica* 8 (4): 657–677
3 Almeida-Toledo LF (1978) *Contribuição à Citogenética dos Gymnotoidei (Pisces, Ostariophysi)*. PhD Thesis, Universidade de São Paulo
4 Oliveira C, Almeida-Toledo LF, Foresti F (1996) Revisão dos estudos citogenéticos em peixes neotropicais de águas continentais. *Abstr VI Simpósio de Citogenética Evolutiva e Aplicada de Peixes Neotropicais*, 104
5 Almeida-Toledo LF, Coutinho-Barbosa AC, Rosenberg C, Daniel MFZ, Toledo-Filho SA (1993) Hibridação *in situ* com sondas de rDNA em cromossomos de *Piaractus mesopotamicus*, *Colossoma macropomum* e de seus híbridos. *Rev Brasil Genet* 16 (3, supl): 141
6 Almeida-Toledo LF, Stocker AJ, Foresti F, Toledo-Filho SA (1996) Fluorescence *in situ* hybridization with rDNA probes on chromosomes of two nucleolus organizer region phenotypes of two species of *Eigenmannia* (Pisces, Gymnotoidei, Sternopygidae). *Chromosome Res* 4: 301–305
7 Haaf T, Schmid M, Steinlein C, Galetti PM, Huntington W (1993) Organization and molecular cytogenetics of a satellite DNA family from *Hoplias malabaricus* (Pisces, Erythrinidae). *Chromosome Res* 1: 77–86
8 Mestriner CA, Valentini SR, Moreira-Filho O, Galetti Jr PM (1996) A distribuição do DNA satélite *Kpn*I em *Astyanax scabripinnis* e a origem do seu cromossomo supernumerário. *Abstr VI Simpósio de Citogenética Evolutiva e Aplicada de Peixes Neotropicais, São Carlos SP*, 8
9 Almeida-Toledo LF, Viegas-Péquignot E, Coutinho-Barbosa AC, Foresti F, Niveleau A, Toledo-Filho SA (1998) Localization of 5-Methylcytosine in heterochromatic regions of metaphase chromosomes in diploid and triploid *Piaractus mesopotamicus* (Pisces, Characiformes). *Cytogenetics and Cell Genet* 83: 21–24
10 Post A (1965) Vergleichende Untersuchungen Chromosomenzahlen bei Süsswasser Teleosten. *Z Zool Syst Evol Forsch* 3: 47–93
11 Scheel JJ, Simonsen V, Gyldenholm AO (1972) The karyotypes and some electrophoretic patterns of fourteen species of the genus *Corydoras*. *Sonderdruck aus Z Zool Syst Evolutiosforsch* 10: 144–152
12 Lande R (1979) Effective deme sizes during long-term evolution estimated from rates of chromosomal rearrangement. *Evolution* 33: 234–251
13 Slatkin M (1981) Estimation of levels of gene flow in natural populations. *Genetica* 99: 323–335
14 Galetti Jr PM, Bertollo LAC, Moreira-Filho O (1994) Trends in chromosome evolution of Neotropical chariciform fishes. *Caryologia* 47: 289–298
15 Vari RP (1983) Phylogenetic relationships of the families Curimatidae, Prochilodontidae, Anostomidae and Chilodontidae (Pisces, Characiformes). *Smithsonian Contrib Zool* 378: 1–60
16 White MJD (1973) *Animal Cytology and Evolution*, 3rd ed. Cambridge University Press, Cambridge
17 Jones RN, Rees H (1982) *B-Chromosomes*. Academic Press, New York
18 Pauls E, Bertollo LAC (1990) Distribution of a supernumerary chromosome system and aspects of karyotype evolution in genus *Prochilodus* (Pisces, Prochilodontidae). *Genetica* 81: 117–123
19 Oliveira C, Almeida-Toledo LF, Mori L, Toledo-Filho SA (1993) Cytogenetic and DNA content in six genera of the family Callichthyidae (Pisces, Siluriformes). *Caryologia* 46 (2–3): 171–188
20 Salvador LB, Moreira-Filho O (1992) B-Chromosomes in *Astyanax scabripinnis* (Pisces, Characidae). *Heredity* 69: 50–56
21 Stange EAR, Almeida-Toledo LF (1993) Supernumerary B-chromosomes restricted to males in *Astyanax scabripinnis* (Pisces, Characidae). *Rev Brasil Genet* 16 (3): 601–615
22 Oliveira C, Almeida-Toledo LF, Mori L, Toledo-Filho SA (1992) Extensive Chromosome Rearrangements and Nuclear DNA content changes in the evolution of the armoured catfishes genus *Corydoras* (Pisces, Siluriformes, Callichthyidae). *J Fish Biol* 40: 419–431
23 Ohno S, Atkin NB (1966) Comparative DNA values and chromosome complements in eight species of fishes. *Chromosoma* 18: 455–456
24 Britski HA (1992) Conhecimento atual das relações filogenéticas de peixes neotropicais. *In*: AA Agostinho, E Benedito-Cecílio (eds): *Situação atual e perspectivas da Ictiologia no Brasil*. Editora da Universidade Federal de Maringá, Maringá, 42–57

25 Carvalho C, Oliveira C, Foresti F (1998) Nuclear DNA content of thirty species of Neotropical fishes. *Genet Mol Biol* 21 (1): 47–54

26 Fauaz G, Vicente VE, Moreira-Filho O (1994) Natural triploidy and B chromosomes in the Neotropical fish genus *Astyanax* (Characidae). *Rev Brasil Genet* 17 (2): 157–163

27 Foresti F, Almeida-Toledo LF, Toledo-Filho SA (1981) Polymorphic nature of nucleolus organizer regions in fishes. *Cytogenet Cell Genet* 31: 137–144

28 Fernandes-Matioli FMC, Almeida-Toledo LF, Toledo-Filho SA (1997) Extensive Nucleolus Organizer Region polymorphism in *Gymnotus carapo* (Gymnotoidei, Gymnotidae). *Cytogenet Cell Genet* 78: 236–239

29 Almeida-Toledo LF, Bigoni APV, Bernardino G, Foresti F, Toledo-Filho SA (1996) Karyotype and NOR conservatism with heterochromatin reorganization in Neotropical Bryconids. *Caryologia* 49: 35–43

30 Almeida-Toledo LF, Foresti F, Toledo-Filho SA, Bernardino G, Ferrari E, Alcântara RCG (1987) Cytogenetic studies in *Colossoma mitrei*, *Colossoma macropomum* and their interspecific hybrids. *In*: K Tiews (ed.): *Selection, Hybridization and Genetic Engeneering in Aquaculture*, vol. 1. Heenemann Verlagsgesellschaft mbH, Berlin, 190–195

31 Daniel-Silva MFZ (1996) *Estudos citogenéticos comparativos em quatro espécies do gênero* Astyanax *(Pisces, Characidae)*. Master's Thesis, Universidade de São Paulo, Brasil

32 Almeida-Toledo LF (1996) Molecular and Immunocytogenetics of Brazilian fishes. *Ciênc Cultur* 48 (5/6): 377–382

33 Daniel MFZ, Almeida-Toledo LF (1995) Bandamento por endonucleases de restrição em quatro espécies do gênero *Astyanax* (Pisces, Characidae). *Rev Brasil Genet* 18 (3, supl.): 451

34 Medrano L, Bernardi G, Couturier Dutrillaux B, Bernardi G (1988) Chromosome banding and genome compartimentalization in fishes. *Chromosoma* 96: 178–183

35 Gold JR, Li YC, Shipley NS, Powers PK (1990) Improved methods for working with fish chromosomes with a review of metaphase chromosome banding. *J Fish Biol* 37: 563–575

36 Almeida-Toledo LF, Viegas-Pequignot E, Foresti F, Toledo-Filho SA, Dutrillaux B (1988) BrdU replication patterns demonstrating chromosome homeologies in two fish species, genus *Eigenmannia*. *Cytogenet Cell Genet* 48: 117–120

37 Almeida-Toledo LF, Foresti F, Toledo-Filho SA (1988) An early stage of sex chromosome differentiation in the fish *Eigenmannia virescens* (Sternopygidae). *Genome* 30 (1, Supl): 132

38 Scheel JJ (1973) Fish chromosomes and their evolution. *Internal Report of Danmarks Akuartium*, Charlottenlund, Denmark

39 Tave D (1993) *Genetics for Fish Hatchery Managers*. 2nd Ed. Van Nostrand Reinhold, New York

40 Bertollo LAC, Fontes MS, Fenocchio AS, Cano J (1997) The X_1X_1Y sex chromosome system in the fish *Hoplias malabaricus*. I. G-, C- and chromosome replication banding. *Chromosome Res* 5: 493–499

41 Feldberg E, Bertollo LAC, Almeida-Toledo LF, Foresti F, Moreira-Filho O, Santos AF (1987) Biological aspects of Amazonian fishes. IX Cytogenetic studies in two species of the genus *Semaprochilodus* (Pisces, Characidae). *Genome* 29: 1–4

42 Galetti Jr PM, Foresti F (1986) Evolution of the ZZ/ZW system in *Leporinus* (Pisces, Anostomidae). Role of constitutive heterochromatin. *Cytogenet Cell Genet* 43: 43–46

43 Galetti PM, Lima NRW, Venere PC (1995) A monophyletic ZW sex chromosome system in *Leporinus* (Anostomidae, Characiformes). *Cytologia* 60: 375–382

44 Koehler MR, Dehm D, Guttenbach M, Nanda I, Haaf T, Molina WF, Galetti Jr PM, Schmid M (1997) Cytogenetics of the genus *Leporinus* (Pisces, Anostomidae). 1. Karyotype analysis, heterochromatin distribution and sex chromosomes. *Chromosome Res* 5: 12–22

45 Falcão JN (1988) *Caracterização cariotípica em peixes do gênero* Triporteus *(Teleostei, Characiformes, Characidae)*. PhD thesis, Universidade de São Paulo

46 Bertollo LAC, Cavallaro ZI (1992) A highly differentiated ZZ/ZW sex chromosome system in a Characid fish *Triportheus guentheri*. *Cytogenet Cell Genet* 60: 60–63

47 Almeida-Toledo LF, Foresti F, Toledo-Filho SA (1984) Complex sex chromosome system in *Eigenmannia* sp. (Pisces, Gymnotiformes). *Genetica* 64: 165–169

48 Moreira-Filho O, Bertollo LAC, Galetti Jr PM (1980) Evidences for multiple sex chromosome system with female heterogamety in *Appareiodon affinis* (Pisces, Parodontidae). *Caryologia* 33: 83–91

49 Moreira-Filho O, Bertollo LAC, Galetti Jr PM (1993) Distribution of sex chromosomes mechanisms in Neotropical fish and description of a ZZ/ZW system in *Parodon hilarii* (Parodontidae).

Caryologia 46: 115–125
50 Haaf T, Schmid M (1984) An early stage of ZW/ZZ sex chromosome differentiation in *Poecilia sphenops* var. *melanistica* (Poeciliidae, Cyprinodontiformes). *Chromosoma* 89: 37–41

Chromosomes Today, Vol. 13
ed. by E. Olmo and C.A. Redi

Repetitive DNA as a tool to study the phylogeny of cold-blooded vertebrates

Teresa Capriglione

Dipartimento di Biologia Evolutiva e Comparata, Università di Napoli, Federico II, Via Mezzocannone 8, I-80134 Napoli, Italy

Introduction

At the end of the 60s, when scientists found out that repetitive DNA sequences were part of the eukaryotic genomes [1, 2], they could not foresee that they would be very useful markers for molecular and cytogenetic analysis and to investigate the phylogenetic relationships between closely related species.

At the beginning, interest was focussed on the study of the peculiar quantitative and compositional characteristics of these repeated DNAs, which sometimes appeared to be a remarkable fraction of the genomes analyzed. Attempts were also made to understand how they originated and whether they might have any genomic function [3, 4].

From the data collected, mainly on invertebrates and mammals, it was clear that repetitive DNA sequences can be accumulated as families in constitutive heterochromatin or be dispersed in the genome as a consequence of recombination and/or replication slippage [5, 6]. They can be tandemly arrayed in clusters with variable degrees of redundancy and array size. Moreover, many of the scattered repeated DNAs consist of classes of transposable elements, sequences able to insert copies of themselves into the genomes via DNA or RNA (retrotransposition) intermediates [7]).

All these repeated DNA families appear to be conserved in closely related species, being, however, free to diverge over time; their evolutionary rate seems to be proportional to the time of interspecific divergence [8]. It has been proposed that repetitive DNA sequences are conserved because their mutagenic activities might contribute to the evolution of the population or be functionally important for host organisms [9].

Although repeated DNA might probably be transcriptionally inactive (or, at least, its transcripts do not seem to have any function), its structural role has been extensively documented. In fact, the repetitive sequences present in heterochromatin seem to be involved in: i) chromosome segregation [10, 11], ii) sister chromatid cohesion [12, 13], and iii) inhibition of recombination [14].

Furthermore, some repetitive families present in centromeric heterochromatin, such as alphoids in mammals, seem to contain a conserved sequence motif, the CENP-B box, important for the attachment of chromosomes to the mitotic spindle [15–17]. All these functions are accomplished through the linkage of particular proteins (i.e. CENP-B) [18, 19], which also appear to be conserved in several organisms.

The evolutionary trend and the function of repetitive DNA have been extensively studied in invertebrates and mammals, but less data are available for other vertebrates. Therefore, over the last few years, we have carried out phylogenetic studies on the repetitive DNA fractions present in the genomes of several heterothermal vertebrates, especially fish and reptiles. Due to their varying degree of conservativeness, they have provided useful information on the phylogeny of these taxa. In addition, sequence analysis has been informative about a possible functional role for at least two of the repetitive families studied.

Repetitive sequences in fish

A nice paper about the low and uniform genome size in birds, published in *Nature* a few years ago, suggested "small genomes for better flyers" [20]. Most bony fish also show small genomes (probably for "better swimmers") ranging from 0.8 pg/nucleus to about 9 pg/nucleus (2C) [21], the organization of which has been deeply studied by Bernardi et al. [23] using density gradient centrifugation. The authors found that low compositional heterogeneity strikingly distinguishes fish DNA from the DNAs of warm–blooded vertebrates. Moreover, fish satellite DNAs are cryptic or poorly resolved [22, 23].

Most information available on repetitive DNA families in fish genomes has recently been acquired and is still scarce if one considers that this taxon represents half of the living vertebrates.

Recently, several repetitive DNA families have been isolated from the genomes of different osteichthyian species. Some of these families are localized in constitutive heterochromatin as typical tandemly arrayed satellites, others are organized as clusters dispersed in the genome [24–30].

The majority of these satellite DNAs are AT-rich and mapped to the centromeres of metaphase chromosomes.

Hartley [27] used the satellite monomer to carry out a phylogenetic analysis at the interspecific level within the genus *Leporinus*. Garrido-Ramos [28] analyzed the conservation of an *Eco*RI satellite DNA in ten species of the family Sparidae, finding a 9 bp motif in the monomeric sequence, which has also been demonstrated to be conserved in other vertebrate satellite DNAs.

Interestingly, the sequence evolution of the centromeric satellite DNA isolated by Haaf in *Hoplias* [24] occurs in a concerted fashion on homologous chromosomes in a way reminiscent of the primate α-satellite.

Transposable elements (TEs) have been also found in fish DNA. In particular, retrotransposition events seem to have worked in shaping salmonid genomes. Several families of short interspersed elements (SINEs), which form a specific retroposon class which has been irreversibly inserted, appear to be differently amplified during evolution within certain salmonid lineages; this has prompted investigators to attempt a phylogenetic tree for the salmonid species [31–33].

The same approach has been successfully used to provide a basis for the elucidation of the phylogenetic framework of cichlid fish in Lake Tanganyika. Analyzing the pattern of conservativeness and divergence of a SINE family, it has been demonstrated that each of the previously identified fish tribes forms a monophyletic group [34].

Another SINE family, DANA, has been isolated from the zebra fish, *Danio rerio*. This family, present only in the genus *Danio,* is made up of four cassettes of alternating conserved and hypervariable sequences. Its massive amplification may have played a role in *Danio* speciation [35].

Though the SINEs so far considered appear to be specific in closely related taxa, Shimoda et al. [36] found a new family of retrotransposons, designated *Mermaid*, composed of highly polymorphic sequences in the Japanese medaka fish. This new family is common to other vertebrates as well. In particular, it appears to be widely diffused in fish, amphibians and primates, but is lacking in mouse. Due to its characteristics, the *Mermaid* detected in the zebra fish could facilitate the genome mapping of this organism, which would be very useful in the study of the early development of vertebrates.

Recently, the occurrence of elements belonging to the Tc1-like transposon family has been detected in several teleostean fish [37–39].

These elements were first identified in invertebrates such as *Caenorhabditis elegans* [40] and *Drosophila melanogaster* [41], and are thought to transpose via a DNA intermediate. Tc1-like transposons are approximately 1600 bp long and share structural similarities, namely terminal inverted repeats with an identical 4-7 bp sequence at or near the ends and an open reading frame encoding a putative transposase (encoding a 273 amino acid sequence) [42, 43].

Another Tc1-like transposon, designated pIF, has been isolated by us in the genome of an Antarctic ice-fish, *Chionodraco hamatus*, belonging to the Notothenioidi, a suborder of Perciformes [26]. This element is only 802 bp, shorter than the other Tc1 elements so far identified, large deletions being present within the terminal regions. However, pIF conserves the 5 bp terminal sequences (CAGTG at the 5' end and GTCAC at the 3' end) found within the ITRs of other Tc1 transposons [42].

The open reading frame, beginning at position 61 and ending at position 504, is also reduced in size (Fig. 1). It potentially encodes a 144 amino acid protein with up to 54% identity to the putative Tc1-like transposase of *Drosophila virilis*, *Caenorhabditis elegans*, and Pacific hagfish.

Figure 1 (See color plate VIII between pages 204 and 205.) Above: consensus sequence and restriction map of pIF, the Tc1-like TE isolated in *C. hamatus*. Below: Shown is a schematic representation of pIF structure. DR: direct repeat; IR: inverted repeat.

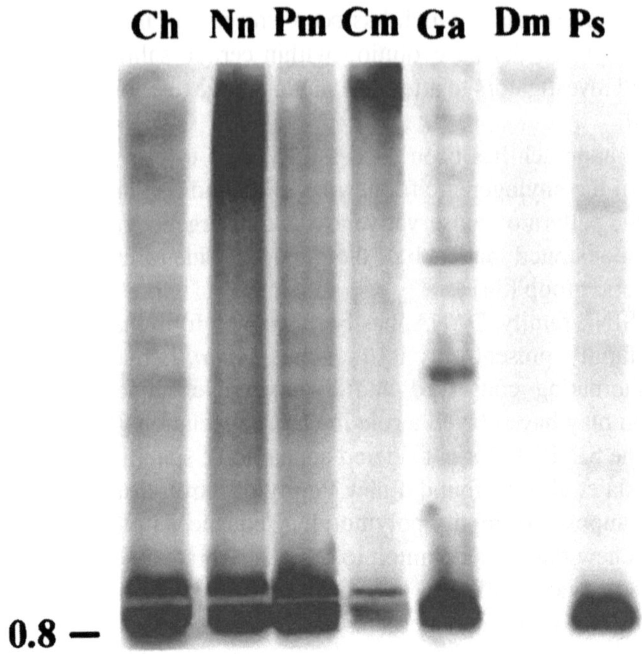

Figure 2 Southern hybridization of *Bgl* II-restricted DNAs of notothenioids probed with DIG-labelled pIF. Ch: *Chionodraco hamatus*, Nn: *Notothenia neglecta*, Pm:*Pagetopsis macropterus*, Cm:*Cygnodraco mawsonii*, Ga:*Gymnodraco acuticeps*, Dm: *Dissosticus mawsonii*, Ps: *Pogonophrine scottii*.

pIF strongly hybridizes, showing slightly different patterns, to all the species of notothenioids studied by us except *D. mawsonii* (Fig. 2). However, when we used primers designed from pIF conserved regions, to carry out polymerase chain reaction amplification on the genome of this last species, we could amplify two bands that, although shorter than pIF, shared up to 58% identity with it (Fig. 3).

Although the efficiency of these widespread transposable elements as phylogenetic markers is still being debated, we can hypothesize that the Tc1-like family isolated in *C. hamatus* appeared early in the genome of notothenioids, and its divergence probably accompanied or paralleled the phylogenetic history of this taxon.

Moreover, *in situ* experiments showed that this Tc1-like element is localized to the heterochromatic regions of *C. hamatus* chromosomes [26] and, sometimes, interstitially. In particular it hybridizes interstitially to the long arm of the *C. hamatus* sex chromosome Y, which is supposed to result from the fusion between a gonosome and an autosome. This suggests that, as hypothesized for other transposable elements, pIF might have been involved in chromosome rearrangements during notothenioid karyotype evolution.

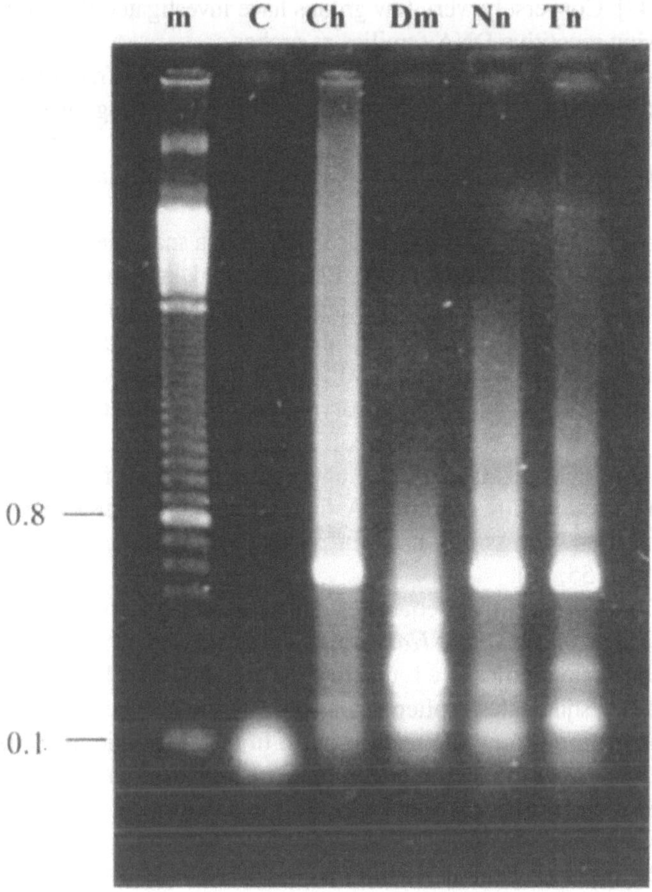

Figure 3 Amplification of DNAs from four notothenioid species with oligopIFA and oligopIFB primers. m: marker, C. blank, Ch: *Chionodraco hamatus,* Dm: *Dissosticus mawsonii*, Nn: *Notothenia neglecta,* Tn: *Trematomus newnesii.*

Whether Tc1-like transposable elements may have played any role in genome evolution during phylogeny is still unknown. However, the fact that they have been conserved for so many millions of years points to the action of selective forces in their favor, probably to guarantee variability of the genome during evolution [38].

Repetitive sequences in reptiles

Over the last decades, the DNA repetitive fractions of the genomes of amphibians, particularly urodeles, have been extensively studied by several research groups [44–47]. Conversely, very few groups have investigated the phylogeny of reptiles using repetitive DNA families as probes.

One of the first studies on the evolution of repetitive DNA fractions in reptiles was performed by Olmo and colleagues in 1985, using reassociation kinetics. They found that, in reptiles, variations in repetitive fractions were modulated by constraints and seemed to accompany the speciation processes [48, 49].

In reptiles, a repeated DNA was first isolated from a snake genome, for taxonomic purposes. This satellite DNA, Bkm (banded krait minor satellite), recovered from the W sex-determining chromosome of a venomous species of elapid snake, *Bungarus fasciatus*, was shown to be made up of tandemly arrayed, simple (GATA)n repeats [50]. In further studies, Bkm-related sequences appeared to occur in all eukaryote genomes, sometimes preferentially amplified in the heterogametic sex [51–54].

Some authors hybridizing a Bkm clone to the genome of Chelonia and Crocodilia obtained well-resolved DNA fingerprints, demonstrating that this clone could also be successfully used for population studies and perhaps taxonomic analysis [55, 56]

Porter [57] used repetitive sequences isolated from genomic libraries of *Cnemidophorus neomexicanus*, *Holbrookia maculata* and *Crotalus atrox* to investigate squamate reptiles. He found that the type and distribution of repetitive sequences in squamates is often taxon specific, and may be useful as characters for elucidating taxonomic relationships. In fact, genomic hybridization of repetitive sequences isolated from *Holbrookia* provided evidence that the three species of earless lizards ascribed to this genus formed a clade distinct from the greater earless lizard, *Cophosaurus*.

A particular restriction endonuclease analysis of the repetitive DNA fraction was used by a Russian group who carried out taxonomic investigations on the genome of reptiles, particularly lizards [58, 59]. Using 5 to 10 short-cutting restriction enzymes they generated species-specific multiple banding patterns, called taxonprints, unique for each enzyme used. They demonstrated that taxonprint bands can be subdivided into bands specific for a single species and those specific for groups of closely related species, genera and even families. The data collected seemed to be of some use for species identification and phylogenetic studies.

Very little is known on the occurrence of transposable elements in reptilian genomes. Flavell et al. [60] isolated Ty1-copia sequences from the genomes of three reptiles. The evolution of these retrovirus-related transposable elements is of interest because of their wide distribution in different phyla, but scarcely informative as regards the phyletic relationships among the different reptilian species in the genome of which they have been found. So far, little information is available on the other TEs recently isolated in reptiles, such as the Bov-b LINEs (considered to be restricted to the genomes of Ruminantia) and the CR1-like elements found in the genomes of turtles [61] and crotaline snakes [62].

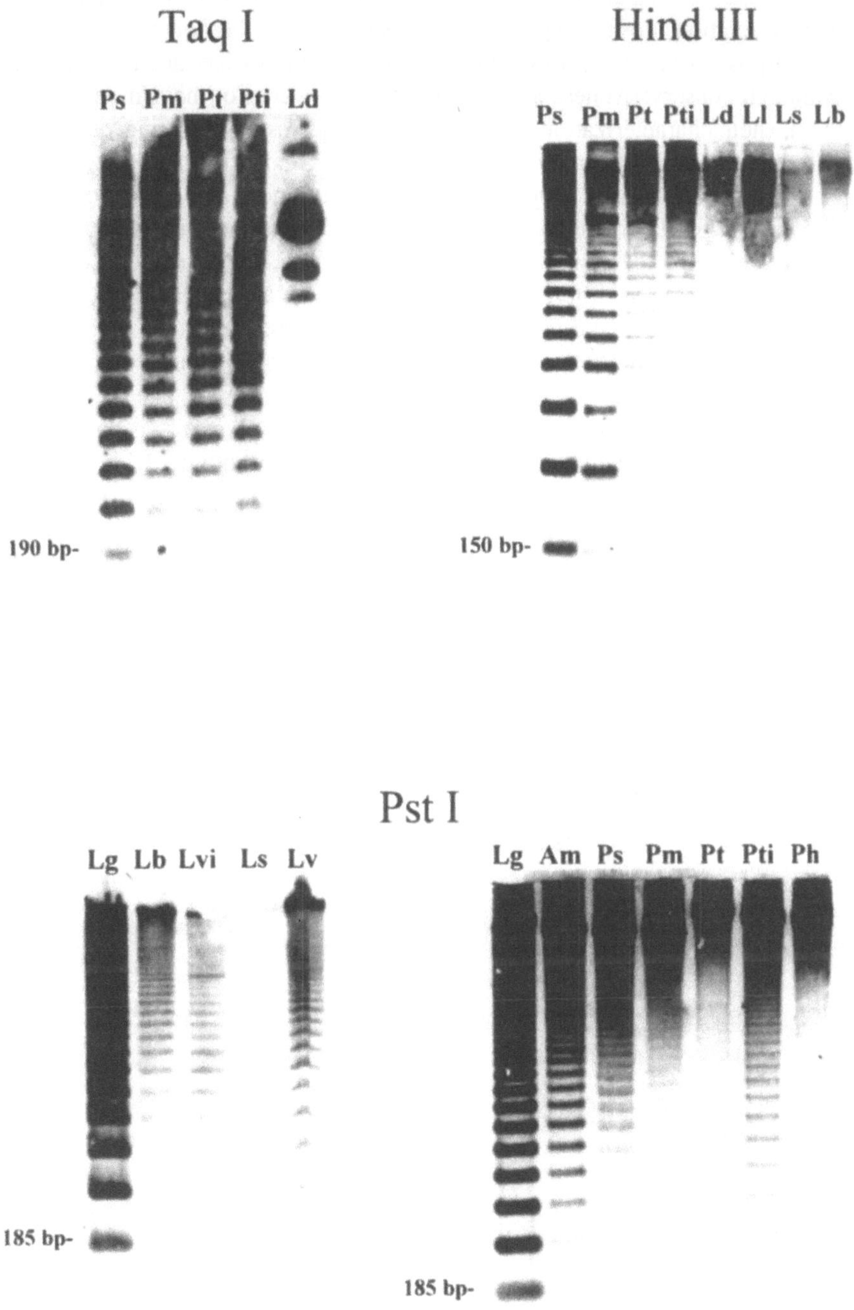

Figure 4 Genomic hybridization of the *Taq* I and *Hind* III repetitive DNA families isolated from *Podarcis sicula* (top), and of *Pst* I isolated from *Lacerta graeca* (bottom). Ps: *P. sicula*, Pm: *P. muralis*, Pt: *P. taurica*, Pti: *P. tiliguerta*, Ph: *P. hispanica* Ld: *L. dugesii*; Ll: *L. lepida*; Ls: *L. saxicola*; Lb: *L. bedriagae*; Lg: *L. graeca*; Lviv: L *vivipara*; Lv: *L. viridis*; Am: *Algyroides moreoticus*.

Over the last years, we have been concerned with the analysis of the phylogenetic relationships among the species belonging to the *Lacerta* group. For our purpose, we isolated some repetitive DNA families from lacertid genomes. Although conservativeness and divergence rates were not the same for all of them, their characterization provided interesting results agreeing quite well with those obtained from morphological and immunological studies.

The most informative repetitive families for taxonomic investigations were *Taq* I and *Hind* III isolated in *P. sicula* [63–65]. The former family, found with few variations in the genome of all the *Podarcis* studied, is also conserved in species belonging to different genera supposedly related to *Podarcis*, such as *Algyroides*, *L graeca* and *L. dugesii* (Fig. 4). The latter, restricted to the genus *Podarcis,* shows a more rapid evolutionary rate and is informative at the interspecific level (Fig. 4).

A different behavior is displayed by the *Pst*I satellite family isolated from the genome of *L. graeca* [66]. In fact, this DNA shows quite a wide distribution, being found in species considered either closely related, or not, to *L. graeca* (Fig. 4). Sequence analysis of the satellite monomeric unit (pGPS) has shown the presence of regions with a high degree of homology with the CDE III centromeric sequence of yeast (72%) [67] and with the mammalian CENP-B box (60%) [18] (Fig. 5). Similarities have also been found comparing pGPS with a data set of alphoid sequences of human chromosomes (Fig. 6). These observations suggest that this repetitive DNA family might play a role in determining the centromeric function in lacertid lizards. Moreover, in

```
pGPS     GTTTCTGAAAAAAC--TTTTTTAGGGATTCGGCGCCCCAGTCTTAAAGTGGGTT---TTAGACTG   60
CHR 6    GTTTCTGAAACACCCTTCTTGTAGGAATTC-TCGAAAC-TTCTTTAAGTTATATACATTCAAGTC

pGPS     CAAGA-CGTTAAGTGAGTTT--TTATAACGGTTTGACAAGATGTTATAGACGACTTTCCAAAC   120
CHR 6    ACAGAGTGGAAACTTCCTTTGGATAGAGCAGTTTGAAACGCTGTGGTTGTAGTATTTCCAAGC
CHR 4             AAGTGTATTTGGATA----GCTTTGAGGATTTGTTGGAAA

pGPS     CGCCACCTTTGTCTTCCGAAAAATAGTATCCCTTTTTTTAGACCGGAGCTGTTGAAATGG      180
CHR 6    GG
CHR 17     CC-CCTTTCAGTTTCTACAAAAAG----AGTGTTT-CAGACCTGAACTAT
```

Figure 5 Comparison of pGPS sequence with alphoid sequences of human chromosome 6 (CHR6), 4 (CHR4), 17 (CHR17).

```
pGPS     GTTTCTGAAAAAACTTTTTTAGGGATTCGGCGCCCCAGTCTTAAAGTGGGTTTTAGACTG   60
CENP                                                  CTTCGTTGGAAACGG

         CAAGACGTTAAGTGAGTTTTTATAACGGTTTGACAAGATGTTATAGACGACTTTCCAAAC   120
         GA                                             CTTCGTTGGAAAC

         CGCCACCTTTGTCTTCCGAAAAATAGTATCCCTTTTTTTAGACCGGAGCTGTTGAAATGG   180
         GGGA                          CTTCGTTGGAAACGGGA
CDE III     TGTTTTTGNTTTCCGAAAANNNAAA
```

Figure 6 Comparison of pGPS sequence with the mammalian CENP-B box and CDEIII of yeast.

view of the phylogenetic relationships between reptiles and mammals, pGPS could be considered a sort of *ancestral alphoid* that during the evolution might have undergone amplification and divergence, though preserving its functional role.

Conclusions

The data which have been accumulating over the last years on the phylogeny of several fish and some reptiles by utilizing clones of repeated DNA as probes for genome hybridization, indicate that this approach is still considered a rapid and reliable method for taxonomic and phylogenetic studies.

Moreover, the increased use of more refined techniques, such as PCR amplification and the analysis of sequence characteristics, has allowed more and more precise data. PCR amplification has contributed to a deeper knowledge of the evolutionary trend of the various elements, allowing the detection of repeated sequences integrated at orthologous loci. Sequence comparison has allowed the identification of sequence motifs that, being functionally active, appear to be conserved in most organisms.

In addition, the identification and the use of interspersed repeated DNA families, isolated from the genomes of heterotherms, might facilitate the genome mapping of these organisms; which can be considered suitable models for the study of early development of vertebrates.

Acknowledgements

We acknowledge the financial support from the Italian National Program for Antarctic Research (P.N.R.A.)

References

1 Britten R, Kohne D (1968) Repeated sequences in DNA. Hundreds of thousands of copies of DNA sequences have been incorporated into the genomes of higher organisms. *Science* 161: 529–540
2 Walker PMB (1971) Repetitive DNA in higher organisms. *Prog Biophys Mol Biol* 23: 145–190
3 John B, Miklos GLG (1977) Functional aspects of satellite DNA and heterochromatin. *In*: G Bourne and J Danielli (eds): *Int Rev Cytol* 58, Academic Press, New York, San Francisco and London, 1–113
4 Doolittle WF, Sapienza C (1980) Selfish genes, the phenotype paradigm and genome evolution. *Nature* 284: 601–603
5 Kurnit DM (1979) Satellite DNA and heterochromatin variants. The case for unequal mitotic crossing over. *Hum Genet* 47: 169–186
6 Smith GP (1976) Evolution of repeated DNA sequences by unequal crossover. *Science* 191: 528–535
7 Berg DE, Howe MM (eds) (1989) *Mobile DNA.* American Society for Microbiology, Washington DC
8 Miklos GLG (1985) Localized highly repetitive DNA sequences in vertebrates and invertebrates

genomes. *In*: RJ MacIntyre (ed.): *Mol Evol Genet*, Plenum Publ Corporation, 241–319
9 Charlesworth B, Sniegowski P, Stephan W (1994) The evolutionary dynamics of repetitive DNA in eukaryotes. *Nature* 371: 215–220
10 Haaf T, Warburton PE, Willard HF (1992) Integration of human alpha-satellite DNA into simian chromosomes: Centromere protein binding and disruption of normal chromosome segregation. *Cell* 70: 681–696
11 Dernburg AF, Sedat JW, Hawley RS (1996) Direct evidence of a role for heterochromatin in meiotic chromosome segregation. *Cell* 86: 135–146
12 Lica LM Narayanswami S, Hamkalo BA (1986) Mouse satellite DNA, centromere structure, and sister chromatid pairing. *J Cell Biol* 103: 1145–1151
13 Warburton PE, Cooke HJ (1997) Hamster chromosomes containing amplified human alpha-satellite DNA show delayed sister chromatid separation in the absence of *de novo* kinetochore formation. *Chromosoma* 106: 149–159
14 Ahmad K, Golik KG (1996) Somatic reversion of chromosomal position effects in *Drosophila melanogaster. Genetics* 144: 657–670
15 Masumoto H, Kenji S, Tuneko O (1989) Alphoid satellite DNA is tightly associated with centromere antigens in human chromosomes throughout the cell cycle. *Exp Cell Res* 181: 181–196
16 Ikeno M, Masumoto H, Okazaki T (1994) Distribution of CENP-B boxes reflected in CREST centromere antigenic sites on long range α–satellite DNA arrays of human chromosome 21. *Hum Mol Genet* 3: 1245–1257
17 Sugimoto K, Furukawa K, Mikeno M (1994) Functional cloning of centromere protein B (CENP-B) box enriched alphoid DNA repeats utilizing the sequence-specific DNA binding activity of human CENP-B *in vitro. Chromosome Res* 2: 453–459
18 Cooke CA, Bernat RL, Earnshaw WC (1990) CENP-B: a major human centromere protein located beneath the kinetochore. *J Cell Biol* 110: 1475–1488
19 Sullivan RF, Glass GA (1991) CENP-B is a highly conserved mammalian centromere protein with homology to the helix-loop-helix family of proteins. *Chromosoma* 100: 360–370
20 Hughes AL, Hughes MK (1995) Small genomes for better flyers. *Nature* 377 (6548): 391
21 Hinegardner RT (1968) Evolution of cellular DNA content in teleost fishes. *Amer Naturalist* 102: 517–523
22 Hudson AP, Cuny G, Cortadas J, Haschemeyer AE, Bernardi G (1980) An analysis of fish genomes by density gradient centrifugation. *Eur J Biochem* 112: 203–210
23 Bernardi G, Bernardi G (1990) Compositional patterns in the nuclear genome of cold-blooded vertebrates. *J Mol Evol* 31: 265–281
24 Haaf T, Schmid M, Steinlein C, Galetti PMJr Willard HF (1993) Organization and molecular cytogenetics of a satellite DNA family from *Hoplias malabaricus* (Pisces, Erythrinidae). *Chromosome Res* 1: 77–86
25 Kubota S, Kuro-o M, Mizuno S, Kohno S (1993) Germ line-restricted repeated DNA sequences and their chromosomal localization in a Japanese hagfish (*Eptatretus okinoseanus*). *Chromosoma* 102: 163–173
26 Capriglione T, Morescalchi A, Olmo E, Rocco L, Stingo V, Manzo S (1994) Satellite DNAs, heterochromatin and sex chromosomes in *Chionodraco hamatus* (Channichthyidae, Perciformes). *Polar Biol* 14: 285–290
27 Hartley SE, Davidson WS (1994) Distribution of satellite DNA sequences isolated from Artic charr, *Salvelinus alpinus*, in the genus *Salvelinus. Can J Fish Aquat Sci* 51: 277–283
28 Garrido-Ramos MA, Jamilena M, Lozano R, Ruiz Rejon C, Ruiz Rejon M (1994) Cloning and characterization of a fish centromeric satellite DNA. *Cytogenet Cell Genet* 65: 233–237
29 Elder JF, Turner BJ (1994) Concerted evolution at the population level: Pupfish HindIII satellite DNA sequences. *Proc Natl Acad Sci USA* 91: 994–998
30 Reed KM, Phillips RB (1995) Molecular characterization and cytogenetic analysis of highly repeated DNAs of lake trout, *Salvelinus namayscush. Chromosome Res* 104: 242–251
31 Murata S, Takasaki T, Saitoh M, Okada N (1993) Determination of the phylogenetic relationships among Pacific salmonids by using short interspersed elements (SINEs) as temporal land marks of evolution. *Proc Natl Acad Sci USA* 90: 6995–6999
32 Murata S, Takasaki T, Saitoh M, Tachida H, Okada N (1996) Details of retropositional genome dynamics that provide a rationale for a generic division: the distinct branching of all the pacific salmon ant trout (*Oncorhynchus*) from the atlantic salmon and trout (*Salmo*). *Genetics* 142: 915–926

33 Takasaki N, Yamaki T, Hamada M, Park L, Okada N (1997) The salmon Sma I family of Short Intersersed Repetitive Elements (SINEs): interspecific and intraspecific variation on the insertion of SINEs in the genomes of chum and pink salmon. *Genetics* 146: 369–380
34 Takahashi K, Terai Y, Nishida M, Okada N (1998) A novel family of Short Intersersed Repetitive Elements (SINEs) from cichlids: The patterns of insertion of SINEs at orthologous loci support the proposed monophyly of four major groups of cichlid fishes in Lake Tanganyika. *Mol Biol Evol* 15: 391–407
35 Izsvak Z, Ivics Z, Garcia-Estefania D, Fahrenkrug S (1996) DANA elements: A family of composite, tRNA-derived short interspersed DNA elements associated with mutational activities in zerbrafish (*Danio rerio*). *Proc Natl Acad Sci USA* 93: 1077–1081
36 Shimoda N, Chevrette M, Ekker M, Kikuchi Y, Hotta Y, Okamoto H (1996) *Mermaid*, a family of short interspersed repetitive elements, is useful for genome mapping. *Biochem Biophys Res Commun* 220: 226–232
37 Goodier J, Davidson W (1994) Tc1 transposon-like sequences are widely distributed in salmonids. *J Mol Biol* 241: 26–34
38 Radice A, Bugaj B, Fitch D, Emmons S (1994) Widespread occurrence of the Tc1 transposon family: Tc1-like transposons from teleost fish. *Mol Gen Genet* 244: 606–612
39 Ivics Z, Izsvak Z, Minter A, Hackett P (1996) Identification of functional domains and evolution of Tc1-like transposable elements. *Proc Natl Acad Sci USA* 93: 5008–5013
40 Rosenzweig B, Liao L, Hirsh D (1983) Sequence of the *C. elegans* transposable element Tc1. *Nucl Acid Res* 16: 4201–4209
41 Hunt JA, Bishop IIIJG, Carson HL (1984) Chromosomal mapping of a middle-repetitive DNA sequence in a cluster of five species of Hawaiian *Drosophila. Proc Natl Acad Sci USA* 81: 7146–7150
42 Brezinsky L, Wang G, Humphreys T, Hunt J (1990) The transposable element Uhu from Hawaiian *Drosophila* – member of the widely dispersed class of Tc1-like transposons. *Nucl Acid Res* 18: 2053–2059
43 Avancini R, Walden K, Robertson H (1996) The genomes of most animals have multiple members of the Tc1 family of transposable elements. *Genetica* 98: 131–140
44 Baldwin L, Macgregor HC (1985) centromeric satellite DNA in the newt *Triturus cristatus karelinii* and related species: its distribution and transcription on lampbrush chromosomes. *Chromosoma* 92: 100–107
45 Barsacchi G (1991) Satellite DNA in the newts (Amphibia, Urodela). *In*: G Ghiara, F Angelini, E Olmo, L Varano (eds): *Symp Evol Terr Vert.* Mucchi Modena,171–198
46 Batistoni R, Pesole G, Marracci S, Nardi I (1995) A tandemly repeated DNA family originated from SINE-related elements in the European Plethodontid salamanders (Amphibia, Urodela). *J Mol Evol* 40: 608–615
47 Vignali R, Nardi I (1996) Unusual features of the urodele genome: do they have a role in evolution and development? *Int J Dev Biol* 40: 637–643
48 Olmo E, Odierna G, Capriglione T, Cobror O (1985) Different trends in the variations of the main genomic components in turtles and scaly-reptiles. *Comp Biochem Physiol* 80B: 441–446
49 Olmo E, Capriglione T, Odierna G (1989) genome size evolution in vertebrates: trends and constraints. *Comp Biochem Physiol* 92B: 447–453
50 Singh L, Purdom IF, Jones KW (1980) Sex chromosome associated satellite DNA. Evolution and conservation. *Chromosoma* 79: 137–157
51 Erickson RP, Ross CE, Gorski Jl Stalvey JR, Drumm MM (1988) Bkm sequences from the human X chromosome contain large clusters of GATA/GACA repeats. *Ann Hum Genet* 52: 167–176
52 Kent MG, Elliston KO, Shroeder W, Guise KS, Wachtel SS (1988) Conserved repetitive DNA sequences (Bkm) in normal equine males and sex-reversed females detected by *in situ* hybridization. *Cytogenet Cell Genet* 48: 99–102
53 Wachtel S, Demas S, Tiersch T, Pechan P, Shapiro D (1991) Bkm satellite DNA and ZFY in the coral reef fish *Anthias squamipinnis. Genome* 34: 612–617
54 Rajyashri KR, Singh L (1995) A Bkm-associated human Y-chromosomal DNA is conserved and transcribed in the testis of mouse. *Chromosoma* 104: 274–281
55 Demas S, Wachtel S (1991) DNA fingerprinting in reptiles: Bkm hybridization patterns in Crocodilia and Chelonia. *Genome* 34: 472–476
56 Aggarwal RK, Majumdar KC, Langh JW, Singh L (1994) Generic affinities among crocodilians as revealed by DNA fingerprint with a Bkm-derived probe. *Proc Natl Acad Sci USA* 91:

10601–10605
57 Porter AC (1994) Organization and chromosomal location of repetitive DNA sequences in three species of squamates reptiles. *Chromosome Res* 2: 263–273
58 Grechko VV, Fedorova LV, Fedorov AN, Slobodyanyuk SY, Ryabinin DM, Melnikova MN, Bannikova AA, Lomov AA, Sheremet'éva VA, Gorshkov VA et al (1997) Restriction endonuclease analysis of highly repetitive DNA as a phylogenetic tool. *J Mol Evol* 45: 332–334
59 Fedorov AN, Fedorova LV, Grechko VV, Ryabinin DM, Shermet'eva VA, Bannikova AA, Lomov AA, Ryskov AP, Darevsky IS (1999) Variable and invariable DNA repeat characters revealed by taxon print approach are useful for molecular systematics. *J Mol Evol* 48: 69–76
60 Flavell AJ, Jackson V, Iqbal MP, Riach I, Waddell S (1995) Ty-copia group retrotransposon sequences in Amphia and Reptilia. *Mol Gen Genet* 246: 65–71
61 Endoh H, Nagahashi S, Okada N (1990) A highly repetitive and transcribable sequence in the tortoise genome is probably a retroposon. *Eur J Biochem* 189 (1): 25–31
62 Nobiusha I, Ogawa T, Deshimaru M, Chijiwa T, Nakashima Ki Chuman Y, Shimoigashi Y, Fukumaki Y, Hattori S, Ohno M (1998) Retrotransposable CR1-like elements in crotaline snake genomes. *Toxicon* 36 (6): 915–20
63 Capriglione T, Olmo E, Odierna G, Smith DI, Miller OJ (1988) Genome composition and tandemly repetitive sequences at some centromeres in the lizard *P. sicula* Raf. *Genetica* 79: 85–91
64 Capriglione T, Cardone A, Olmo E, Odierna G (1991) Evolution of a centromeric satellite and phylogeny of lacertid lizards. *Comp Biochem Physiol* 100B: 641–645
65 Capriglione T, Cardone A, Olmo E, Odierna G (1994) Further data on the occurence and evolution of satellite DNA families in the lacertid genome. *Chromosome Res* 2: 327–330
66 Capriglione T, De Santo MG, Olmo E, Odierna G (1998) An alphoid–like satellite DNA sequence is present in the genome of a lacertid lizard. *J Mol Evol* 46: 240–244
67 Clark L, Carbon I (1985) the structure and function of yeast centromeres. *Annu Rev Genet* 19: 29–56

Chromosomes Today, Vol. 13
ed. by E. Olmo and C.A. Redi

Comparative gene mapping permits new insights into karyotype evolution in mammals

Günther Rettenberger[1,2], Walther Vogel[1], Horst Hameister[1]

[1] *Abteilung Medizinische Genetik, Universität Ulm, 89069 Ulm, Germany*
[2] *Labor Drs. Mehnert, Wegener Straße 15, 89231 Neu-Ulm, Germany*

Introduction

One of the most fascinating themes in biology concerns the evolutionary history of todays mankind. One would like to know which genetic mechanisms made possible the extraordinary development to human. To enlighten this fundamental question the description of the differences between human and species which are most closely related to human is highly informative. In todays molecular world such comparisons are performed for specific genes at the DNA or amino acid sequence level or – when the entire genome is to be compared – on the level of the whole karyotype. In this review we will concentrate on the latest achievements concerning karyotype comparisons.

Early data from the prebanding area

More than 1000 mammalian species have been studied cytogenetically. Enormous variability between karyotypes has been observed. This concerns the number of chromosomes, which varies between 6 [1] and 102 [2], and the chromosome structure, which can be mainly acrocentric or a combination of metacentric, submetacentric and acrocentric chromosomes. During the early cytogenetic studies the most prominent characteristics of a karyotype were the number of chromosomes, $2n$, and the shape of the chromosomes as determined by their length and the position of the centromere. In the light of these early observations different hypotheses were developed about karyotype evolution in mammals. S. Ohno favored the idea of an ancestral karyotype with $2n = 96$ acrocentric chromosomes, which gave rise to the different karyotypes of present day species by a variable number of metacentric fusions ("fusion hypothesis") [3]. N.B. Todd, who was influenced by karyotype studies in marsupials, took the opposite position and claimed the ancestral karyotype to consist of $2n = 14$ large chromosomes ("fission hypotheses") [4]. R. Matthey, who had studied for himself several hundreds of species, suggested that the founder karyotype most likely should consist of $2n = 48$ meta-, submeta-, and acrocentric chromosomes [5].

These different views represent not only the respective field of interest with its particular set of observations but focus also on different aspects of mammalian evolution. Ohno was fascinated by the tetraploidization event observed between different fish species and mammals [3]. The duplication of $2n = 48$ acrocentric chromosomes in fish to $2n = 96$ acrocentric chromosomes was seen as the starting point for mammalian evolution; although this event may have happened long before the mammalian lineage split off. After fusion of these acrocentric chromosomes a karyotype with $2n = 48$ meta- and submetacentric chromosomes arose. This is in remarkable agreement with the hypothetical founder karyotype suggested by Matthey [5].

Application of chromosomal banding techniques

The introduction of chromosomal banding techniques allows us to compare directly chromosomes from different species. Similarity in the banding pattern is evident only between closely related species and therefore with these techniques the ancestral karyotype can be reconstructed only within orders, e.g. primates [6, 7], carnivores [8]. The variation in banding patterns on homologous chromosomal segments is quite surprising because it provides evidence for a "microevolution" within chromosomes which is accounted for by non-coding sequences. Our knowledge about these sequence differences is still rather limited. They invaded the respective genome after radiation into the different genera as has been proven for the 300 bp long repetive Alu sequence elements [9].

Application of gene mapping data

Another level of comparison between genomes is offered by the application of gene mapping data. First apparent for the X chromosome [10], and later confirmed by the mapping data in human and mouse, it was shown that the ordering of the genes along chromosomes is highly conserved. As early as 1984, Nadeau and Taylor [11] calculated, on the basis of 13 linkage groups of conserved synteny between human and mouse, a maximum of approximately 180 segments with the same gene order conserved in both species. This first calculation is proven to be completely correct when the number of conserved segments is recalculated today on a much firmer basis of nearly 2000 genes mapped both in human and in mouse [12, 13].

Another prominent example of species comparison on the gene mapping level was provided in 1982 by O'Brien and Nash, who compared human and cat [14]. With only 31 loci mapped both in human and cat at that time the conservation of synteny relationships was at once apparent. Human and cat offer

Table 1. Current status of ongoing mapping projects (Sept. 98)

	gene number	human homology
human	9632 (6541 mapped)	-
mouse	9184	2968
rat	775	530
cattle	362	268
sheep	237	133
pig	312	117
cat	100	63

a further advantage since also for both species homology of the chromosomal banding pattern can be shown [8, 15, 16].

Since that time comparative gene mapping has evolved as a new discipline. Though the genomes of different mammalian species differ according to size by a factor of two, the number of genes is identical for all species. For some prominent species for which mapping initiatives are underway the recent data are listed in Table 1. Wakefield and Graves [17] have recently prepared a comparative map for all vertebrates for which enough mapping data are available. Linkage for some gene groups has been conserved since the divergence of birds from the lineage leading to mammals more than 300 million years ago. Comparative gene mapping is a powerful tool for transferring the gene mapping data from map rich species – human and mouse – to all other mammalian species [18]. Furthermore, these data are valuable to understanding chromosome structure and function, to defining basic units of gene order which are common to all mammals and for helping to describe animal models for human genetic diseases.

These data also provide new insight into karyotype evolution. This is especially evident when the position of the centromere is considered. When comparing the human and mouse gene map [19] it is obvious that there are several examples of conserved gene order across the centromere but the centromere itself is repositioned, e.g. *Homo sapiens* (HSA) 20 on *Mus musculus* (MMU) 2, HSA 17 on MMU 11, HSA X and MMU X. The centromeres and also the telomeres and nucleolar organizer regions (NOR's) are reshuffled in the mammalian genome independently of the conserved gene order and at a higher rate than the genes themselves. This observation has been confirmed by gene mapping data in other species. A prominent example is pig chromosome 7. The short arm and the proximal long arm *Sus scrofa* (SSC) 7 are homologous to

Figure 1. (See color plate IX between pages 204 and 205.) Chromosome painting of the pig karyotype with human chromosome specific DNA probes (ZOO-FISH). The human Y chromosome probe gave no signal. There are a total of 48 segments of conserved synteny between human and pig. On the left of each chromosome is listed the human homologous chromosome segment. The color-code of the human chromosomes is shown on the left.

HSA 6p (Fig. 1) [20]. The pig homologue of the human HLA complex is called SLA and maps just to the centromere of SSC 7. The class I and III genes of the SLA are localized in the p arm and the class II genes in the q arm [21]. The centromere is integrated into the SLA complex separating the class II genes from the rest of the SLA complex. One learns from this example that centromeres are moved through the genomes independently of a conserved gene order. In addition, centromeres and the pericentromeric heterochromatin do not influence the gene function of neighbouring genes in the form of a position effect.

These examples provide compelling evidence for the notion that it is necessary to reevaluate our ideas of karyotype evolution which in former times was biased by the position of the centromere. Centromeres form a characteristic chromosomal landmark; however, they are reshuffled in the mammalian genome at a higher rate than the genes themselves. The gene order relationships are far more informative and allow insight into karyotype arrangements which took place very far back in evolutionary times.

Application of chromosomal painting techniques (ZOO-FISH analysis)

To perform comparative mapping in an economic manner, O'Brien [18] has described 300 anchor loci which, when mapped in species other than human, should allow the transfer of mapping information from human to the species in question. However, it takes a lot of time and efforts to map these 300 loci precisely. A powerful new approach was provided by a chromosomal painting technique [22] which is also known as a ZOO-FISH analysis when applied to a non-human species[23–25]. This technique allows us to directly visualize the segments into which a specific human chromosome is split. We have applied this technique for the first time to paint the whole karyotype of another species with all human chromosomes [26]. In Figure 1 the ZOO-FISH data are summarized for all human chromosomes when hybridized to metaphase chromosomes from pig, *Sus scrofa domestica* (SSC).

Several observations are at once evident: (1) There are only 48 large segments of conserved synteny. This means a 3–4 times higher conservation of gene order than calculated for the human/mouse comparison. Even large chromosomes such as HSA 3 and HSA 4 are conserved as a single segment or chromosome. (2) As already mentioned above, the centromeres have been repositioned irrespective of conserved gene order, e.g. HSA 13 on SSC 11, HSA 20 on SSC 17, HSA 3 and 21 on SSC 13. (3) Several intrachromosomal inversions are apparent, e.g. HSA 18 and 15 on SSC 1q, HSA 1 and 18 on SSC 6q, HSA 14 and 15 on SSC 7q. However, chromosomal painting techniques cannot reveal intrachromosomal inversions within a segment of conserved synteny. By ZOO-FISH analysis HSA 4 is conserved in total on SSC 8 (Fig. 1). Independent mapping data have revealed that indeed the gene order is interrupted by one inversion when HSA 4 and SSC 8 are compared [27]. It is to be expected that other examples of intrachromosomal inversion will be observed when more independent gene mapping data are available. However, this uncertainty does not compromise the main message, that conservation of gene order is the rule in mammals and that only a limited number of rearrangements have occurred since separation of artiodactyles and primates 70 to 100 million years ago in mesozoic times. This finding is also confirmed when chromosomal painting is performed in the reverse order with pig chromosome-specific DNA probes hybridized to human metaphase chromosomes [28].

In the meantime, chromosomal painting has been extended to many other primate (e.g. [23–32]) and non-primate species [33–41]. In this review we will concentrate on non-primate species. In Table 2 the ZOO-FISH data obtained in mammals thus far are summarized. Not included are the data for the rodents mouse and rat since chromosome painting has not been successful in spite of the wealth of mapping data.

Several conclusions can be derived from the combined data. In non-primate species the number of segments, into which the human autosomes have been split, varies and ranges from 30 in cat and more than 52 in cattle (Tab. 2). This number of segments is considerably lower than the number of segments calculated for the human/mouse comparison [11–13]. Karyotype and genome evolution in rodents and especially mouse is a special situation and occurs at a higher rate than in other species [41]. Surprising is the low number of segments determined for three species out of the order of carnivores. Chromosomal banding studies have already proven that karyotype evolution in some species of the carnivores is very conservative [8, 15, 16]. The dolphin, *Tursiops truncatus*, out of the order of Cetacea has a gene arrangement very similar to the human arrangement [40].

The ZOO-FISH data for the gibbon, *Hylobates concolor* [28], are included in Table 2. *Hylobates concolor* belongs to the hominoids and has separated from human some 15 million years ago. In *Hylobates concolor* the karyotype is found rearranged in 64 segments, which is the highest number of segments determined so far. When compared with the afore-mentioned carnivore species and the dolphin, which separated from human more than 60 million years ago, this difference in the rate of rearrangements forms a strong argument that the rate with which chromosomal rearrangements occur is a species specific [42]. There are enormous differences within the rate at which rearrangements occur. From the number of rearrangements separating two species it is not possible to derive the time when both these species diverged from each other. In other words, the "molecular clock" runs at a different rate in each species [42, 43].

In Table 2 some examples of chromosome conservation are highlighted. (1) HSA 13, 17, and 20 have been found to be conserved in most species in the form of a single segment (one chromosome or part of a larger chromosome). As mentioned earlier, chromosome 20 is also totally conserved in mouse on MMU 2 as is most of HSA 17 on MMU 11 [19]. (2) It is well established that

Table 2. Summary of painting data with human chromosome specific DNA libraries

	Primates		Artiodact.			Perissod.	Carnivor.			Insectiv.	Chiropt.	Cetacea
human painting probe	*Pan troglodytes*	*Hylobates concolor*	*Sus scrofa*	*Bos taurus*	*Muntjacus muntjak*	*Equus caballus*	*Felis catus*	*Mustela vison*	*Phoca vitulina*	*Sorex araneus*	*Glossoph soricina*	*Tursiops truncatus*
1	1	4	6	4	3	3	2	2	2	2	3	1
2	2	7	2	2	3	3	2	2	2	1	2	2
3	1	4	1	2	4	2	2	1	1	3	2	2
4	1	3	1	3	2	2	2	2	2	2	3	1
5	1	4	2	5	2	2	1	1	1	2	2	1
6	1	5	2	2	3	2	1	1	1	1	2	2
7	1	3	3	2	2	1	1	2	1	1	2	2
8	1	2	2	3	2	1	2	2	2	2	2	1
9	1	2	1	2	3	2	1	1	1	1	1	1
10	1	3	2	3	2	2	1	2	2	2	2	1
11	1	3	2	2	1	1	1	1	1	2	2	1
12	1	3	2	3	3	4	2	2	1	2	2	2
13	1	2	1	1	1	1	1	1	1	1	1	1
14	1	1	3	2	3	2	1	1	1	1	2	1
15	1	2	4	3	2	1	1	1	1	1	2	1
16	1	3	2	2	2	2	2	2	2	2	2	2
17	1	5	1	1	1	1	1	1	1	1	1	1
18	1	1	3	2	1	1	1	1	1	1	1	1
19	1	3	2	2	2	2	2	2	2	1	2	2
20	1	2	1	2	2	1	1	1	1	1	1	1
21	1	1	1	1	2	1	1	1	1	1	1	1
22	1	1	2	3	2	3	2	2	2	1	2	2
total segments	**23**	**64**	**46**	**52 +**	**48**	**40**	**31**	**32**	**30**	**32+**	**40**	**30+**

HSA 2 is a very recent fusion product of two chromosomes, which are represented in chimpanzee by PTR 12 and 13 [6]. At the fusion point in HSA 2q13 some telomere sequences are left which are reminders of this evolutionary late fusion event [44]. It is, however, interesting to note that in most other non-primate species the HSA 2 homologs are found conserved in just these two segments which are known from chimpanzee. Therefore, it becomes evident that the arrangement in two segments seems to be the ancestral configuration (Fig. 2).

In a similar way several other ancestral chromosomal configurations are apparent when the ZOO-FISH data from different species are compared [33, 36, 45]. There are several typical fusions of human chromosomes or parts of chromosomes. These fusions are observable in several other species and they are believed to represent the ancestral karyotype configuration.

The most obvious fusion concerns HSA 14 and 15. This fusion was already observed during the first complete ZOO-FISH analysis, which was performed in pig [26]. On two different chromosomes, SSC 1 and 7, segments from HSA 14 and 15 localize adjacent to each other. It was already suggested at that time that in a more ancestral configuration there should be a simple fusion of both chromosomes. Later on this ancestral configuration has been observed in cat, *Felis catus* (FCA) [33, 43], but also in several other species.

Another fusion represents SSC 13 where HSA 3 and 21 are found fused (Figs 1, 2). Again, this fusion has been repeatedly observed in other species and therefore this configuration is thought to be the more ancestral one.

ZOO-FISH analysis in pig revealed two chromosomes, SSC 5 and 14, on which homologous segments from HSA 12 and 22 are fused. Again this fusion has been repeatedly observed in all mammalian species analyzed so far. Mostly there is a tandem fusion of segments from both human chromosomes. In muntjak [36] and horses [39] the segments are found on the same chromosome but separated by another interspersed segment from a different chromosome.

A similar fusion is obvious for part of HSA 16 and 19 which is seen in pig on SSC 6. Independent gene mapping data from other species show that HSA 16q and 19q are fused, the p arms of the respective chromosomes forming independent segments.

The respective human chromosomes and the most common arrangement of the homologs of these chromosomes in other mammalian species is depicted in Figure 2.

To the best of our knowledge these most common arrangements are thought to represent the ancestral configuration which has been conserved independ-

Figure 2. (See color plate X between pages 204 and 205.) Schematic illustration of some recurrent associations and dissociations of human chromosomes observed in different mammalian species. The human chromosomes are shown left and a representative ancestral arrangement from another species on the right. HSA *Homo sapiens*, FCA *Felis catus*, MVI *Mustela vison*, ECA *Equus caballus*, BTA *Bos taurus*.

ently in the different mammalian orders since radiation some 60 to 80 million years ago. Based on these preliminary data Chowdhary et al. [45] derived a potential pattern of the ancestral mammalian founder karyotype. This karyotype is described with $2n = 48$.

At an earlier stage of ZOO-FISH analysis there was strong evidence to suggest that the cat karyotype with $2n = 38$ might be a karyotype most similar to the ancestral founder karyotype [33]. Now on the broader basis of much more painting data Chowdhary et al. [45] have reevaluated the painting data and come up with the karyotype formula: 1p, 1q, 2pter-q13, 2q13-qter, 3/21, 4, 5, 6, 7, 8, 9, 10, 11, 2 × 12/22, 13, 14/15, 16q/19q, 16p, 17, 18, 19p, 20, X, Y ($2n = 48$). This is exactly the karyotype formula which was suggested 25 years ago by R. Matthey [5]. In addition, much to our surprise, we learn that the present day human karyotype is very similar to this ancestral configuration. In other words, no gross chromosomal changes during the last 100 million years of evolution have occured. The ancestral mammalian founder karyotype turns out to be an already well balanced gene arrangement which meets all requirements of mammalian evolution. Genome evolution, visible as a chromosomal karyotype was already in its final stage 100 million years ago. To explain the extraordinary and very efficient evolution to present-day *Homo sapiens*, karyotype studies are not very informative. Evolution on the basis of chromosomal changes is far too slow to explain the rapid evolution to human.

It is unquestionable that the mystery of speciation and evolution has a genetic basis. All mammals have the same number of genes and nearly the same gene order and a very similar karyotype. Hence, evolution in mammals is brought back to the allelic variation of the genes themselves. This is a situation which was encountered already more than 60 years ago by Dobzhansky [46]. Allelic differences or specific allele combinations are involved in the process of the evolution of subpopulations. One of the mechanisms for building up effective reproductive barriers between specific subpopulations is the introduction of chromosomal rearrangements [47]. The remarkable stability and conservation of the mammalian founder karyotype casts doubt on a major role of these chromosomal rearrangement on phenotype development. The species differ from each other in the alleles they possess and in which way these alleles interact with each other during development. These interactions are difficult to study, but with the completion of the human genome project we will know the location of all genes. To understand the function of the genome we need to know how different alleles interact with each other. This is an intellectual challenge rather similar to the genetic resolution of multifactorial diseases.

References

1 Wurster DH, Benirschke K (1970) Indian muntjac, *Muntiacus muntjak*: A deer with a low diploid chromosome number. *Science* 168: 1363–1366

2 Contreras LC, Torres-Mura JC, Spotorno AE (1990) The largest known chromosome number for

a mammal, in a south American desert rodent. *Experimenta* 46: 506–508
3 Ohno S (1970) *Evolution by gene duplication.* New York, Springer
4 Todd NB (1970) Karyotypic fissioning and canid phylogeny. *J Theor Biol* 26: 445–480
5 Matthey R (1973) The chromosome formulae of eutherian mammals. *In*: AB Chiarelli, E Capanna (eds): *Cytotaxonomy and Vertebrate Evolution.* Academic Press, London, 531–616
6 Yunis JJ, Prakash O (1982) The origin of man: a chromosomal pictorial legacy. *Science* 215: 1525–1529
7 Dutrillaux B, Couturier J, Viegas-Péquignot E (1981) Chromosomal evolution in primates. *Chromosome Today* 7: 176–191
8 Dutrillaux B, Couturier J (1983) The ancestral karyotype of carnivora: comparison with that of platyrrhine monkeys. *Cytogenet Cell Genet* 35: 200–208
9 Deininger PL (1991) SINEs: Short interspersed repeated DNA-elements in higher eukaryotes. *In*: DE Berg, MM Howe (eds): *Mobile DNA.* American Society of Microbiology, Washington D.C
10 Ohno S (1973) Ancient linkage groups and frozen accidents. *Nature* 244: 259–262
11 Nadeau JH, Taylor BA (1984) Lengths of chromosomal segments conserved since divergence of man and mouse. *Proc Natl Acad Sci USA* 81: 814–818
12 DeBry RW, Seldin MF (1996) Human/mouse homology relationships. *Genomics* 33: 337–351
13 Nadeau JH, Sankoff D (1998) The lengths of undiscovered conserved segments in comparative maps. *Mamm Genome* 9: 491–495
14 O'Brien SJ, Nash WG (1982) Genetic mapping in mammals: chromosome map of domestic cat. *Science* 216: 257–265
15 Nash WG, O'Brien SJ (1982) Conserved regions of homologous G-banded chromosomes between orders in mammalian evolution: carnivores and primates. *Proc Natl Acad Sci USA* 79: 6631–6635
16 von Kiel K, Hameister H, Somssich IE, Adolph S (1985) Early replication banding reveals a strongly conserved functional pattern in mammalian chromosomes. *Chromosoma* 93: 69–76
17 Wakefield MJ, Graves JAM (1996) Comparative maps of vertebrates. *Mamm Genome* 7: 715–734
18 O'Brien SJ, Womack JE, Lyons LA, Moore KJ, Jenkins NA, Copeland NG (1993) Anchored reference loci for comparative genome mapping in mammals. *Nat Genet* 3: 103–112
19 http://www.informatics.jax.org/homtools.html
20 Frönicke L, Chowdhary BP, Scherthan H, Gustavsson I (1996) A comparative map of the porcine and human genomes demonstrates ZOO-FISH and gene mapping-based chromosomal homologies. *Mamm Genome* 7: 285–290
21 Smith TPL, Rohrer GA, Alexander LJ, Troyer DL, Kirby-Dobbels KR, Janzen MA, Cornwell DL, Louis CF, Schook LB, Beattie CW (1995) Directed integration of the physical and genetic linkage maps of swine chromosome 7 reveals that the SLA spans the centromere. *Genome Res* 5: 259–271
22 Lichter P, Cremer T, Borden J, Manuelidis L, Ward DC (1988) Delineation of individual human chromosomes in metaphase and interphase cells by *in situ* suppression hybridization using recombinant DNA libraries. *Hum Genet* 80: 224–234
23 Jauch A, Wienberg J, Stanyon N, Arnold N, Tofanelli S, Ishida T, Cremer T (1992) Reconstruction of genomic rearrangements in great apes and gibbons by chromosome painting. *Proc Natl Acad Sci USA* 89: 8611–8615
24 Wienberg J, Stanyon R, Jauch A, Cremer T (1992) Homologies in human and *Macaca fuscata* chromosomes revealed by *in situ* suppression hybridization with human chromosome specific DNA libraries. *Chromosoma* 101: 265–270
25 Scherthan H, Cremer T, Arnason U, Weier HU, Lima-de-Faria A, Frönicke L (1994) Comparative chromosome painting discloses homologous segments in distantly related mammals. *Nat Genet* 6: 342–347
26 Rettenberger G, Klett C, Zechner U, Kunz J, Vogel W, Hameister H (1995) Visualization of the conservation of synteny between humans and pigs by heterologous chromosomal painting. *Genomics* 26: 372–378
27 Ellegren H, Fredholm M, Edfors-Lilja I, Winter AK, Andersson L (1993) Conserved synteny between pig chromosome 8 and human chromosome 4 but rearranged and distorted linkage maps. *Genomics* 17: 599–603
28 Goureau A, Yerle M, Schmitz A, Riquet J, Milan D, Pinton P, Frelat G, Gellin J (1996) Human and porcine correspondence of chromosome segments using bidirectional chromosome painting. *Genomics* 36: 252–262
29 Koehler U, Bigoni F, Wienberg J, Stanyon R (1995) Genomic reorganization in the Concolor gibbon (*Hylobates concolor*) revealed by fluorescence *in situ* hybridization. *Genomics* 30: 287–292

30 Consigliere S, Stanyon R, Koehler U, Agoramoorthy G, Wienberg J (1996) Chromosome painting defines genomic rearrangements between red howler monkey subspecies. *Chromosome Res* 4: 264–270
31 Sherlock JK, Griffin DK, Delhanty JDA, Parrington JM (1996) Homologies between human and Marmoset (*Callithrix jacchus*) chromosomes revealed by comparative chromosome painting. *Genomics* 33: 214–219
32 Koehler U, Arnold N, Wienberg J, Tofanelli S, Stanyon R (1995) Genomic reorganization and disrupted chromosomal synteny in the Siamang (*Hylobates syndactylus*) revealed by fluorescence *in situ* hybridization. *Amer J Phys Anthropol* 97: 37–47
33 Rettenberger G, Klett Ch, Zechner U, Bruch J, Just W, Vogel W, Hameister H (1995) ZOO-FISH analysis: Cat and human karyotypes closely resemble the putative ancestral mammalian karyotype. *Chromosome Res* 3: 479–486
34 Hayes H (1995) Chromosome painting with human chromosome-specific DNA libraries reveals the extent and distribution of conserved segments in bovine chromosomes. *Cytogenet Cell Genet* 71: 168–174
35 Solinas-Toldo S, Lengauer C, Fries R (1995) Comparative genome map of human and cattle. *Genomics* 27: 489–496
36 Yang F, Muller S, Just R, Ferguson-Smith MA, Wienberg J (1997) Comparative chromosome painting in mammals: Human and the Indian muntjac (*Muntiacus muntjak vaginalis*). *Genomics* 39: 396–401
37 Hameister H, Klett C, Bruch J, Dixkens C, Vogel W, Christensen K (1997) ZOO-FISH analysis: The American mink (*Mustela vison*) closely resembles the cat karyotype. *Chromosome Res* 5: 5–11
38 Frönicke L, Müller-Navia J, Romanakis K, Scherthan H (1997) ZOO-FISH maps of the harbor seal (*Phoca vitulina*) and the putative ancestral carnivore karyotype. *Chromosoma* 106: 108–113
39 Raudsepp T, Frönicke L, Scherthan H, Gustavsson I, Chowdhary BP (1996) ZOO-FISH delineates conserved chromosomal segments between horse and man. *Chromosome Res* 4: 218–225
40 Bielec PE, Gallagher DS, Womack JE, Busbee DL (1998) Homologies between human and dolphin chromosomes detected by heterologous chromosome painting. *Cytogenet Cell Genet* 81: 18–25
41 Dixkens C, Klett C, Bruch J, Kollak A, Serov OL, Zhdanova N, Vogel W, Hameister H (1998) ZOO-FISH analysis in insectivores: "Evolution extols the virtue of the *status quo*". *Cytogenet Cell Genet* 80: 61–67
42 Gibbons A (1995) When it comes to evolution, humans are in the slow class. *Science* 267: 1907–1908
43 Wienberg J, Stanyon R, Nash WG, O'Brien PCM, Yang F, O'Brien SJ, Ferguson-Smith MA (1997) Conservation of human vs. feline genome organization revealed by reciprocal chromosome painting. *Cytogenet Cell Genet* 77: 211–217
44 Ijdo JW, Baldini A, Ward DC, Reeders ST, Wells RA (1991) Origin of human chromosome 2: an ancestral telomere-telomere fusion. *Proc Natl Acad Sci USA* 88: 9051–9055
45 Chowdhary BP, Raudsepp T, Frönicke L, Scherthan H (1998) Emerging patterns of comparative genome organization in some mammalian species as revealed by ZOO-FISH. *Genome Res* 8: 577–589
46 Dobzhansky T (ed) (1937) *Genetics and the origin of species*. Columbia University Press, New York
47 Coyne JA (1991) Genetics and speciation. *Nature* 355: 511–515

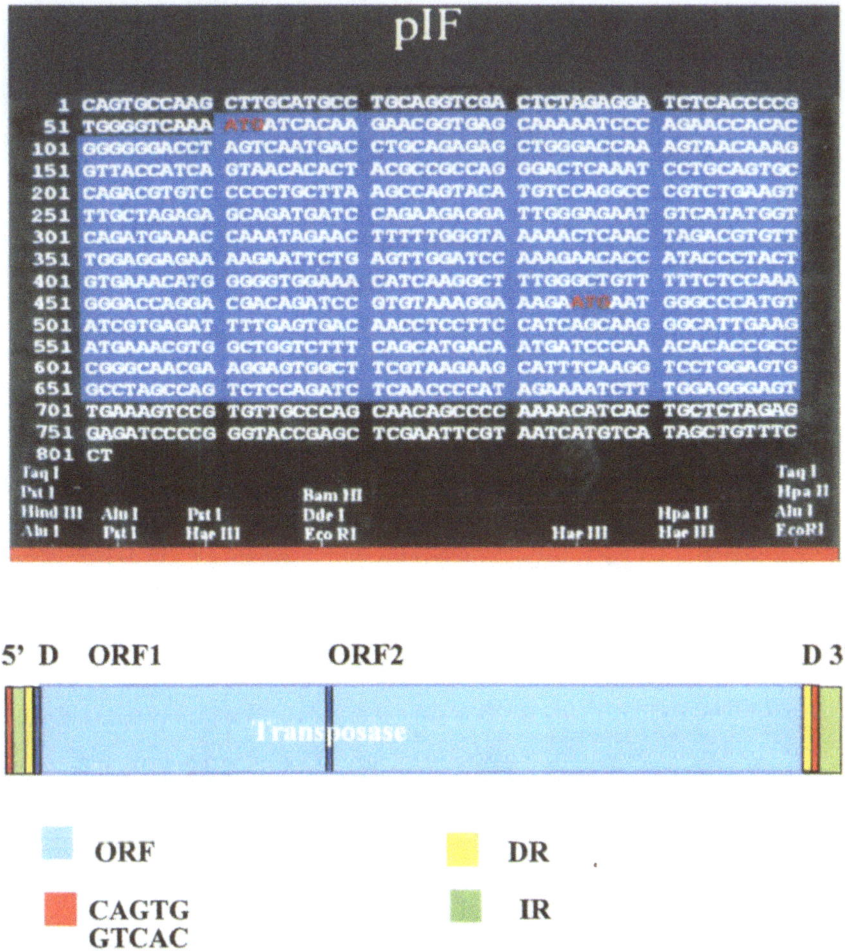

Figure 1. Legend see p. 185

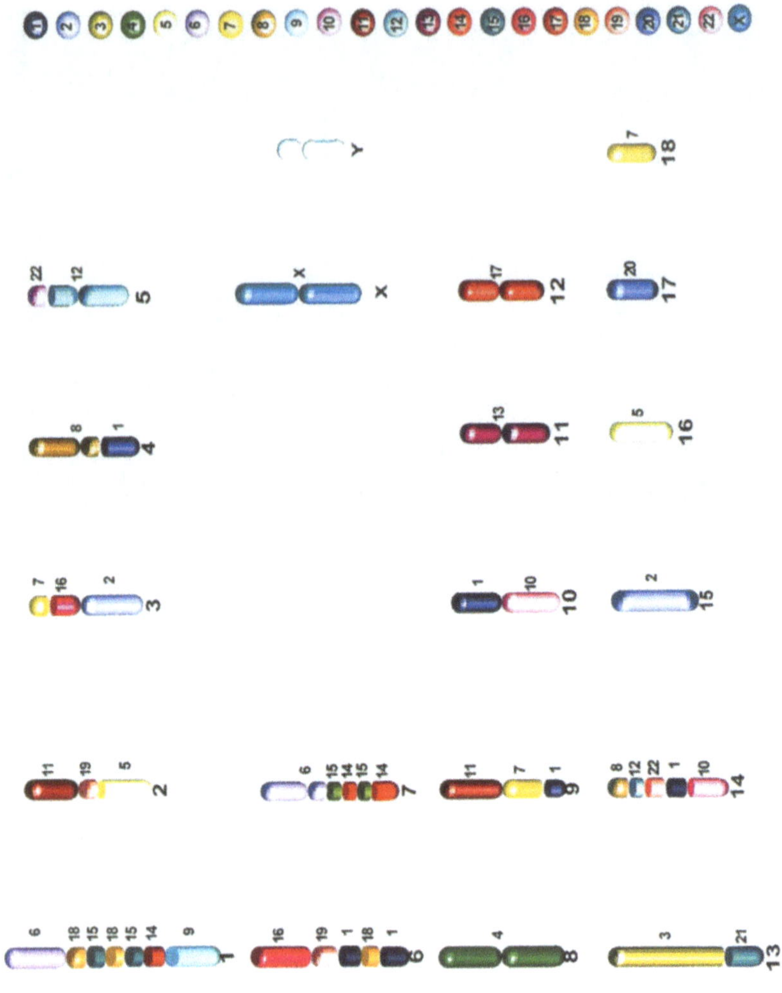

Figure 1. Legend see p. 197

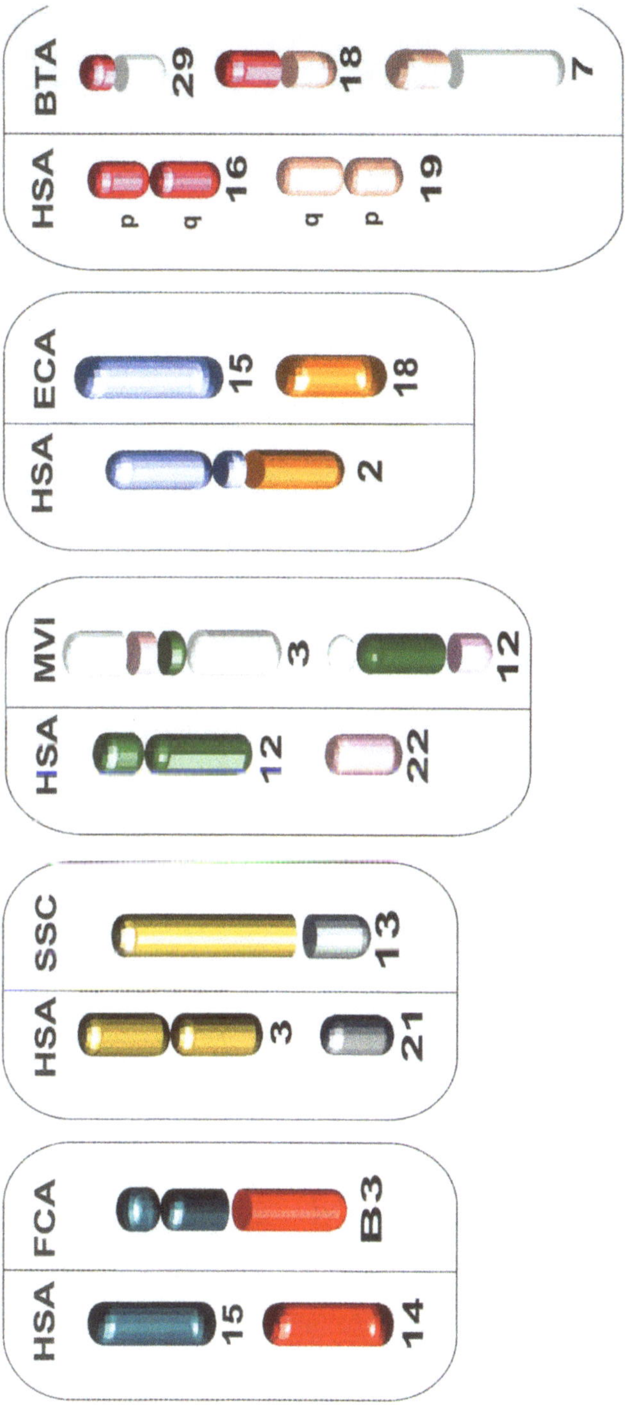

Figure 2. Legend see p. 201

Part five

Medical and cancer cytogenetics

Bernard Dutrillaux

Brunella Franco

Jennifer M. Varley

Chromosomes Today, Vol. 13
ed. by E. Olmo and C.A. Redi

Chromosome and gene alterations in human cancers in relation to aging

Bernard H. Dutrillaux

UMR 147, Institut Curie-CNRS, 26 rue d'Ulm, 75248 Paris Cedex 05 and Département de Radiobiologie et Radiopathologie, CEA/DSV, BP N° 6 Fontenay-aux-Roses, France

Introduction

Cancer cytogenetics constitutes an important component of all cytogenetics literature. How representative of diseases are the few hundred papers appearing each year is an interesting question. A brief survey of the matter they cover indicates that there is an extraordinary bias of publication. Cancer registers, confirming all clinical data, indicate that cancer is a pathology of senescence. For instance, considering the register for males from England and Wales [1], taken as reference, the annual incidence by the age of 80 is about 330-fold that by the age of 10 years. Cancer is not a single disease. It is comprised of some 200 distinct pathologies, which cytogenetic and molecular studies have contributed to individualize. These pathologies can be rougthly grouped into three unequal categories: hematological malignancies (leukemias and lymphomas), carcinomas (cancers of the epithelium, i.e. lung, breast, skin, colon, oesophagus, stomach, etc...) and other solid tumours (sarcomas, germ cell tumours, tumours of neuronal origin...). The incidence of almost all of these pathologies increases with age, but that of carcinomas much more than others. As a consequence, carcinomas represent about 95% of all malignancies by the age of 70–80 years and the average percentage is above 90% during the whole life span, while the other two categories hardly reach 5% each. Considering the papers which appeared in the two most specialized journals (Cancer Genetics and Cytogenetics, and Genes, Chromosomes and Cancer, 1989–90, 1992, 1995 and 1998) it appears that, on average, hematological malignancies, carcinomas and other solid tumours accounted for about 50%, 20% and 30%, respectively.

Thus, a case of leukemia or lymphoma has about 50-fold more chance than a carcinoma of being reported in a publication. The difference is yet more marked between sarcomas and carcinomas. The reason for this bias is obvious, and it is not without importance to point it out. The high reporting of non-carcinoma tumours is related to the simple fact that specific rearrangements, reciprocal translocations in particular, activating oncogenes exist in a high proportion (above 50%) of leukemias and sarcomas, but are exceptional in carcinomas. Thus, most publications deal with specific translocations and their consequences: fusion (onco-) gene formation or proto-oncogene activation. This extraordinary bias led some investigators to believe that all tumours should carry specific chromosome rearrangements, and that such rearrangements will be found when accurately looked for. As a matter of fact, specific translocations are found in about 5% of all malignancies and it is very likely that this percentage will only slightly increase following technical improvements and increasing studies.

Gene alterations in cancer

Cytogenetics aided by molecular studies of cancers, has proven to be a powerful means to discover new genes involved in cell cycle, differentiation, apoptosis, signal transduction and cell migration control. It is not our purpose to describe them, but let's just recall that almost 200 genes involved in these functions are known. About half of them have a positive function and are tightly regulated. To simplify, they may be considered as proto-oncogenes. Their function may be altered by either point mutation or translocation. Specific translocations can be detected on metaphase chromosomes. They result in an activation: the affected proto-oncogene becomes an oncogene. In almost all tumours studied, the other allele being normal, the oncogene activation can be regarded as a dominant mutation. The second half of the group of cancer genes displays a negative control. Their alteration results in a loss of their function, but the involvement of a single allele is generally insufficient. In tumour cells where they are involved, their function is completely abolished by the alteration of both alleles. Thus, their mutations have a recessive expression. Initially called anti-oncogenes, they are now generally called tumour suppressor genes.

Mutations activating oncogenes are targeted. In the case of a point mutation, only the alteration of a very limited number of codons of a given gene is effective. A classical example is given by RAS genes, whose activating mutations are almost limited to codons 12, 13 and 61 [2]. In the case of translocation, the point of breakage can vary (not at the cytogenetic scale) along each chromosome, but the need to recombine two distinct sequences probably makes the event unfrequent. *In vitro* experiments using clastogenic agents such as radiations failed to reproduce such rearrangements (personal observations).

In contrast, molecular studies indicate that tumour suppressor genes can be inactivated by a large variety of alterations [2]. This simply means that any alteration of coding sequences has more of a chance to suppress than to increase or modify their function.

Chromosome alterations and cytogenetic patterns

As shown by the numerous publications collected in reviews or catalogs [3, 4], all known types of chromosome alterations are observed in cancer cells. They can be both numerical (monosomies, trisomies, endoreduplication) and structural (balanced and unbalanced translocations, deletions, isochromosomes, inversions, various complex changes, amplifications as homogeneously staining regions and double minutes, etc.). These are consensus data, which will not be the case for the following cytogenetic patterns. A major difficulty, in addition to the publication bias discussed above, is the heterogeneity of reported data. This heterogeneity could be due to variations of either the populations studied or of diagnoses. In my opinion, they are mostly related to technical biases. Tumours are heterogeneous tissues composed of cancerous, inflammatory, stromal, epithelial, and vascular cells whose growth capacities may strongly differ *in vitro* and *in vivo*. Cytogenetic studies are performed after cell culture under a large variety of conditions. Control of ploidy of the original tumour by flow cytometry is only exceptionally performed and that of tumorigenic capacities of cultured cells, by xenografting on nude mice, still less frequently done in association with cytogenetic studies. Consequently, a proportion of published karyotypes from tumours may refer rather to non-cancerous lines which grow better *in vitro* than the cancerous lines, especially when cultures are prolonged for more than 2–3 days. A typical example is given by gliomas, for which large series of cases are published. In low grade gliomas, 45, X, -X or 45, X,-Y karyotypes are frequently described. In high grade gliomas, which supposedly derive from low grade gliomas, sex chromosome losses are rare, but autosomal alterations frequent. In our series of xenografted gliomas, there is never any sex chromosome loss, and always autosomal alterations. Thus, most of the 45, X karyotypes described are likely to correspond to non-cancerous cells.

Therefore, to understand chromosome evolution of tumours and establish their characteristic cytogenetic patterns, it is necessary to interpret published data, which can be highly criticized and can only reasonably be performed in fields covered by a personal experience. The existence of six main patterns can be proposed.

1) cancers with a normal karyotype: normal type

For the reasons described above, the incidence of tumours with a normal karyotype is impossible to evaluate. Our personal data, with xenografting controls, indicate:

- their existence, at least in colorectal and breast cancer;
- their occurrence in about 5–7% of colorectal cancer, probably lower in breast cancer;

- their strong association with a mutator phenotype (micro-satellite instability) related to the alteration of one of the mismatch repair genes [5];
- their exceptional association with a *TP53* mutation.

2) cancers with a single chromosome alteration

Here again, many descriptions from the literature may not refer to cancer cells. Excluding sex chromosome losses, two types of alterations occur:

- autosomal gains (+7 in particular), whose meaning is uncertain, since they were described in a variety of non-cancerous cells;
- deletions or complete autosome losses, rarely observed as a single change in cancer cells.

3) cancers with a haploid or near-haploid karyotype

In addition to a variety of leukemias (acute lymphoblastic and chronic myeloid leukemias), near haploid karyotypes are rarely observed in solid tumours.

A drastic loss of chromosomes occurs in the absence of structural rearrangements making this rare chromosome evolution quite distinct from that of the monosomic type which will be described below.

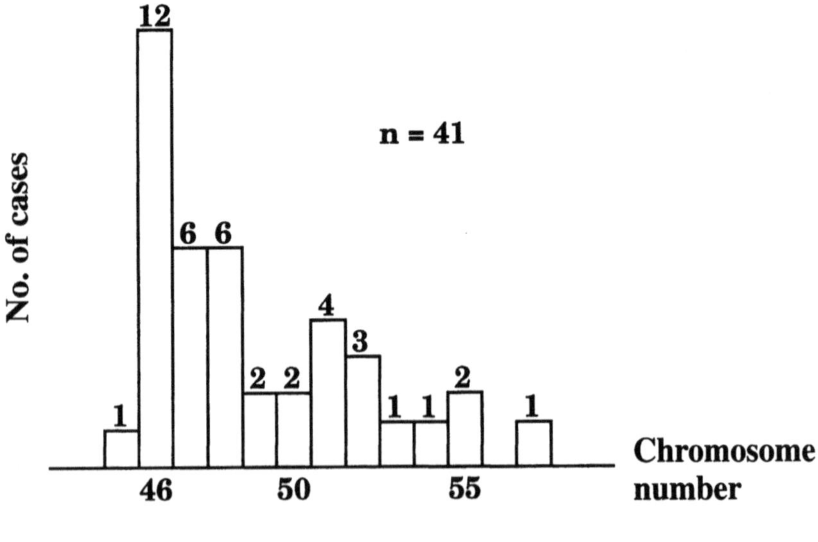

Figure 1. Distribution of ploidies in a series of published cases of Ewing sarcoma. Recurrent chromosome gains lead to hyperdiploid formulas in 68% (28/41) of cases.

4) cancers with a reciprocal translocation (or inversion): translocation type (tlt) (Fig. 1)

In tumours with a specific (highly recurrent) translocation, the rearrangement, frequently observed as the sole anomaly, has been associated with oncogene activation or formation in many instances [6]. Occasional additional anomalies can occur. They are most frequently recurrent chromosome gains which leads to hyperdiploidy. This applies to both hematological malignancies and sarcomas (Tabs 1 and 2).

Table 1. Distribution of ploidies in hematological malignancies with a specific balanced rearrangement

Specific rearrangements	Pathology	Ploidy (%)				Nb. cases
		Hypo-	Diplo-	Hyper-	Poly-	
t(1;19)(q23;p13)	ALL	6	58	35		144
t(2;8)(p11;q24)	ALL, NHL	0	65	35		23
t(3;3), or inv(3)(q21q26)	AML, MDS	45	52	3		67
t(4;11)(q21;q23)	ALL	2	79	18		267
t(6;9)(p23;q34)	AML	5	83	12		66
t(7;11)(p15;p15)	AML	0	100	0		21
t(8;14)(q24;q32)	ALL, NHL	5	62	33	<1	329
t(8;14)(q24;q11)	ALL	0	78	22		27
t(8;16)(q11;p13)	AML	0	87	13		24
t(8;21)(q22;q22)	AML	47	44	9		681
t(8;22)(q24;q21)	ALL, NHL	5	42	53		57
t(9;11)(p21;q21)	AML	0	76	24		102
t(9;22)(q34;q11)	CML	10	32	58		1475
	ALL	17	60	23		448
t(11;14)(p13;q11)	LL	4	82	14		51
t(11;14)(q13;q32)	LL, MM	24	51	22	3	148
t(11;19)(q23;p13)	AML, ALL	2	70	28		87
t(14;14), or inv(14)(q11q34)	ALL	32	47	19	2	81
t(14;18)(q32;q21)	NHL	5	27	63	5	346
t(15;17)(q22;q21)	AML	2	81	17	<1	438
inv(16)(p13q22)	AML	3	69	28		305

Except for rearrangements involving only chromosome 3 or 14, and the t(8;21) frequently associated with sex chromosome loss, hypodiploid formulas are rare, compared to pseudo- or hyperdiploid ones (calculated data collected in [3])

ALL: Acute Lymphoblastic Leukemia; AML: Acute Myeloid Leukemia; NHL: Non-Hodgkin Lymphoma; CML: Chronic Myeloid Leukemia; MDS: MyeloDysplastic Syndrome; LL: Lymphoblastic Leukemia; MM: Multiple Myeloma.

Table 2. Age of sarcoma onsets in relation to karyotypic data

Tumour type	Mean age of onset (years) ±SD	Number of cases without/with endoreduplication	Mean number of chromosomes without/with endoreduplication
Tumours with a specific balanced translocation (TLT)			
Askin tumour, t(11;22)	21 ± 8	5/0	46 ± 3/–
Clear cell sarcoma, t(12;22)	28 ± 11	7/2	48 ± 3/ ± 4n
Congenital fibrosarcoma t(12;15)	1 ± 1	9/0	48 ± 2/–
Ewing sarcoma, t(11;22)	16 ± 7	72/3	48 ± 3/77 ± 11
Ewing sarcoma, t(21;22)	11 ± 6	3/0	47 ± 1/–
Extrasqueletal mixoid chondrosarcoma t(9;22)	66 ± 12	5/1	46 ± 2/92
IDSRCT, t(11;22)	17 ± 4	3/0	48 ± 2/–
Liposarcoma, mixoid, t(12;16)	48 ± 14	30/0	46 ± 1/–
Liposarcoma, mixoid, t(12;22)	33 ± 9	4/0	47 ± 1/–
Peripheral neuroepithelioma, t(11;22)	19 ± 13	12/2	48 ± 2/91–95
Rabdomyosarcoma, t(1;13)	4 ± 2	4/2	48 ± 2/91–92
Rabdomyosarcoma, t(2;13)	15 ± 5	11/14	48 ± 2/93 ± 5
Synovial sarcoma, t(X;18)	32 ± 16	28/2	47 ± 3/90
Mean	24 ± 17	85%/15%	47 ± 1/90 ± 5

Table 2. (continued)

Tumour type	Mean age of onset (years) ±SD	Number of cases without/with endoreduplication	Mean number of chromosomes without/with endoreduplication
Tumours with a monosomic type karyotype (MT)			
Angiosarcoma	61 ± 12	3/1	44 ± 4/–
Chondrosarcoma (grade I or II)	63 ± 11	9/3	41 ± 5/60 ± 7
Chordoma	69 ± 7	8/1	42 ± 3/72
Fibrosarcoma (adult)	34–39	2/1	43–45/hypotetraploid
Leiomyosarcoma	54 ± 13	22/16	44 ± 3/70 ± 8
Malignant fibrous histiocytoma	68 ± 13	Very dispersed chromosome numbers and multiple rearrangements	
Malignant mixed mesodermal tumours	67 ± 9	1/8	44/62 ± 9
Mesothelioma	63 ± 10	64/36	44 ± 6/79 ± 8
Osteosarcoma	21 ± 16	11/10	42 ± 5/75 ± 9
Pleomorphic liposarcoma	68 ± 10	2/7	41–45/63–210
Mean	57 ± 16	60%/40%	43 ± 1/69 ± 7

Mean ages of tumour onset and cytogenetic data of sarcomas with either a recurrent balanced translocation (TLT) or a Monosomic Type (MT) karyotype: MT tumours have lower mean chromosome numbers, higher rates of endoreduplications and occur later than TLT tumours. IDSRCT: Intra-abdominal Desmoplastic Small Round-Cell Tumours.

5) cancers with chromosome gains: trisomic type (tt) (Fig. 2)

Some tumour karyotypes are characterized by multiple chromosome gains, whereas structural rearrangements are rare or non-existent. Frequent in childhood tumours, this pattern leads to near triploidy, and is often associated with a fairly favourable prognosis. Some of these near triploid tumours may even undergo spontaneous regression.

The TT pattern is proportionally rarer in adult tumours, especially in carcinomas, where it is observed in endometrial cancers and about 20% of colorectal cancer.

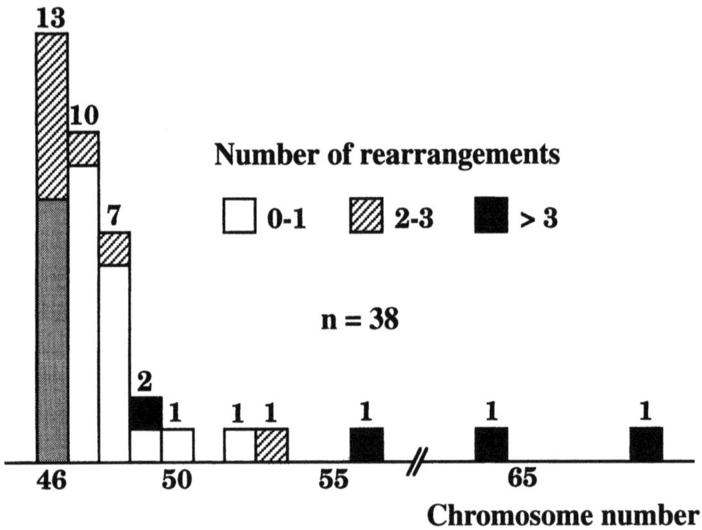

Figure 2. Distribution of ploidies in a series of 38 human endometrial carcinomas [20]: a typical Trisomic Type (TT) is observed, after recurrent gains of chromosomes 2 and 10 in particular. Structural rearrangements of chromosomes are infrequent compared to other carcinomas. Euploidy (46 normal chromosomes) is indicated in grey.

6) cancers with chromosome deletions and losses: monosomic type (mt) (Fig. 3)

This is by far the most frequent pattern of chromosome evolution both in cancers and during cell transformation *in vitro*. It is characterized by the occurrence of unbalanced rearrangements which lead to deletions. The most frequently deleted chromosome arms are 1p, 3p, 5q, 6q, 8p, 9p, 10q, 11p, 11q, 13q, 16q, 17p, 18p and q and 22q. The association of some of these deletions, which accumulate during tumour progression, is quite characteristic of a given tumour type. For instance, deletions of 17p, 18p and q arms are associated with

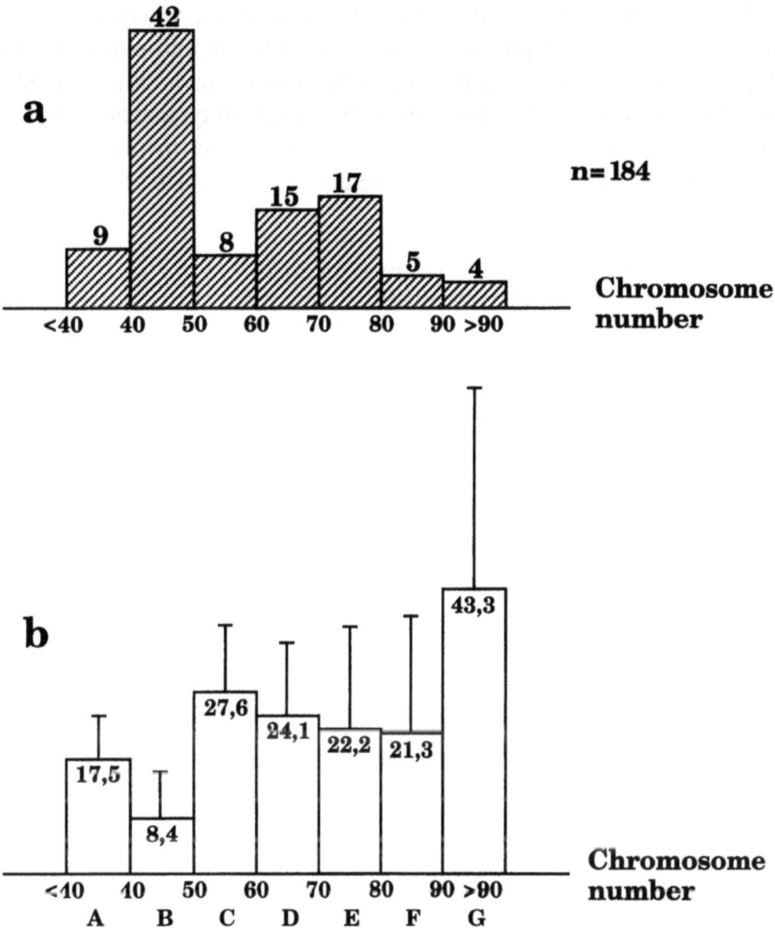

Figure 3. Distribution of ploidies in a series of 182 breast carcinomas (a) and mean numbers of rearranged chromosomes in relation to ploidy (b). The increase in the numbers of rearrangements indicates the progression in the following order: B → A (chromosome losses) → F (after endoreduplication) → E → D → C (progressive chromosome losses) → G (after a second endoreduplication).

about 70% of colorectal cancers in which these deletions are also associated with those of 1p and 8p arms in about 40–50% of cases. In breast cancer, there are concerted deletions of 6q, 8p, 11p, 13q, 16q and 17p arms etc… [7, 8]. This leads to a progressive reduction in the number of chromosomes, by formation of derivatives composed of two arms from different chromosome origins. As the number of chromosome dicreases, down to 35, endoreduplications become frequent, leading to hypotetraploid clones, which frequently acquire a selective advantage and replace the original hypodiploid cell population. These hypotetraploid tumours continue to lose chromosome material, and we could show that karyotypes with 50–60 chromosomes could have passed through endoreduplication [9]. At this stage, the karyotypes which carry multiple unbalanced structural rearrangements (50–90% of chromosomes may be abnormal) can easily be distinguished from those with similar numbers of chromosomes derived by a trisomic or translocation type evolution, in which almost all chromosomes are normal alongside those involved in balanced translocations. Most carcinomas can be classified as MT tumours (Tab. 3).

Table 3. Annual incidences and cytogenetic types of carcinomas

Site	Age (years)				Cytogenetic types
	20–24	40–44	60–64	80–84	
Lung, bronchus	0.3	13	301	764	MT
Prostate	0	0.5	64	539	MT
Colon, rectum	0.5	12	120	427	MT >> TT
Skin (melanoma excluded)	0.9	17	131	395	MT, others?
Breast	1	92	192	249	MT
Stomach	0.1	2	66	242	MT
Bladder	0.4	5.5	70	239	MT
Pancreas	0	2.9	29	99	MT
Oesophagus	0.1	2.1	28	79	MT
Ovary	1.5	12	45	51	MT/TT
Kidney	0.1	3.2	23	39	TRT?/TT/MT
Corpus uteri	0.1	4.5	37	37	TT
Head and neck	0.2	3.8	22	36	MT >> others
Larynx	0	2.2	20	29	MT
Cervix uteri	2.1	24	32	20	MT >> TT
Liver	0.1	0.7	7	16	MT?
Thyroid	0.3	0.8	2	4.1	TRT, MT, TT

Annual incidences per 100 000 population at various ages of cancers (mostly carcinomas) from different organs calculated from [1] and cytogenetic types (personal observations and collected in [3]): MT is proportionally much more frequent in tumours whose incidence strongly increases with aging.

Gene alterations associated with these deletions were intensively investigated. It is now generally accepted that these deletions unmask recessive mutations of tumour suppressor genes. Some of the genes have been characterized:

3p: VHL and FHIT; 5q: APC; 6q: SOD2; 9p: p16; 10q: PTEN; 11p: WT1; 13q: RB1, BRCA2; 16q: ECAD; 17p: TP53; 22q: NF2 [1, 10–12]. Many others remain to be discovered. A number of these genes (VHL, APC, p16, WT1, RB1, BRCA2, ECAD, TP53, NF2) can be altered in germ cells: their mutation constitutes a predisposition to cancer which may be familial. Cytogenetic studies of tumours developed in predisposed patients are infrequently reported. Data exist for renal cell carcinoma after VHL mutation. In these tumours, the deletion of the 3p arm is almost constant, frequently associated with other

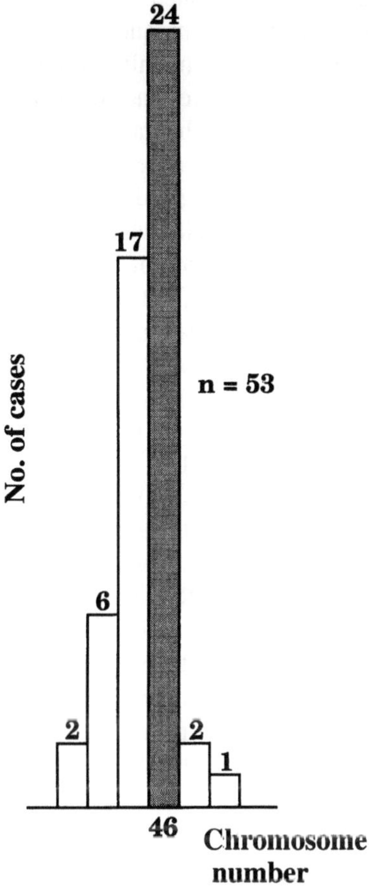

Figure 4. Distribution of ploidies in non-papillary renal cell carcinomas developed in patients affected by a Von Hippel Lindau syndrome, determined by the recessive mutation of the VHL gene mapped on the 3p arm. In addition to the 3p arm deletion, most aneuploid karyotypes are hypodiploid (25/28), indicating a MT evolution. Data from literature [3]. Pseudodiploidy is indicated in grey.

deletions, and a hypodiploid formula (Fig. 4). We have studied a few tumours with APC, BRCA1 or TP53 germinal mutations (unpublished observations). Here again, a monosomic type karyotype was observed.

Initiating gene alterations and tumour karyotypes

Tumours developed in genetically predisposed patients are particularly interesting because the first genetic alteration is known. Except for tumours caused by mismatch repair gene alterations, their karyotype is highly rearranged and carries not only the deletion unmasking the normal allele of the altered gene, but other deletions, as in similar tumours developed in non-predisposed patients. Thus, a reasonable generalization is that when a tumoral process is initiated by the recessive mutation of a tumour suppressor gene, chromosome evolution follows a Monosomic Type pattern.

The situation is quite different when an oncogene-related specific translocation exists. Such a chromosome alteration is frequently unique in the tumour karyotype, and when other alterations occur, they are mainly chromosome gains or other balanced translocations. Deletions are rare, and never involve the normal homologues of the chromosomes involved in the translocation. The oncogene activation is clearly dominant, and this can be generalized to all known examples. Thus, chromosome evolution following a dominant activation of an oncogene initiating a cancerous process recalls what we called a Trisomic Type pattern. The nature of the initiating mutation of Trisomic Type tumours without specific translocation is more uncertain, but some data suggest it may be a dominant mutation. A first indirect indication is that trisomic type evolution is not (or rarely?) observed in tumours from genetically predisposed patients. A second indication is provided by molecular studies of a number of leukemias in which there were chromosome gains, but not the expected translocation, such as the t(9;22)(q34;q11) in chronic myeloid leukemia. It was shown that the chimeric gene Abelson leukemya (ABL-BCR) usually formed by the t(9;22) was present, even in the absence of cytogenetically visible translocation [13]. Another element is provided by colon adenomas (not cancer) where both a point mutation of K-rat sarcoma (K-RAS) and a trisomic type evolution are quite frequent [14]. More convincing is the study of congenital fibrosarcomas, which display a typical trisomic type evolution, with recurrent gains of chromosomes 8, 11, 17 and 20. Molecular studies demonstrated the presence of a ETV6-NTRK3 fusion, and this fusion was shown to derive from a t(12;15)(p13;q25) which had not been detected by classical cytogenetics [15].

Finally, a direct demonstration was provided by mammary tumours from WAP-rat sarcoma (WAP-RAS) transgenic mice [16], in which all the cytogenetically abnormal tumours developed after H-RAS activation were of the trisomic type (Fig. 5). Thus, at least a proportion of Trisomic Type tumours are formed after the dominant activation of an oncogene.

Somatic gene mutation and age of tumour onset

That dominantly inherited predisposition to cancer is caused by recessive mutations, at the cellular level, was proposed by Knudson [17] and Commings [18] on the basis of epidemiological data. The germinal mutation is recessive because the second allele must be disabled to enable tumour development. This “two hits” mechanism explains why, as for retinoblastoma, the predisposed develop their tumour earlier than non-predisposed patients, the first mutation being present since conception in predisposed patients.

Our purpose is to generalize this relationship between the age of tumour onset and the nature of the initiating mutations to all cancers developed in non-predisposed patients. Dominant-activating mutations can initiate the tumoral

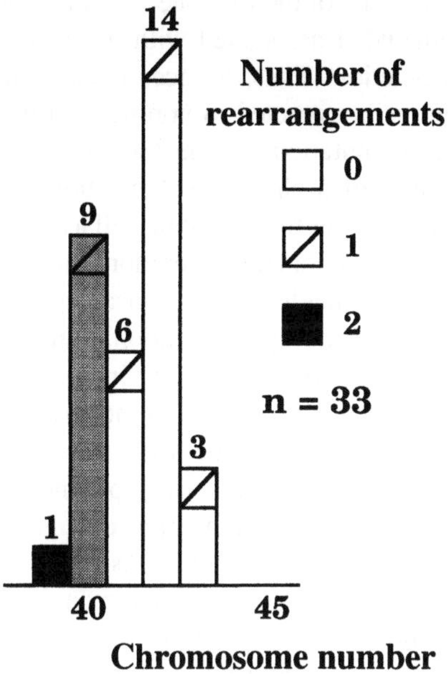

Figure 5. Distribution of ploidies in mammary cancers developed in transgenic WAP-RAS mice [16], in which the H-RAS gene is strongly activated when mice are breast feeding. These cancers, initiated by a dominant oncogene, follow a typical TT evolution. Euploidy (40 chromosomes) is indicated in grey.

process without delay allowing tumour progression to occur. Tumour diagnosis may however be delayed for months or years, depending on various parameters, such as tumour type, cell proliferation, associated alterations, genomic instability etc... As discussed above, dominant mutations are assumed to only rarely occur because they must be precisely targeted. Depending on a single hit, they can occur at any moment during a lifetime, being not strictly dependent on other gene alterations.

Recessive activating mutations can initiate the tumoral process only when both alleles are involved. As for dominant mutations, the mutation of a single allele is a stochastic event which can occur at any age. When formed, the mutation, which is silent, can remain in the affected cell and its descendents. At the level of a cell population, it means that both the proportion of mutant cells and the number of recessive mutations per cell increase with aging. This will increase the probability of an alteration to the second allele of one of the altered genes, and hence, the expression of a recessive mutation: the more altered alleles a cell carries, the higher is the probability of a second hit. This means that when a mutant cell initiates a tumour process after invalidation of both alleles of a tumour suppressor gene, there is a high probability that other genes have one altered allele. Such recessive mutations can be further unmasked during tumour progression by deletions, and this explains the occurrence of concerted deletions in the Monosomic Type pattern.

Thus, tumours initiated by recessive mutations should be strongly dependent in aging, in contrast to those initiated by dominant mutations. The nature of the second mutation may be different from that of the first one. The first one must not be deleterious: it may be eliminated if associated with other gene deficiencies, such as those caused by a deletion. Thus, an accumulation of point mutations or small intragenic deletions is expected in non-transformed or non-malignant senescent cells. The second mutation, in contrast, has a phenotypic consequence, which will be the loss of a check point function if a tumour suppressor gene is affected. This may confer to the mutant cell a strong selective advantage which may counterbalance associated alterations such as those determined by large deletions or chromosome losses. This also explains the occurrence of multiple deletions in Monosomic Type evolution, unmasking recessive mutations accumulated either before or after tumour initiation.

Because accumulation of single allele mutations is needed to increase the probability of bi-allelic mutations, tumours initiated by recessive mutations should occur in old adults, and their incidence should increase supralinearly when several tumour suppressor genes may be involved in the same oncogenic process. Such tumours should be quite rare in young people. As a consequence, tumours of the young should be principally those initiated by a dominant mutation whose incidence should not strongly increase with aging, except for cases with a hereditary predisposition where the first mutation is germinal.

Tumour karyotype and aging

We have seen that cancers initiated by a dominant mutation are those characterized by a specific translocation (TLT), and perhaps by a trisomic type evolution (TT). Those initiated by a recessive mutation should be of the monosomic type (MT), i.e. with a highly rearranged karyotype and multiple deletions. Thus, if the above assertions are correct, TLT and TT should affect young and MT cancers old patients.

In the classification of sarcomas in relation to their cytogenetic type, performed on published data [3, 4], we evaluated the mean ages of tumour onset (Tab. 2). An obvious difference exists when MT are compared to either TT or TLT, but not when TT and TLT are compared. MT sarcomas obviously occur in older patients, which fits with the hypothesis that they were initiated by recessive mutations. As shown in Figure 6, the slight increase in sarcomas with aging is related to those with MT karyotype.

Such classification could be done for carcinomas, but almost all in them are classified as MT. Indeed, this fits with the strong increase in their incidence with aging (Tab. 3). Exceptions are interesting to consider. Firstly, endometrial cancer is a hormonal-dependent tumour, with a typical TT karyotype which is rare for carcinomas. Compared to another hormonal-dependent, but typically MT tumour such as breast cancer, its incidence only slightly increases with aging (Figs 6, 7).

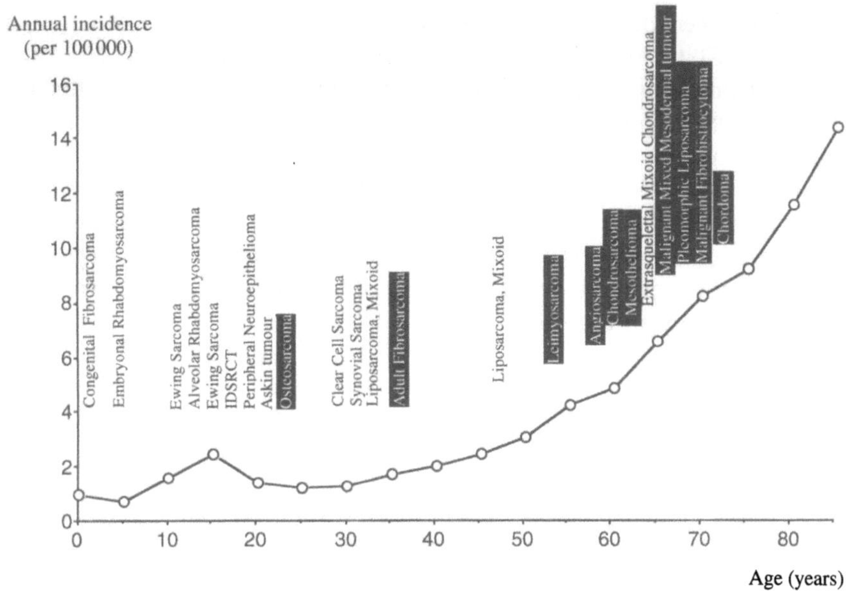

Figure 6. Annual incidence of bone and soft tissue tumours, principally sarcomas [1], and mean age of occurrence of Translocation Type (TLT, dark letters) and Monosomic Type (MT: white letters) malignant tumours (data from Tab. 3). Most MT tumours occur in old adults, which causes the overall increase in incidence of bone and soft tissue tumours with aging. For abbreviations, see legend to Table 2.

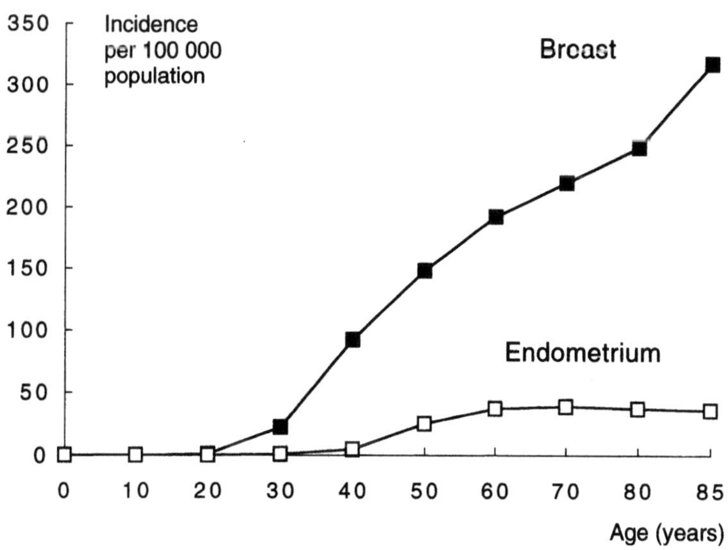

Figure 7. Annual incidences of two human hormone-dependent carcinomas exhibiting the higher incidence of breast (MT) than endometrial (TT) cancer (calculated from [1]).

Ovarian cancer is another hormonal-dependent tumour in which various karyotypes were described. Considering the data from Thompson et al. [19], a subset of TT tumours exists in addition to a large proportion of MT karyotypes. Comparing the ages of tumour onset, there is a significant difference ($p = 0.017$): TT tumours occur 12 years earlier than MT tumours (48 versus 60 years) on average.

Indeed, the hypothesis that initiating mutations, karyotype evolution pathways and age of tumour onset are tightly related requires further examination, and the study must be extended to hematological malignancies. It is proposed as a heuristic hypothesis which may help in the search for genes involved in the cause of tumour initiation.

References

1 Parkin DM, Muir CS, Whelan SL, Gao YT, Ferlay J, Powell J (eds) (1992) *Cancer incidence in five continents*, vol. VI (IARC Scientific Publications, N° 120) Lyon, International Agency for Research on Cancer

2 Hesketh R (1997) *The oncogene and tumour suppressor gene factsbook* (2nd edition). Academic Press, San Diego

3 Mitelman F (1994) *Catalog of chromosome aberrations in cancer* (5th edition), Wiley-Liss, New York

4 Sandberg AA, Bridge JA (1995) *The cytogenetics of bone and soft tissue tumours*. Medical Intelligence Unit, Springer, New York

5 Remvikos Y, Vogt N, Muleris M, Salmon RJ, Malfoy B, Dutrillaux B (1995) DNA-repeat instability is associated with colorectal cancers presenting minimal chromosome rearrangements. *Gene Chromosome Cancer* 12: 272–276

6 Rabbitts TH (1994) Chromosomal translocations in human cancer. *Nature* 372: 143–149

7 Muleris M, Salmon RJ, Dutrillaux B (1990) Cytogenetics of colorectal adenocarcinomas. *Cancer Genet Cytogenet* 46: 143–156

8 Dutrillaux B, Gerbault-Seureau M, Zafrani B (1990) Characterization of chromosomal anomalies in human breast cancer. A comparison of 30 paradiploid cases with few chromosome changes. *Cancer Genet Cytogenet* 49: 203–217

9 Dutrillaux B, Gerbault-Seureau M, Remvikos Y, Zafrani B, Prieur M (1991) Breast cancer genetic evolution: I. Data from cytogenetics and DNA content. *Breast Cancer Res Treat* 19: 245–255

10 Bravard A, Sabatier L, Hoffschir F, Ricoul M, Luccioni C, Dutrillaux B (1992) SOD2: A new type of tumour-suppressor gene? *Int J Cancer* 51: 476–480

11 Ohta M, Inoue H, Cotticelli MG, Kastury K, Baffa R, Palazzo J, Siprashvili Z, Mori M, McCue P, Druck T et al (1996) The FHIT gene, spanning the chromosome 3p14.2 fragile site and renal carcinoma-associated t(3;8) breakpoint, is abnormal in digestive tract cancers. *Cell* 84: 587–597

12 Guilford P, Hopkins J, Harraway J, McLeod M, McLeod N, Harawira P, Taite H, Scoular R, Miller A, Reeve AE (1998) E-cadherin germline mutations in familial gastric cancer. *Nature* 392: 402–405

13 Fitzgerald PH, Beard MEJ, Morris CM, Heaton DC, Reeve AE (1987) Ph-negative chronic myeloid leukemia. *Brit J Haematol* 66: 311–314

14 Muleris M, Zafrani B, Validire P, Girodet J, Salmon RJ, Dutrillaux B (1994) Cytogenetic study of 30 colorectal adenomas. *Cancer Genet Cytogenet* 74: 104–108

15 Knezevich SR, McFadden DE, Tao W, Lim JF, Sorensen PHB (1998) A novel ETV6-NTRK3 gene fusion in congenital fibrosarcoma. *Nat Genet* 18: 184–187

16 Le Roy H, Ricoul M, Ogata H, Apiou F, Dutrillaux B (1993) Chromosome anomalies in mammary carcinoma from transgenic WAPRAS Mice: comparison with human data. *Genes Chromosome Cancer* 6: 156–160

17 Knudson AG (1971) Mutation and cancer: statistical study of retinoblastoma. *Proc Natl Acad Sci USA* 68: 820–823
18 Commings DE (1973) General theory of carcinogenesis. *Proc Natl Acad Sci USA* 70: 3324–3328
19 Thompson FH, Liu Y, Emerson J, Weinstein R, Makar R, Trent JM, Taetle R, Alberts DS (1994) Simple numeric abnormalities as primary karyotype changes in ovarian carcinoma. *Genes Chromosome Cancer* 10: 262–266
20 Dutrillaux B (1995) Pathways of chromosome alteration in human epithelial cancers. *Adv Cancer Res* 67: 59–81

Chromosomes Today, Vol. 13
ed. by E. Olmo and C.A. Redi

Xp contiguous gene syndromes: from clinical observation to disease gene identification

Brunella Franco and Andrea Ballabio

Telethon Institute of Genetics and Medicine (TIGEM), San Raffaele Biomedical Science Park, Via Olgettina 58, I-20132, Milan, Italy

Introduction

The X chromosome is currently among the best characterized of all human chromosomes. With the exception of genes located in the pseudoautosomal region [1], and a few others which have active copies on both the X and Y chromosomes, most genes on the X chromosome display a haploid status in males. As a consequence, recessive mutations in X-linked disease genes result in hemizygous affected males and heterozygous, and usually asymptomatic, carrier females. the easily-recognizable inheritance pattern of X-linked diseases allows the instant assignment of the corresponding genes to the X chromosome, narrowing down their position from anywhere in 3,000 Mb, which is the size of the human genome, to 150 Mb, the size of the human X chromosome. Furthermore, the haploid status of the X chromosome in males simplifies the analysis with genetic markers. Deletions on the X result in nullisomy in males which can be easily identified using common techniques such as Southern blotting or polymerase chain reaction (PCR). For all these reasons genetic analysis and mapping have been greatly facilitated on the X chromosome.

The evolution of the distal portion of the short arm of the X chromosome (Xp22.3) is complex and still rich with unresolved issues. The most telomeric 2.6 Mb of Xp22.3 is referred to as the pseudoautosomal region (PAR). Genes in the PAR have homologous and functional alleles on the Y chromosome, they escape X-inactivation and, therefore, show absence or partial sex linkage (for a review, see [2]). There is evidence that this segment of identity between the X and the Y chromosome short arms is necessary to allow pairing and chiasma formation of sex chromosomes during meiosis and subsequent proper segregation [3]. An obligatory crossing-over event occurs within the PAR in any functional male meiotic division [4]. It is known that a number of genes located on the short arm of the human X chromosome map on at least two autosomal clusters in marsupials and monotremes. Therefore, it has been suggested that Xp was originally autosomal and was translocated to the X early in eutherian lineage [5, 6]. The homology between the distal Xp and the Y chromosome is not restricted to PAR but it extends beyond the pseudoautosomal boundary into the more proximal, X-specific segment of Xp22.3. The three genes located within this segment, STS, KAL-1, and GS1, share high sequence identity (88–96%) with a region of the Y chromosome long arm (Yq11.2). It has been proposed that this segment was originally part of an ancestral pseudoautosomal region that was disrupted by a pericentric inversion which occurred on the Y chromosome in recent evolution [7–9].

The distal part of Xp is an exception to the rule that all of the loci mapped to the human X chromosome have murine homologs that are also located on the X chromosome, within clusters of conserved order [10]. The high evolutionary divergence of this region in mammals is probably due to the occurrence of an obligatory crossing-over in the pseudoautosomal region in male meiosis, which leads to a higher frequency of double strand breakage and repair, and subsequently to a higher mutation rate. None of the human distal Xp genes has been shown to have a homolog on the murine X chromosome [11, 12]. In most cases, a mouse homolog could not be identified at all, due to the complete lack of cross-hybridization to mouse DNA. Recently, the murine Sts gene has been cloned, after intensive efforts, and found to be pseudoautosomal [13].

Furthermore, the distal short arm of the X chromosome is characterized by a very high frequency of chromosomal rearrangements [14]. We have constructed a detailed physical map of this region (Xpter to the DMD locus). The backbone of the map is represented by a single oriented contiguous stretch of 585 overlapping yeast artificial chromosome (YAC) clones covering the entire region [15]. In addition, the map displays the precise localization of 85 chromosomal breakpoints from patients with different phenotypes [15].

The presence of numerous patients with deletions and translocations has been of considerable help for the mapping of X-linked disease genes in this region. An accurate phenotype/genotype correlation in these cases can lead to both the identification of disease genes and the understanding of the pathogenetic mechanisms involved in the disease phenotype.

Chromosomal rearrangements

Distal Xp rearrangements include 5 types of abnormalities: interstitial deletions, terminal deletions, X/Y translocations, X/autosomal translocations and Xp duplications. A summary of the distal Xp abnormalities and of their causative molecular mechanism is given in Table 1.

Interstitial deletions, usually not detectable by chromosome analysis, are the most common type of abnormality found in the Xp22-pter region. In the large majority of the patients these deletions involve a region of approximately 2 Mb [16], including the steroid sulfatase (STS) locus, and resulting in isolated X-linked ichthyosis (XLI)[17–23]. These chromosomal abnormalities are due to homologous recombination (unequal crossing over) between

Table 1. Common molecular mechanisms of distal Xp abnormalities

Abnormality	Mechanism
Interstitial deletions	Homologous recombination (unequal crossing over) between elements of a family of repeats clustered in Xp22.3
X/translations	Homologous recombination between regions of the X and Y chromosomes sharing high sequence identity (Xpter/Ypter or Xp22.3/Yq11.2)
X/autosomal translocations	unknown
Terminal deletions	unknown
Xp duplications	unknown

Modified from Ballabio and Andria [14].

elements of a family of repeats clustered in Xp22.3 on either side of the STS gene [16]. In some instances the deletion is not limited to the STS locus but also involves adjacent disease genes, causing contiguous gene syndromes [24–26].

Terminal deletions are usually detected both by cytogenetic and molecular analysis [27–33]. The name "terminal deletion" may not be used in a strict sense in these cases, since the deleted Xp arm presumably retains a functional telomere, which may or may not correspond to the normal Xp telomere. However, chromosomes in which Xp subtelomeric markersare deleted are usually referred to as terminal deletions. The position of the breakpoints in the various patients with terminal deletions, and consequently the sizes of the deleted regions, are very variable (anywhere from 3 to 15 Mb). Therefore, the complexity of the phenotype of these patients is also variable and depends on both the number of disease genes involved and the sex of the affected individual.

X/Y translocations involving the distal short arm of the X chromosome can be divided into two major groups: Xp/Yp and Xp/Yq translocations. X/Y translocations are largely due to homologous recombination between regions of the X and Y chromosomes sharing high sequence identity (Xpter/Ypter or Xp22.3/Yq11.2). Both types of translocation events cause a terminal Xp deletion. Therefore, as in "pure" terminal deletions, the length of the deleted region and the complexity of the phenotype depend on both the position of the breakpoint on the X chromosome and on the sex of the affected individual.

X/autosomal translocations with an Xp22 breakpoint have been found in only a few cases [34–36]. In balanced translocations the phenotype is presumed to be due to the disruption of a putative disease gene on the X chromosome at the site of the translocation breakpoint. However, the presence of a complex (translocation-deletion) rearrangement or a positional effect of gene dosage cannot be ruled out in X/autosomal translocations. In unbalanced translocations the phenotype is due to both the monosomy of the distal Xp region in affected females and to the trisomy of the autosomal region involved in the translocation. However, spreading of X-inactivation into the autosome, and preferential survival of cells with the normal X active, could prevent the effects of the trisomy.

Clinical findings

Male patients with deletions in Xp22.3 are nullisomic for this region and consequently they show contiguous gene syndromes characterized by different combinations of phenotypes according to the length of the deletion [25] (see Fig. 1). These contiguous gene syndromes are characterized by the association of up to six of the following diseases[37]: short stature (SS; MIM312865), X-linked recessive chondrodysplasia punctata (CDPX; MIM302950), mental retardation (MRX; MIM309530), X-linked ichthyosis due to STS deficiency (XLI; MIM308100), Kallmann syndrome (KAL; MIM308700) and Ocular

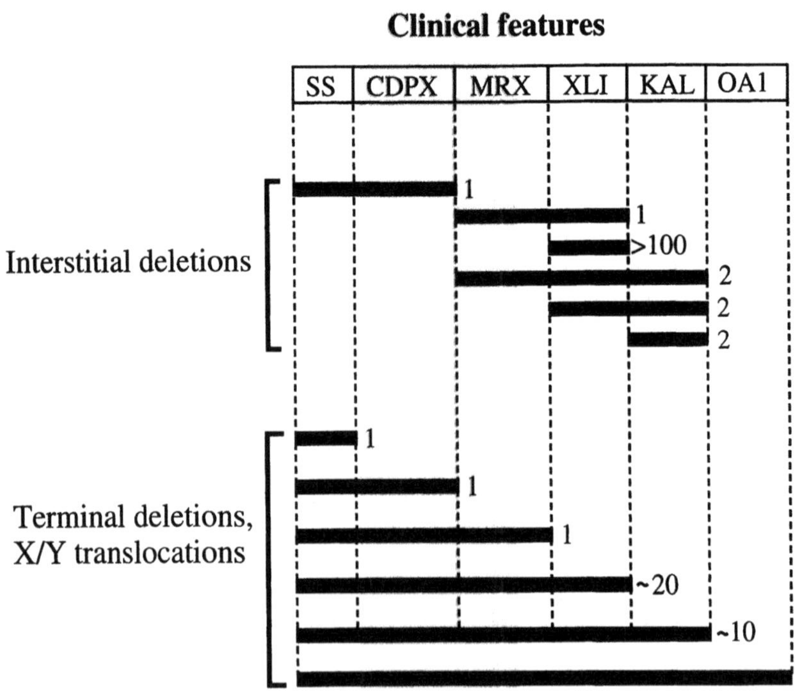

Figure 1. Clinical features of male patients with Xp22 deletions and translocations. SS, short stature; CDPX, X-linked recessive chondrodysplasia punctata; MRX, mental retardation; XLI, X-linked ichthyosis; KAL, X-linked Kallmann syndrome; OA1, ocular albinism type 1. Bold lines indicate presence of a disease phenotype. Numbers beside the bold lines indicate the approximate number of unrelated patients observed in whom the chromosomal rearrangement was demonstrated either by cytogenetic or molecular analysis. Modified from Ballabio and Andria [90].

albinism type 1 (OA1; MIM300500). The phenotype can be predicted in each patient from the results of molecular analysis. A description of the Xp22.3 CGS associated disorders is reported in Table 2.

Table 2. Xp22.3 contiguous gene syndomes

Disease entities	Main features
Short stature (MIM312865)	Short stature
X-linked recessive chondrodysplasia punctata (MIM302950)	Nasal hypoplasia, focal calcification of cartilage, distal phalangeal hypoplasia
X-linked mental retardation (MIM309530)	Mental retardation, usually mild to moderate
X-linked ichthyosis (MIM308100)	Scaly skin, steroid sulfatase deficiency
X-linked Kallmann syndrome (MIM308700)	Hypogonadotropic hypogonadism and anosmia
X-linked ocular albinism (MIM300500)	Impaired visual acuity, nystagmus, strabismus, photophobia

Female patients with terminal Xp deletions, X/Y translocations or X/autosomal translocations typically do not show any of the recessive diseases described in affected males because of the presence of a normal X chromosome. However, they almost invariably show short stature. This observation is consistent with the presence of a gene affecting height in the pseudoautosomal region, two active copies of which are needed for normal growth [25, 38].

A paradigmatic example of a contiguous gene syndrome involving this region is the patient described by Meindl et al., in whom each of the six genetic disorders previously described was observed [39]. Molecular analysis of the patients carrying Xp22 chromosomal abnormalities has allowed us to construct a detailed deletion map of the distal part of the short arm of the X-chromosome and to assign each disease locus to a precise position within the map [40]. This effort allowed us to define the order of the six disease loci constituting the Xp22.3 contiguous gene syndrome: TEL-SS-CDPX-MRX-XLI-KAL-OA1-CEN (Fig. 2) [40].

Short stature (SS)

Growth is a fundamental aspect of the development of an organism. A role for the human sex chromosomes in growth has been suggested and at least two different loci controlling growth have been described [41]. Cytogenetic studies have provided further evidence that terminal deletions of the short arms of

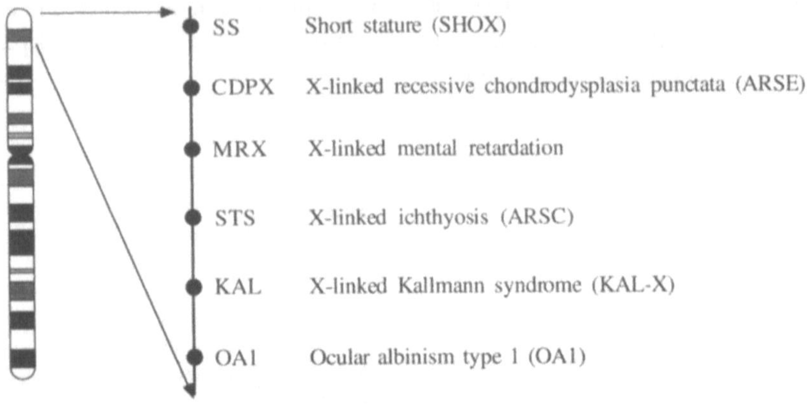

Figure 2. The position of disease loci in the deletion map of the distal short arm of the X chromosome. Deletions result in nullisomy in male individuals who show "recessive" traits. SS, short stature; CDPX, X-linked recessive chondrodysplasia punctata; MRX, mental retardation; XLI, X-linked ichthyosis; KAL, X-linked Kallmann syndrome; OA1, ocular albinism type 1. The genes responsible for these diseases are indicated in parenthesis. MRXis the only Xp22.3 CGS associated disorder for which the gene responsible has not been identified yet.

either the X or the Y chromosome consistently lead to short stature [25, 27, 40, 42]. Physical mapping studies on 19 individual with partial monosomy of the pseudoautosomal region and correlation of the height of those individuals with their deletion breakpoints allowed the definition of a 170 kb critical region [38, 43–45]. This region was subsequently cloned in overlapping cosmid clones which were used for cDNA selection and exon trapping experiments. These gene identification techniques led to the isolation of a homeobox-containing gene (SHOX). A point mutation in SHOX was identified in one out of 91 individuals with idiopathic short stature. This mutation introduces a termination codon at amino acid position 195 of the predicted protein product resulting in a truncated protein deleted for the 3' region which is highly conserved between human and mouse [45]. No additional mutations have been reported to date and further studies will be needed to understand the role of SHOX in normal growth.

Recently the phenotypic spectrum associated with functional haploinsufficiency of SHOX has been expanded to include Leri-Weill Dyschondrosteosis (LWD). LWD is a dominantly inherited skeletal dysplasia characterized by variable short stature with predominantly mesomelic limb shortening. Submicroscopic PAR1 deletions encompassing SHOX and point mutations leading to truncated SHOX proteins have been reported by two different groups [46, 47].

X-linked ichthyosis (XLI)

Steroid sulfatase deficiency is an inborn error of metabolism causing X-linked ichthyosis, a skin disorder inherited as an X-linked trait. The condition affects between 1 in 2000 and 1 in 6000 males in many different populations from a range of geographical locations and racial and ethnic background (see [48] for review). The phenotype of STS deficiency is characterized by the presence of dark scaly skin starting between birth and 4 months of age. In fetal life, placental deficiency of STS causes adiminished estrogen biosynthesis by the maternal-fetal-placental pregnancy unit. This often results in prolonged labor due to difficulty in cervical effacement, which leads to cesarean section in many cases. Patients with STS deficiency have increased levels of cholesterol sulfate in both plasma and stratum corneum. The increase of cholesterol sulfate levels in the stratum corneum appears to be responsible for the ichthyotic changes observed in patients with STS deficiency. A gene necessary for the expression of human STS was assigned by somatic cell hybrids studies, deletion mapping and linkage analysis to the Xpter-Xp22.3 [49–51]. The STS protein was purified from human placenta and polyclonal antibodies were raised in rabbits [52, 53] and used to screen a λgt11 expression library. This effort allowed the identification of the STS gene, which is composed of 10 exons spanning 140 kb of genomic DNA, by different groups [17, 21]. Mutation analysis demonstrated that 85–90% of STS patients have submicroscopic deletions detectable by either Southern blotting or polymerase chain reaction (PCR) analysis [22, 23]. Most of the deletions span the entire STS gene and flanking sequences on Xp22.3. A minority of patients with STS deficiency (10–15%) do not show any abnormality by Southern blotting. In some of these patients point mutations in the coding region of the STS gene have been identified [54].

X-linked recessive chondrodysplasia punctata (CDPX)

The term "chondrodysplasia punctata" refers to a group of skeletal dysplasias characterized by abnormal calcium deposition in regions of enchondral bone formation. This abnormality results in a peculiar radiological finding, commonly referred to as "stippling" of epiphyses or "paint-spattered" calcifications. These calcifications tend to disappear within the first few years of life as bone development progresses. These skeletal dysplasias are genetically heterogeneous and one of them has been localized to the Xp22.3 region by deletion mapping. X-linked recessive chondrodysplasia punctata (CDPX) is characterized by aberrant bone mineralization, severe underdevelopment of nasal cartilage and distal phalangeal hypoplasia. A virtually identical phenotype is observed in the warfarin embryopathy, which is due to the teratogenic effects of coumarin derivatives during pregnancy. The CDPX critical region was defined by the breakpoints of two X/Y translocation male patients only one of which was affected by CDPX. Overlapping yeast artificial chromosome (YAC) clones spanning this entire critical region (600 kb) were isolated and characterized. Groups of cosmids covering the YAC clones were isolated and used for exon trapping experiments. This effort allowed us to isolate three adjacent genes showing highly significant homology to the sulfatase gene family (Fig. 3) [55, 56].

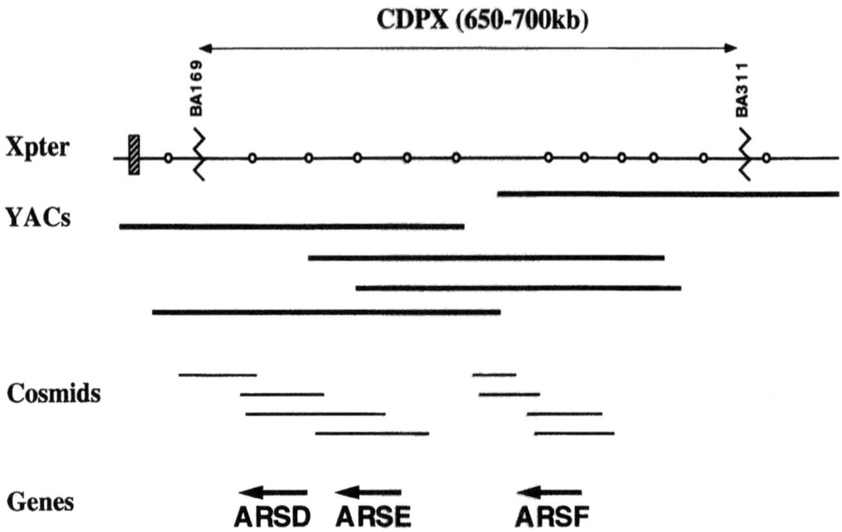

Figure 3. Physical mapping of the CDPX critical region. The CDPX critical region is defined by the breakpoints of patients BA311 and BA169. YAC clones are shown as thick bars. Dots show the position of single copy probes used for YAC identification and for assembly of the YAC contig. Thin bars indicate cosmid clones used for exon amplification experiments. The position and orientation of transcription of the three new sulfatase genes isolated from the region are shown at the bottom. Modified from Franco et al. [55].

The sulfatases are a group of enzymes which hydrolyze sulfate ester bonds in a wide variety of structurally different compounds [48]. Nine different human sulfatases have been described and characterized biochemically. The importance of these proteins in human metabolism is emphasized by the presence of seven distinct inherited disorders resulting from specific sulfatase deficiencies ([48, 57, 58] for reviews). Three of the previously described human sulfatases, known as arylsulfatases A, B and C (ARSA, ARSB and ARSC), are able to hydrolyze sulfated artificial substrates containing a phenolic ring, suchas p-nitrocatechol sulfate or 4-methylumbelliferyl sulfate (4-MU sulfate) [57]. ARSC is more specifically known as steroid sulfatase (STS) because of its ability to hydrolyze steroid sulfates. While ARSA, ARSB and ARSC each act on a distinct natural substrate, they all have a "promiscuous" activity against artificial substrates (arylsulfatase activity).

The newly-identified transcripts showed the highest homology with the arylsulfatase gene family, and were thus named arylsulfatase D (ARSD), arylsulfatase E (ARSE) and arylsulfatase F (ARSF). Transient expression of the three genes in COS cells demonstrated that both ARSE and ARSF have a heat-labile arylsulfatase activity, while ARSD does not hydrolyze sulfated artificial substrate [55, 56]. Point mutations in ARSE were identified in six patients with CDPX, demonstrating that ARSE is the gene responsible for CDPX. Interestingly, expression of ARSE in COS cells resulted in a heat-labile arylsulfatase activity which is inhibited by warfarin and a deficiency of a heat-labile arylsulfatase activity was demonstrated in patients with deletions spanning the CDPX region [55]. On the contrary the ARSF arylsulfatase activity is not inhibited by warfarin[56]. These data indicate that CDPX is caused by an inherited deficiency of ARSE and suggest that warfarin embryopathy might involve drug-induced inhibition of the same enzyme.

Characterization of the genomic structure revealed a conserved genomic organization shared by ARSD, ARSE, ARSF and ARSC, but completely different from that of all the other sulfatase genes [56, 59]. Sequence analysis and pairwise comparisons revealed that ARSD, ARSE, and ARSF are more similar to each other than to the ARSC protein, and more similar toARSC than to the other members of the human sulfatase gene family. Furthermore, ARSD, ARSE and ARSF have the same direction of transcription from centromere to telomere [55, 56]. As reported for other genes from the Xp22.3 region, ARSD and ARSE displayed homology to the Y-chromosome. The Y linked homologs were identified and sequence analysis indicate that they represent truncate pseudogenes. Sequence identity values between the X and Y copies of each gene is on average 91%, significantly higher than the values obtained by comparing different members of the family. Fluorescent *in situ* hybridisation (FISH) mapping experiments performed in several primate species revealed an identical localization of the X-linked copies to that in man, but different localizations of the Y homologs. Together our data indicate that the cluster of sulfatase genes on human Xp22.3 was created through duplication events and support the view that PAR has undergone multiple changes during recent mammalian evolution [59].

Kallmann syndrome (KS)

Kallmann syndrome (KS) is an inherited disorder defined by the association of hypogonadism and inability to smell (anosmia). Families with autosomal dominant, autosomal recessive and X-linked recessive inheritance patterns have been described, indicating genetic heterogeneity of the defect.

The clinical hallmark of the disease is the characteristic association of delayed gonadal development and inability to smell. Endocrinological tests in patients with KS show the presence of hypogonadotropic hypogonadism of hypothalamic origin, due to a deficiency of gonadotropin-releasing hormone (Gn-RH)[60]. Anosmia is due to hypoplasia or aplasia of olfactory bulbs and tracts, as proved by anatomical [61] and neuroimaging studies [62]. In addition to these symptoms, some individuals with Kallmann syndrome display mirror movements, eye movement abnormalities, cerebellar dysfunction, horizontal nystagmus, *pescavus*, unilateral renal agenesis and cleft palate [63–66].

Several lines of evidence indicate that KS is a neuronal migration defect. The first hint came from the observation that Gn-RH secreting neurons and olfactory neurons share a common origin during development [67, 68]. Both Gn-RH and olfactory neurons originate in the olfactory placode, a discrete thickening of the head ectoderm which will later form the olfactory epithelium. From this peripheral location, olfactory neurons project their axons through the cribriform plate and through the meningeal tissue to the olfactory bulb, where they form synapses with dendrites of mitral cells. Olfactory neurons have a half-life in the range of weeks and they are continuously replaced by new neurons that differentiate from a stem cell population present in the olfactory epithelium [69]. Therefore, migration and targeting of olfactory axons are required throughout life. During development, Gn-RH neurons migrate along olfactory nerves and across the olfactory bulb to eventually reach their definitive location in the hypothalamus. Thus, Gn-RH neurons and olfactory axons also share a common migratory pathway.

Both axonal and neuronal migration are dependent on a number of guidance molecules. The hypothesis that one of these molecules might be deficient in KS was therefore formulated and subsequently reinforced by the histopathological findings in a 19-week-old human fetus with X-linked KS. In this fetus, olfactory axons developed normally and started their migration towards the forebrain, but arrested prematurely within the meninges between the cribriform plate and the forebrain [70].

By taking advantage of the availability of several patients with Xp22.3 rearrangements we assigned KS to a specific deletion interval [14, 40]. Positional cloning efforts in this region led to the isolation, by our group and others, of a candidate gene for X-linked KS (KAL-X) (Fig. 4) [71, 72]. This gene was found to be deleted in patients with KS having Xp22.3 contiguous gene syndromes. In addition to deletions, several point mutations in the KAL gene have been identified in patients with isolated KS ([73] and T. Meitinger, unpublished observations). Furthermore, we identified a patient with KS having a 3 kb intragenic deletion within KAL, thus providing final evidence that KAL is the Kallmann syndrome gene [74].

Sequence analysis revealed that KAL encodes a 680 amino acids protein, sharing homology with a number of molecules involved in neuronal migration and axonal pathfinding. The amino-terminus of the protein contains a cysteine-rich domain, referred to as the "four-disulfide core" domain, while a fibronectin type III repeat is present at the carboxy-terminal. Recently, by

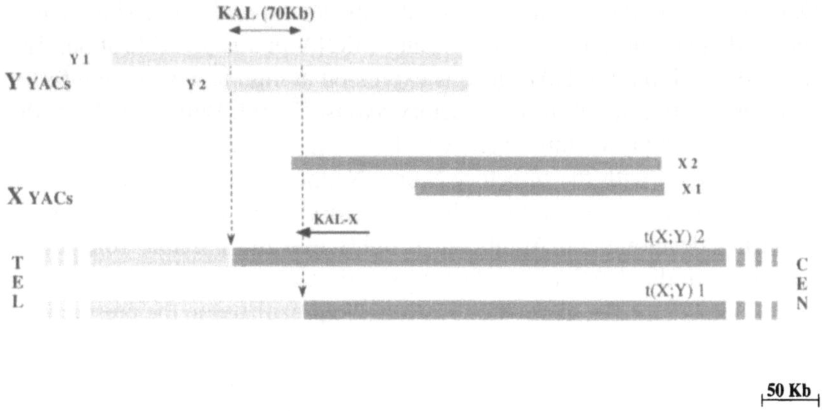

Figure 4. Physical mapping and overlap cloning of the KAL region. The physical map of the four YAC clones and their position with respect to the Kallmann's syndrome gene deletion interval is shown. A different pattern has been used for Y (light) and X (dark) derived sequences. X/Y translocations defining the 70 kb Kallmann's syndrome critical interval are shown at the bottom: t(X;Y)1 is affected by KS, t(X;Y)2 is not affected, respectively. The localization and direction of transcription of the KAL-X gene is also indicated. Modified from Franco et al. [71].

overexpressing both the human and chick KAL cDNAs in eukaryotic cells, we have provided evidence that KAL is a glycosylated peripheral membrane protein with an apparent molecular weight of approximately 100 kDa [75]. We have also shown that this 100 kDa component is proteolytically processed to yield a 45 kDa diffusible component, which is detectable with an antisera against the C-terminal part of the protein and binds tightly to cell surfaces [75]. These data suggest that KAL might be either a substrate molecule or a diffusible chemoattractant for olfactory axons and may be required for target invasion.

Characterization of the KAL gene structure revealed the presence of 14 exons spanning approximately 210 kb on Xp22.3 [9, 76]. The gene was shown to escape X-inactivation (i.e. it is expressed both by the active and by the inactive X) [71] and to have a closely related non-functional homologue on the Y chromosome [9, 76].

We isolated and sequenced the chicken homologue of the KAL gene [77]. The overall protein sequence identity between human and chicken genes is 77%, and ranges between 91% and 94% within the putative functional domains, emphasizing their functional importance.

Low stringency Southern blot analysis and screening of cDNA libraries have so far failed to identify any sequence homologous to the KAL gene in the mouse (A. Ballabio, unpublished observations). This is intriguing, in view of the high degree of homology between the human and chicken genes. This lack of homology with the mouse seems to be a peculiarity of the region of the X chromosome (Xp22.3) in which KAL is located.

Our group and others have analyzed the spatiotemporal expression pattern of the KAL gene in chick embryos using mRNA *in situ* hybridization. The results indicated that the KAL gene is expressed at high levels in the olfactory bulb, the central target of the olfactory axons[77, 78]. Within the bulb, the gene is expressed in the mitral cells which are the secondary sensory neurons forming synapses with the olfactory axons. Neither the olfactory epithelium, nor the meningeal tissue through which olfactory axons migrate, express the gene during development. Additional insights into the possible role of the KAL protein in the developing olfactory system were provided by the timing of expression [77]. KAL expression is strongly upregulated in the presumptive olfactory bulb region at day 7, whereas expression is not detectable above background between day 4 and 6.5, which corresponds to the time when the first contacts between the olfactory axons and the forebrain are actually established. These findings suggest that the KAL protein is not required for the initial migration of olfactory axons toward the bulb, but rather for later processes. This is in agreement with the neuropathological analysis in a KS fetus which showed some evidence of normal migration of olfactory axons into the forebrain [70]. Moreover, elegant experiments have demonstrated that the olfactory axons retain the ability to migrate toward the forebrain after removal of the olfactory bulb, indicating that the projections of these neurons to the brain are independent from the target, and therefore unlikely to depend on a molecule secreted by the target [69]. The expression studies in the chick have also revealed other multiple sites of expression of the gene during development, both within and outside the nervous system. Prominent expression has been found in the Purkinje cells of the cerebellum. This pattern of expression could be related to the occurrence of cerebellar symptoms in KS patients.

All these data have contributed to the development of the model represented in Figure 5. We postulate that the KAL protein is a substrate adhesion molecule mediating interactions between dendrites of mitral cells and olfactory axons within the olfactory bulb. In the absence of KAL, these interactions could not be established and this would cause a regression of the olfactory bulb after an initial phase of normal development. The migration defect of Gn-RH neurons would be indirect, caused by the lack of contacts between the olfactory nerves and the forebrain and therefore by the lack of a migration route. The

---

Figure 5. (See color plate XI between pages 256 and 257.) Model for KAL function and KS pathogenesis. In normal individuals, axons of olfactory neurons (ON) traverse the cribriform plate (CP) to reach the olfactory bulb. Within the glomerular layer of the bulb (GL), they make synapses with dendrites of mitral cells (M), whose axons will form the olfactory tracts (OT). A tentative model is proposed in which the KAL protein (green area) is secreted by the mitral cells and is required in the glomerular layer for the establishment/maintenance of proper interactions with olfactory axons. In KS, KAL protein is absent; therefore, olfactory axons cannot interact properly with their target, ending their migration between the cribriform plate and the forebrain. The migration defect of GnRH neurons in KS would be a secondary effect caused by lack of contact between olfactory nerves and forebrain, resulting in the absence of a "migration route". From Rugarli and Ballabio [90]. (Used by permission of JAMA).

genes involved in the autosomal types of Kallmann syndrome may encode additional factors participating in this fascinating developmental system.

Ocular albinism type 1 (OA1)

Ocular albinism of the Nettleship-Falls type (OA1; MIM 300500) is one of the most common forms of ocular albinism (http://www.Tigem.Tt/LOCAL/OMIM/omim.Ttml). This disorder is transmitted as an X-linked recessive trait, with affected males showing the complete phenotype and heterozygous females showing only minor signs of disease [79]. Affected males with OA1 have an extremely severe reduction in visual acuity, which represents a major handicap. They also manifest horizontal and rotary nystagmus, strabismus, and marked photophobia. Ophthalmological examination reveals foveal hypoplasia, hypopigmentation of the retina, and iris translucency [80–82].

Cutaneous changes are minimal in OA1, and mild hypopigmentation of the skin is found in rare cases. Microscopic examination of melanocytes, in both retinal pigment epithelium and in the skin, reveals the presence of large pigment granules called giant melanosomes or macromelanosomes [80, 83–85]. This finding indicates that the underlying defect in OA1 is an abnormality in melanosome formation. The finding of macromelanosomes also helps distinguish between OA1 and other types of albinism.

The OA1 locus has been assigned to the distal short arm of the X chromosome (Xp22.3) by linkage analysis. Additional evidence on the location of the OA1 locus came from the analysis of patients with chromosomal rearrangements involving the distal short arm of the X chromosome [40]. Moreover, a 700 kb submicroscopic deletion encompassing the OA1 critical region was identified in a patient with isolated OA1 [86].

Molecular analysis of these chromosomal rearrangements enabled us to refine the OA1 critical region to approximately 110 kb of genomic DNA. Subsequently, we assembled a cosmid contig spanning the entire OA1 critical region using a conventional cosmid walking strategy. The minimum number of cosmid clones spanning the entire OA1 critical region were used in strategies aimed at the identification of transcribed sequences from the region. Two novel genes representing strong candidates for OA1 were identified in the region by exon amplification experiments (Fig. 6). The first gene we isolated encodes a 1616 amino acid protein sharing significant similarities with the Apical Protein from *Xenopus laevis* (APX), which is implicated in amiloride-sensitive sodium channel activity. The transcript of this newly identified gene, named APXL (for Apical Protein *Xenopus* Like), measures approximately 7.5 kb and was found to be abundant in retina and melanoma and was also detected in brain, placenta, lung, kidney, and pancreas. The APXL gene covers about 160 kb of genomic DNA and is organized in ten exons, being distributed over 80 of the 110 kb of the OA1 critical region. Although mapping, expression, and patient

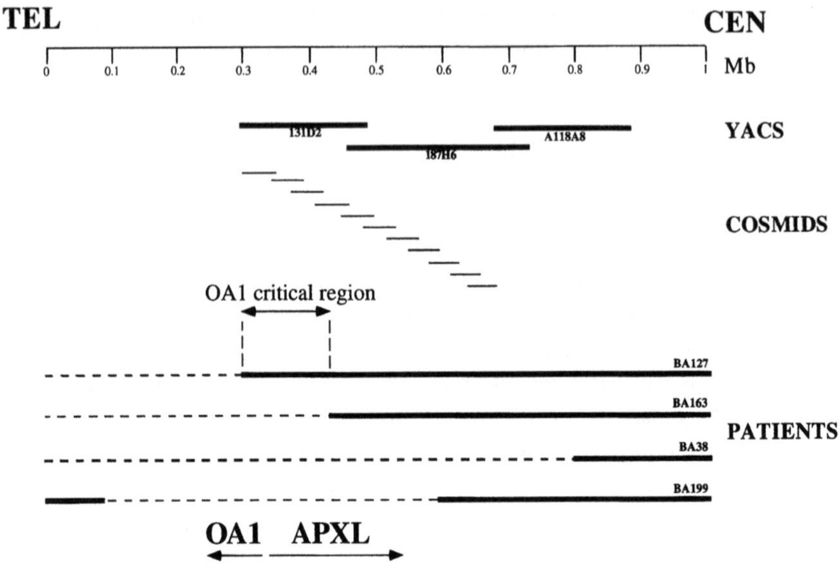

Figure 6. Map of the OA1 critical region. YAC and cosmid contigs spanning the OA1 critical region are shown at the top. A schematic representation of the deletions in patients defining the OA1 critical region is shown at the bottom. Patient BA127 does not have OA1; patients BA163 and BA38 have OA1 as part of the Xp22.3 contiguous gene syndrome; and patient BA199 has isolated OA1. The OA1 critical region is indicated, as well as the position and orientation of the OA1 and APXL transcripts. Modified from Schiaffino et al. [87].

analysis data led us to consider the APXL gene a strong candidate for OA1, mutation analysis of the coding region in 57 unrelated patients compelled us to exclude a direct involvement of this gene in OA1 [87].

Five small intragenic deletions and a CG insertion in exon 8, were found in patients with OA1 in the second transcript isolated from the critical region, providing formal genetic evidence that mutations in this gene, consequently named OA1, cause the disease [88]. The OA1 transcript measures about 1.6 kb and was readily detected by Northern blot exclusively in RNA preparations from pigment cells, such as melanocytes, melanoma, and retinal pigment epithelium. The OA1 predicted protein product is 404 aminoacids long and displays several putative transmembrane domains, sharing no significant similarities with any previously identified molecules [88].

To characterize the OA1 gene protein product, polyclonal antibodies against a recombinant protein expressed in *E. coli* were generated. The use of these antibodies allowed us to identify the endogenous OA1 protein in retinal pigment epithelial cells, in normal human melanocytes, and in various melanoma cell lines [89]. Upon subcellular fractionation and phase separation with the non-ionic detergent TritonX-114, the OA1 protein segregated into the melanosome-rich fraction (LargeGranule Fraction, LGF) and behaved as an authentic integral membraneprotein. Immunofluorescence and immunogold analyses on normal human melanocytes confirmed the melanosomal membrane localization of the endogenous OA1 protein, consistent with its possible involvement in melanosome biogenesis [89].

Conclusions

The availability of a large number of patients with complex phenotypes due to various types of chromosomal rearrangements made the distal short arm of the human X-chromosome an attractive region for positional cloning of disease genes. A close interaction between the clinicians, the cytogeneticist and the molecular biologist was essential for the isolation of the genes responsible for short stature [45], X-linked recessive chondrodysplasia punctata [55], Kallmann syndrome [71], and ocular albinism[88] which are part of the Xp22.3 contiguous gene syndrome.

The successful cloning of these disease genes was the first step towards the understanding of the pathogenetic mechanism underlying these genetic disorders and could possibly provide new diagnostic and therapeutic tools.

Acknowledgements

We wish to thank all the people who contributed to the data described in this paper.

References

1 Ellis N, Goodfellow PN (1989) The mammalian pseudoautosomal region. *Trends Genet* 5: 406–410
2 Rappold GA (1993) The pseudoautosomal regions of the human sex chromosomes. *Hum Genet* 92: 315–324
3 Mohandas TK, Speed RM, Passage MB, Yen PH, Chandley AC, Shapiro LJ (1992) Role of the pseudoautosomal region in sex-chromosome pairing during male meiosis: meiotic studies in a man with a deletion of distal Xp. *Amer J Hum Genet* 51: 526–533
4 Rouyer F, Simmler MC, Johnsson C, Vergnaud G, Cooke HJ, Weissenbach J (1986) A gradient of sex linkage in the pseudoautosomal region of the human sex chromosomes. *Nature* 319: 291–295
5 Watson JM, Spencer JA, Riggs AD, Marshall Graves JA (1990) The X chromosome of monotremes shares a highly conserved region with the eutherian and marsupial X chromosomes despite the absence of X chromosome inactivation. *Proc Natl Acad Sci USA* 87: 7125–7129
6 Watson JM, Spencer JA, Riggs AD, Marshall Graves JA (1991) Sex chromosome evolution: platypus gene mapping suggests that part of the human X chromosome was originally autosomal. *Proc Natl Acad Sci USA* 88: 11256–11260
7 Yen PH, Marsh B, Allen E, Tsai SP, Ellison J, Connolly L, Neiswanger K, Shapiro LJ (1988) The human X-linked steroid sulfatase gene and a Y-encoded pseudogene: Evidence for an inversion of the Y chromosome during primate evolution. *Cell* 55: 1123–1135
8 Fraser N, Ballabio A, Zollo M, Persico G, Craig I (1987) Identification of incomplete coding sequences for steroid sulphatase on the human Y chromosome: evidence for an ancestral pseudoautosomal gene? *Development* 101 Suppl: 127–132

9 Incerti B, Guioli S, Pragliola A, Zanaria E, Borsani G, Tonlorenzi R, Bardoni B, Franco B, Wheeler D, Ballabio A et al (1992) Kallmann syndrome gene on the X and Y chromosomes: implications for evolutionary divergence of human sex chromosomes. *Nat Genet* 2: 311–314
10 Ohno S (1972) An argument for the genetic simplicity of man and other mammals. *J Hum Evol* 1: 651–662
11 Disteche CM, Brannan CI, Larsen A, Adler DA, Schorderet DF, Gearing D, Copeland NG, Jenkins NA, Park LS (1992) The human pseudoautosomal GM-CSF receptor α subunit gene is autosomal in mouse. *Nat Genet* 1: 333–336
12 Milatovich A, Kitamura T, Miyajima A, Francke U (1993) Gene for the α-subunit of the human interleukin-3 receptor (1L3RA) localized to the X-Y pseudoautosomal region. *Amer J Hum Genet* 53: 1146–1153
13 Salido EC, Li XM, Yen PH, Martin N, Mohandas TK, Shapiro LJ (1996) Cloning and expression of the mouse pseudo autosomal steroid sulphatase gene *(Sts)*. *Nat Genet* 13: 83–86
14 Ballabio A, Andria G (1992) Deletions and translocations involving the distal short arm of the human X chromosome: review and hypotheses. *Hum Mol Genet* 1: 221–227
15 Ferrero GB, Franco B, Roth EJ, Firulli BA, Borsani G, Delmas-Mata J, Weissenbach J, Halley G, Schlessinger D, Chinault AC et al (1995) An integrated physical and genetic map of a 35 Mb region on chromosome Xp22.3-Xp21.3. *Hum Mol Genet* 4: 1821–1827
16 Yen PH, Li X-M, Tsai S-P, Johnson C, Mohandas T, Shapiro LJ (1990) Frequent deletions of the human X chromosome distal short arm result from recombination between low copy repetitive elements. *Cell* 61: 603–610
17 Ballabio A, Parenti G, Carrozzo R, Sebastio G, Andria G, Buckle V, Fraser N, Craig I, Rocchi M, Romeo G et al (1987) Isolation and characterization of a steroid sulphatase cDNA clone: Genomic deletions in patients with X-chromosome-linked ichthyosis. *Proc Natl Acad Sci USA* 84: 4519–4523
18 Bonifas JM, Morley BJ, Oakey RE, Kan YW, Epstein EHJ (1987) Cloning of a cDNA for steroid sulfatase: Frequent occurrence of gene deletions in patients with recessive X chromosome-linked ichthyosis. *Proc Natl Acad Sci USA* 84: 9248–9251
19 Conary J, Lorkowski G, Schmidt B, Pohlmann R, Nagel G, Meyer HE, Krentler C, Cully J, Hasilik A, von Figura K (1987) Genetic heterogeneity of steroid sulfatase deficiency revealed with cDNA for human steroid sulfatase. *Biochem Biophys Res Commun* 144: 1010–1017
20 Gillard EF, Affara NA, Yates JRW, Goudie DR, Lambert J, Aitken DA, Ferguson-Smith MA (1987) Deletion of a DNA sequence in eight of nine families with X-linked ichthyosis (steroid sulfatase deficiency). *Nucl Acid Res* 15: 3977–3985
21 Yen PH, Allen E, Marsh B, Mohandas T, Wang N, Taggart RT, Shapiro LJ (1987) Cloning and expression of steroid sulfatase cDNA and the frequent occurrence of deletions in STS deficiency: Implication for X-Y interchange. *Cell* 49: 443–454
22 Ballabio A, Carrozzo R, Parenti G, Gil A, Zollo M, Persico MG, Gillard E, Affara N, Yates J, Ferguson-Smith MA et al (1989) Molecular heterogeneity of steroid sulfatase deficiency: a multicenter study on 57 unrelated patients, at DNA and protein levels. *Genomics* 4: 36–40
23 Shapiro LJ, Yen P, Pomerantz D, Martin E, Rolewic L, Mohandas T (1989) Molecular studies of deletions at the human steroid sulfatase locus. *Proc Natl Acad Sci USA* 86: 8477–8481
24 Schmickel RD (1986) Contiguous gene syndromes: a component of recognizable syndromes. *J Pediat* 109: 231–241
25 Ballabio A, Bardoni B, Carrozzo R, Andria G, Bick D, Campbell L, Hamel B, Ferguson-Smith MA, Gimelli G, Fraccaro M et al (1989) Contiguous gene syndromes due to deletions in the distal short arm of the human X chromosome. *Proc Natl Acad Sci USA* 86: 10001–10005
26 Ballabio A (1991) Contiguous deletion syndromes. *Curr Opin Genet Develop* 1: 25–29
27 Curry CJR, Magenis E, Brown M, Lanman JT Jr, Tsai J, O'Lague P, Goodfellow P, Mohandas T, Bergner EA, Shapiro LJ (1984) Inherited chondrodysplasia punctata due to a deletion of the terminal short arm of an X chromosome. *N Engl J Med* 311: 1010–1014
28 Bick D, Curry CJ, McGill JR, Schorderet DF, Bux RC, Moore CM (1989) Male infant with ichthyosis, Kallmann syndrome, chondrodysplasia punctata, and an Xp chromosome deletion. *Amer J Med Genet* 33: 100–107
29 Temple IK, Hurst JA, Hing S, Butler L, Baraitser M (1990) De novo deletion of Xp22.2-pter in a female with linear skin lesions of the face and neck, microphthalmia, and anterior chamber eye anomalies. *J Med Genet* 27: 56–58
30 Allanson J, Richter S (1991) Linear skin defects and congenital microphthalmia: a new syndrome

at Xp22.2. *J Med Genet* 28: 143–144
31 Ballabio A, Zollo M, Carrozzo R, Caiulo A, Zuffardi O, Cascioli CF, Viggiano D, Strisciuglio P (1991) Deletion of the distal short arm of an X chromosome (Xp) in a patient with short stature, chondrodysplasia punctata, and X-linked ichthyosis due to steroid sulfatase deficiency. *Amer J Med Genet* 41: 184–187
32 Thies U, Gopal Rao VVN, Engel W, Schmidtke J (1990) Physical mapping of two Xp markers DXS16 and DXS143. *Hum Genet* 86: 418–420
33 Naritomi K, Izumikawa Y, Nagataki S, Fukushima Y, Wakui K, Niikawa N, Hirayama K (1992) Combined Goltz and Aicardi syndromes in a terminal Xp deletion: are they a contiguous gene syndrome? *Amer J Med Genet* 43: 839–843
34 Ropers H-H, Zuffardi O, Bianchi E, Tiepolo L (1982) Agenesis of the corpus callosum, ocular, and skeletal anomalies (X-linked dominant Aicardi's syndrome) in a girl with balanced X/3 translocation. *Hum Genet* 61: 364–368
35 Donnenfeld AE, Coyne MD, Beauregard LJ (1990) Microphthalmia and chorioretinal lesions in a girl with an Xp22.2-pter deletion and partial 3p trisomy: clinical observations relevant to Aicardi syndrome gene localization. *Amer J Med Genet* 37: 182–186
36 Lindor NM, Michels VV, Hoppe DA, Driscoll DJ, Leavitt JA (1992) Xp22.3 microdeletion syndrome with microphthalmia, sclerocornea, linear skin defects, and congenital heart defects. *Amer J Med Genet* 44: 61–65
37 McKusick VA (1995) Catalogs of autosomal dominant, autosomal recessive, and X-linked phenotypes. *In*: McKusick (ed.): *Mendelian Inheritance in Man*. Johns Hopkins University Press, Baltimore, MD
38 Henke A, Wapenaar M, van Ommen GJ, Maraschio P, Camerino G, Rappold G (1991) Deletions within the pseudoautosomal region help map three new markers and indicate a possible role of this region in linear growth. *Amer J Hum Genet* 49: 811–819
39 Meindl A, Hosenfeld D, Brückl W, Schuffenhauer S, Jenderny J, Bacskulin A, Oppermann H-C, Bouloux P, Meitinger T (1993) Analysis of a terminal Xp22.3 deletion in a patient with six monogenic disorders: implications for the mapping of X-linked ocular albinism. *J Med Genet* 30: 838–842
40 Schaefer L, Ferrero GB, Grillo A, Bassi MT, Roth EJ, Wapenaar MC, van Ommen GJB, Mohandas TK, Rocchi M, Zoghbi HY et al (1993) A high resolution deletion map of human chromosome Xp22. *Nat Genet* 4: 272–279
41 Ogata T, Matsuo N (1993) Sex chromosome aberrations and stature: deduction of the principal factors involved in the determination of adult height. *Hum Genet* 91: 551–562
42 Zuffardi O, Maraschio P, Lo Curto F, Müller U, Giarola A, Perotti L (1982) The role of Yp in sex determination: new evidence from X/Y translocations. *Amer J Med Genet* 12: 175–184
43 Ogata T, Goodfellow P, Petit C, Aya M, Matsuo N (1992) Short stature in a girl with a terminal Xp deletion to *DXYS15*: Localisation of a growth gene(s) in the pseudoautosomal region. *J Med Genet* 29: 455–459
44 Ogata T, Yoshizawa A, Muroya K, Matsuo N, Fukushima Y, Rappold G, Yokoya S (1995) Short stature in a girl with partial monosomy of the pseudoautosomal region distal to DXYS15: further evidence for the assignment of the critical region for a pseudoautosomal growth gene(s). *J Med Genet* 32: 831–834
45 Rao E, Weiss B, Fukami M, Rump A, Niesler B, Mertz A, Muroya K, Binder G, Kirsch S, Winkelmann M et al (1997) Pseudoautosomal deletions encompassing a novel homeobox gene cause growth failure in idiopathic short stature and Turner syndrome. *Nat Genet* 16: 54–63
46 Shears DJ, Vassal HJ, Goodman FR, Palmer RW, Reardon W, Superti-Furga A, Scambler PJ, Winter RM (1998) Mutation and deletion of the pseudoautosomal gene SHOX cause Leri-Weill dys chondrosteosis. *Nat Genet* 19: 70–73
47 Belin V, Cusin V, Viot G, Girlich D, Toutain A, Moncla A, Vekemans M, Le Merrer M, Munnich A, Cormier-Daire V (1998) SHOX mutations in dyschondrosteosis (Leri-Weill syndrome). *Nat Genet* 19: 67–69
48 Ballabio A, Shapiro LJ (1995) STS deficiency and X-linked ichthyosis. *In*: CR Scriver, AL Beaudet, WS Sly, D Valle (eds): *The Metabolic and Molecular Bases of Inherited Disease*. McGraw-Hill, New York, Vol. II, 2999–3022
49 Mohandas T, Shapiro LJ, Sparkes RS, Sparkes MC (1979) Regional assignment of the steroid sulfatase-X-linked ichthyosis locus: Implications for a non-inactivated region on the short arm of human X chromosome. *Proc Natl Acad Sci USA* 76: 5779–5783

50 Muller CR, Westerveld A, Migl B, Franke W, Ropers HH (1980) Regional assignment of the gene locus for steroid sulfatase. *Hum Genet* 54: 201–204
51 Tiepolo L, Zuffardi O, Fraccaro M, di Natale D, Gargantini L, Müller CR, Ropers H-H (1980) Assignment by deletion mapping of the steroid sulfatase X-linked ichthyosis locus to Xp22.3. *Hum Genet* 54: 205–206
52 van der Loos CM, van Breda AJ, van den Berg FM, Walboomers JMM, Jöbsis AC (1984) Human placental steroid sulphatase – purification and monospecific antibody production in rabbits. *J Inherit Metab Dis* 7: 97–103
53 Epstein EH, Bonifas JM (1985) Recessive X-linked ichthyosis: lack of immunologically detectable steroid sulfatase enzyme protein. *Hum Genet* 71: 201–205
54 Basler E, Grompe M, Parenti G, Yates J, Ballabio A (1992) Identification of point mutations in the steroid sulfatase gene of three patients with X-linked ichthyosis. *Amer J Hum Genet* 50: 483–491
55 Franco B, Meroni G, Parenti G, Levilliers J, Bernard L, Gebbia M, Cox L, Maroteaux P, Sheffield L, Rappold GA et al (1995) A cluster of sulfatase genes on Xp22.3: mutations in chondrodysplasia punctata (CDPX) and implications for Warfarin embryopathy. *Cell* 81: 15–25
56 Puca A, Zollo M, Repetto M, Andolfi G, Guffanti A, Simon G, Ballabio A, Franco B (1997) Identification by shotgun sequencing, genomic organization, and functional analysis of a fourth arylsulphatase gene (ARSF) from the Xp22.3 region. *Genomics* 42: 192–199
57 Kolodny EH, Fluharty AL (1995) Metachromatic leukodystrophy and multiple sulfatase deficiency: sulfatide lipidosis. *In*: CR Scriver, AL Beaudet, WS Sly, D Valle (eds): *The Metabolic and Molecular Bases of Inherited Disease*. McGraw-Hill, New York, Vol. I, 2693–2741
58 Neufeld EF, Muenzer J (1995) The mucopolysaccharidoses. *In*: CR Scriver, AL Beaudet, WS Sly, D Valle (eds): *The Metabolic and Molecular Bases of Inherited Disease*. McGraw-Hill, New York, vol. I, 2465–2494
59 Meroni G, Franco B, Archidiacono N, Messali S, Andolfi G, Rocchi M, Ballabio A (1996) Characterization of a cluster of sulfatase genes on Xp22.3 suggests gene duplications in an ancestral pseudoautosomal region. *Hum Mol Genet* 5: 423–431
60 Naftolin F, Harris GW, Bobrow M (1971) Effect of purified luteinizing hormone releasing factor on normal and hypogonadotropic anosmic men. *Nature* 232: 496–497
61 de Morsier G (1954) Etudes sur les dysraphies cranio-encephaliques. 1. Agenesie des lobes olfactifs (telencephaloschizis lateral) et des commissures calleuse et anterieure (telencephaloschizis median). La dysplasie olfacto-genitale. *Schweiz Arch Neurol Neurochir Psychiat* 74: 309–361
62 Truwit CL, Barkovich AJ, Grumbach MM, Martini JJ (1993) MR imaging of Kallmann syndrome, a genetic disorder of neuronal migration affecting the olfactory and genital systems. *Amer J Neuroradiol* 14: 827–838
63 Sunohara N, Sakuragawa N, Satoyoshi E, Tanae A, Shapiro LJ (1986) A new syndrome of anosmia, ichthyosis, hypogonadism, and various neurological manifestations with deficiency of steroid sulfatase and arylsulfatase C. *Ann Neurol* 19: 174–181
64 Schwankhaus JD, Currie J, Jaffe MJ, Rose SR, Sherins RJ (1989) Neurologic findings in men with isolated hypogonadotropic hypogonadism. *Neurology* 39: 223–226
65 Wegenke JD, Uehling DT, Wear JB Jr, Gordon ES, Bargman JG, Deacon JSR, Herrmann JPR, Opitz JM (1975) Familial Kallmann syndrome with unilateral renal aplasia. *Clin Genet* 7: 368–381
66 White BJ, Rogol AD, Brown SK, Lieblich JM, Rosen SW (1983) The syndrome of anosmia with hypogonadotropic hypogonadism: a genetic study of 18 new families and a review. *Amer J Med Genet* 15: 417–435
67 Schwanzel-Fukuda M, Pfaff DW (1989) Origin of luteinizing hormone-releasing hormone neurons. *Nature* 338: 161–164
68 Wray S, Grant P, Gainer H (1989) Evidence that cells expressing luteinizing hormone-releasing hormone mRNA in the mouse are derived from progenitor cells in the olfactory placode. *Proc Natl Acad Sci USA* 86: 8132–8136
69 Monti Graziadei GA, Graziadei PPC (1979) Studies on neuronal plasticity and regeneration in the olfactory system: morphologic and functional characteristics of the olfactory sensory neuron. *In*: E Meisami, MAB Brazier (eds): *Neural Growth and Differentiation*. Raven Press, New York, 373–396
70 Schwanzel-Fukuda M, Bick D, Pfaff DW (1989) Luteinizing hormone-releasing hormone (LHRH)-expressing cells do not migrate normally in an inherited hypogonadal (Kallmann) syndrome. *Mol Brain Res* 6: 311–326

71 Franco B, Guioli S, Pragliola A, Incerti B, Bardoni B, Tonlorenzi R, Carrozzo R, Maestrini E, Pieretti M, Taillon-Miller P et al (1991) A gene deleted in Kallmann's syndrome shares homology with neural cell adhesion and axonal path-finding molecules. *Nature* 353: 529–536
72 Legouis R, Hardelin J-P, Levilliers J, Claverie J-M, Compain S, Wunderle V, Millasseau P, Le Paslier D, Cohen D, Caterina D et al (1991) The candidate gene for the X-linked Kallmann syndrome encodes a protein related to adhesion molecules. *Cell* 67: 423–435
73 Hardelin J-P, Levilliers J, Blanchard S, Carel J-C, Leutenegger M, Pinard-Bertelletto J-P, Bouloux P, Petit C (1993) Heterogeneity in the mutations responsible for X chromosome-linked Kallmann syndrome. *Hum Mol Genet* 2: 373–377
74 Bick D, Franco B, Sherins RJ, Heye B, Pike L, Crawford J, Maddalena A, Incerti B, Pragliola A, Meitinger T et al (1992) Intragenic deletion of the *KALIG-1* gene in Kallmann's syndrome. *N Engl J Med* 326: 1752–1755
75 Rugarli EI, Ghezzi C, Valsecchi V, Ballabio A (1996) The Kallmann syndrome gene product expressed in COS cells is cleaved on the cell surface to yield a diffusible component. *Hum Mol Genet* 5: 1109–1115
76 del Castillo I, Cohen-Salmon M, Blanchard S, Lutfalla G, Petit C (1992) Structure of the X-linked Kallmann syndrome gene and ist homologous pseudogene on the Y chromosome. *Nat Genet* 2: 305–310
77 Rugarli EI, Lutz B, Kuratani SC, Wawersik S, Borsani G, Ballabio A, Eichele G (1993) Expression pattern of the Kallmann syndrome gene in the olfactory system suggests a role in neuronal targeting. *Nat Genet* 4: 19–25
78 Legouis R, Ayer-Le Lievre C, Leibovici M, Lapointe F, Petit C (1993) Expression of the KAL gene in multiple neuronal sites during chicken development. *Proc Natl Acad Sci USA* 90: 2461–2465
79 Lang GE, Rott H-D, Pfeiffer RA (1990) X-linked ocular albinism. Characteristic pattern of affection in female carriers. *Ophthalmic Paediat Genet* 11: 265–271
80 O'Donnell FE Jr, Hambrick GW Jr, Green WR, Iliff WJ, Stone DL (1976) X-linked ocular albinism: an oculocutaneous macromelanosomal disorder. *Arch Ophthalmol* 94: 1883–1892
81 Witkop CJ Jr, Quevedo WC Jr, Fitzpatrick TB, King RA (1989) Albinism. *In*: CR Scriver, AL Beaudet, WS Sly, D Valle (eds): *The Metabolic Basis of Inherited Disease*. 2905–2947
82 King RA, Summers CG (1988) Albinism. *Dermatol Clinics* 6: 217–228
83 Garner A, Jay BS (1980) Macromelanosomes in X-linked ocular albinism. *Histopathology* 4: 243–254
84 Wong L, O'Donnell FE Jr, Green WR (1983) Giant pigment granules in the retinal pigment epithelium of a fetus with X-linked ocular albinism. *Ophthalmic Paediat Genet* 2: 47–65
85 Yoshiike T, Manabe M, Hayakawa M, Ogawa H (1985) Macromelanosomes in X-linked ocular albinism (XLOA). *Acta Dermatol Venereol (Stockh)* 65: 66–69
86 Bassi MT, Bergen AAB, Wapenaar MC, Schiaffino MV, vanSchooneveld M, Yates JRW, Charles SJ, Meitinger T, Ballabio A (1994) A submicroscopic deletion in a patient with isolated X-linked ocular albinism (OA1). *Hum Mol Genet* 3: 647–648
87 Schiaffino MV, Bassi MT, Rugarli EI, Renieri A, Galli L, Ballabio A (1995) Cloning of a human homologue of the *Xenopus laevis* APX gene from the ocular albinism type 1 critical region. *Hum Mol Genet* 4: 373–382
88 Bassi MT, Schiaffino MV, Renieri A, De Nigris F, Galli L, Bruttini M, Gebbia M, Bergen AAB, Lewis RA, Ballabio A (1995) Cloning of the gene for ocular albinism type 1 from the distal short arm of the X chromosome. *Nat Genet* 10: 13–19
89 Schiaffino MV, Baschirotto C, Pellegrini G, Montalti S, Tacchetti C, De Luca M, Ballabio A (1996) The Ocular Albinism type1 (OA1) gene product is a membrane glycoprotein localized to melanosomes. *Proc Natl Acad Sci USA* 93: 9055–9060
90 Rugarli EI, Ballabio A (1993) Kallmann syndrome: from genetics to neurobiology. *JAMA* 270: 2713–2716

Chromosomes Today, Vol. 13
ed. by E. Olmo and C.A. Redi

Genetic and cytogenetic studies in inherited cancer: Li-Fraumeni syndrome

Jennifer M. Varley[1], John M. Boyle[1], Louise A. James[1], Gail McGown[1], Mary Thorncroft[1], Martin Greaves[1], Anna M. Kelsey[2], D. Gareth R. Evans[3] and Jillian M. Birch[4]

[1] *CRC Section of Molecular Genetics, Paterson Institute for Cancer Research, Wilmslow Road, Manchester M20 4BX, UK*
[2] *Department of Histopathology, Royal Manchester Children's Hospital, Pendlebury, Manchester M27 4HA, UK*
[3] *Department of Medical Genetics, St Mary's Hospital, Hathersage Road, Manchester M13 0JH, UK*
[4] *CRC Paediatric and Familial Research Group, Royal Manchester Children's Hospital, Pendlebury, Manchester M27 1HA, UK*

Li-Fraumeni syndrome (LFS)

In 1969, an inherited cancer predisposition syndrome was proposed by Li and Fraumeni on the basis of four families in which there were at least two cases of soft tissue sarcomas in early life. Other cancers noted at an increased frequency in these families were premenopausal breast cancer and other sarcomas [1], and the same group [2] subsequently further defined the syndrome as a proband with sarcoma diagnosed under the age of 45 years, with a first-degree relative with any cancer under 45, plus another first- or second-degree relative with either any cancer under 45 years or a sarcoma at any age. In addition to sarcomas and premenopausal breast cancer, an excess of brain tumours, adrenocortical tumours and leukaemias was noted [2]. Birch et al. [3] described additional families which did not conform precisely to the criteria of classic LFS, and which were termed Li-Fraumeni like (LFL). LFL families were defined on the basis of a proband with any childhood cancer or sarcoma, brain tumour or adrenocortical tumour diagnosed under the age of 45 with one first- or second-degree relative with a typical LFS cancer at any age, plus a first- or second-degree relative in the same lineage with any cancer under the age of 60. These clinical criteria have become well established, with component tumours comprising predominantly bone and soft-tissue sarcomas, premenopausal breast cancer, brain tumours, adrenocortical tumours and, less frequently, gastric and oesophageal tumours and pancreatic carcinoma (for review see [4]). All these tumours occur at strikingly young ages.

The causative gene – *TP53*

Families conforming to the strict clinical definitions given above are rare. In addition, there is a high mortality precluding formal genetic linkage studies. No constitutional chromosomal abnormality has been identified in any family with LFS. Taken together these factors complicated a straightforward identification of the causative gene. Malkin et al. [5] therefore adopted a candidate gene approach whereby they evaluated genes that could be implicated in LFS for mutation in the germline of affected families. The *TP53* gene was one of the most obvious candidates in that somatic mutations were the most commonly reported genetic lesions in human malignancy, including those tumours that comprised LFS. Furthermore, studies with a transgenic mouse model had shown that a germline *TP53* mutation in the mouse was associated with an elevated incidence of sarcomas, lymphoid malignancies and lung tumours [6]. In their original study [5] all five LFS families examined were reported to have germline mutations, and all the mutations were clustered in one small region of the gene. The *TP53* gene maps to human chromosome band 17p13.1, it spans around 20 kilobase pairs (kb) of DNA, and comprises eleven exons, the first of which is non-coding.

The p53 protein functions as a transcriptional activator or repressor via sequence-specific DNA-binding upstream of target genes. The DNA-binding domain of the protein is encoded largely by exons 5-8, and the majority of mutations in sporadic tumours have been found within these exons [7]. There are 5 highly conserved domains within the *TP53* gene [8], four of which (domains II-V) reside within the DNA-binding region [9]. The fifth conserved domain (domain I) is within the N-terminal region which can transactivate target genes when the protein is interacting with DNA through the central sequence-specific binding region [10]. Sequences within the C-terminus are responsible for tetramerisation of the protein [11, 12]. p53 molecules which are unable to tetramerise are essentially non-functional [13]. As p53 functions as a tetramer, there is the potential for mutant and wild-type p53 monomers to interact if they are expressed in the same cell, and indeed some p53 mutants have demonstrated such a dominant-negative phenotype by blocking the normal function of wild-type p53 [14]. Other mutations can show a gain of function, such as the ability to cooperate with *ras* in the transformation of fibroblasts [15, 16]. Finally, mutations may result in a completely non-functional protein. Such mutations would include nonsense and frameshift mutations, insertions or deletions. These types of mutation and their clinical consequences will be discussed in more detail below.

TP53 mutations in Lf families

There are now over 150 reports of germline *TP53* mutations in the literature, around half of these in families conforming to the definition of LFS or LFL and others in patients presenting with tumours which are typical of the syndrome (see [4] for review). Using the clinical criteria for classic LFS, in our own series of families 79% (19/24) have germline mutations, as do 41% (7/17) of LFL families. These figures are higher than those in other series, and reflect the stringency and accuracy of our methods for mutation detection [4, 17]. In contrast to the assumptions made by many groups studying *TP53* mutations, over one quarter of the mutations in our series have been detected outside the hot-spot domains within the DNA-binding region of the protein [17]. The range of mutations include classic dominant-negative and gain-of-function mutations (eg codons 175, 245, 248, 273) as well as nonsense and splicing mutations. Some of the mutations are novel, and in these cases it is essential to carry out functional studies of the mutant sequence to confirm that it is indeed the causative defect in a family (see below).

Genotype-phenotype correlations

The detailed molecular analysis carried out in LF families in our series has allowed us to analyse genotype-phenotype correlations [18]. We have been rigorous in our ascertainment criteria as defined above, and we have screened for mutations in *TP53* by direct sequencing of all exons (coding and non-coding), all intron-exon boundaries, the promoter and the 3' untranslated region of the gene [17]. Families with germline missense mutations in the core DNA binding-domain showed a more highly penetrant cancer phenotype than families with other mutations, or with no detectable *TP53* mutation. Cancer phenotype in families with mutations in the core DNA-binding domain was characterised by an overall higher cancer incidence, but more specifically a higher incidence of breast and central nervous system (CNS) tumours occurring under the age of 45 years, compared with families carrying protein truncating or other inactivating mutations. In addition, all cases of childhood adrenocortical tumours occurred in individuals with mutations within the DNA-binding domain [3, 17]. Cancers in probands with missense mutations in the DNA-binding domain showed significantly younger ages at diagnosis than those with protein inactivating mutations. In individuals with the former type of mutation there was a significantly lower proportion of tumours that showed loss of the wild-type allele [19]. These results are consistent with observations in experimental systems which demonstrate that certain mutations exhibit gain of function and/or dominant-negative properties. Our results support an enhanced oncogenic potential for such mutations in human populations. It will now be important to extend and refine these studies so that such information can become incorporated into estimated risks in patients with germline *TP53* mutations.

Functional studies of cells carrying a mutant *TP53* allele

In parallel with the molecular analysis in LFS, we are carrying out an extensive programme of cell biology studies. All the work has been carried out on fibroblast or lymphoblastoid cell lines derived from patients with known *TP53* mutation status. Initial studies indicated that fibroblasts from patients with germline defects showed an increased resistance to the effects of low dose-rate ionising radiation, and further data have confirmed that this is a frequent trait [20, 21]. This is paralleled by an increased chromosomal radiosensitivity [22] indicating that cells carrying germline *TP53* mutations are avoiding the normally lethal effects of DNA damage. LF fibroblast strains that are maintained in culture and not subjected to DNA damaging agents develop an excess of chromosomal abnormalities (aneuploidy and structural aberrations) that increase with time in cell culture. Indeed we have shown that this property is the most common trait in fibroblasts derived from patients with germline *TP53* mutations [20]. Increased levels of aneuploidy were also seen in cells from some LF patients in whom no germline mutation had been found. Although lymphoblastoid cell lines carrying germline mutations do not show any defect in G_1/S arrest, fibroblast strains frequently have defects in both transient and permanent G_1 arrest [23–25]. Finally, cell strains from patients with germline *TP53* defects show increased longevity compared to normal control strains.

Among mutation-carrying fibroblast cell strains, all those with missense mutations within the DNA-binding domain showed reduced G_1 arrest. However a number of strains in which there are nonsense or splicing mutations gave responses within the normal range. We have also tested the ability of p53 in LF cells to transactivate a target gene in yeast cells (FASAY). Similar findings to the G_1 arrest data were obtained, with certain splicing and nonsense mutant alleles failing to be detected as mutant in the assay [25, 26].

These data parallel our findings of a correlation between genotype and phenotype as described above. Mutations within the DNA-binding domain show a more highly penetrant and severe phenotype and also more marked abnormalities in a range of *in vitro* assays.

Germline *TP53* mutations in children with adrenocortical tumours

Childhood adrenocortical tumours are rare malignancies, with an incidence of 0.3/million children/year. Adrenocortical carcinoma (ACC) shows an apparent bimodal age distribution, with the median ages of onset of childhood disease of three years, and of adult disease of 55–59 years of age (data based on information from the Manchester Children's Tumour Registry and the Office of National Statistics, England and Wales, respectively). Childhood ACC is known to occur at a considerably increased frequency in association with certain syndromes including Beckwith-Wiedemann Syndrome (BWS), and in families with Li-Fraumeni Syndrome [2, 27]. As discussed above, in a high proportion of the latter, the cancer-prone condition is associated with inheritance of a germline mutation in the *TP53* gene [3, 5, 17]. In LFS families in whom childhood ACC is present, the disease appears to be highly penetrant, with other carriers of the germline *TP53* mutation presenting with typical LFS tumours (eg bone and soft-tissue sarcomas, breast and brain tumours) at exceptionally young ages [3, 17]. In a series of families studied by our own group, all of those in whom there was an adrenocortical tumour carried a germline *TP53* mutation [17].

We were therefore interested to determine the proportion of children (aged less than 15 years) with adrenocortical tumours who had germline *TP53* mutations. We examined a panel of fourteen cases, all except three were ascertained as a population-based series of children with adrenocortical tumours (Manchester Children's Tumour Registry). The children were unselected for a family history. We were able to analyse normal DNA from eleven of these patients, and detected germline *TP53* mutations in 9/11 cases (82%). In a further two cases we were unable to obtain normal DNA from the patients themselves, but detected germline mutations in cancer-affected relatives. This is a remarkably high frequency, and represents the highest frequency of a germline mutation associated with any malignancy reported to date. The mutations detected were somewhat unusual in that a number occurred at codons 152 and 158 [28]. A germline mutation has been reported previously at codon 152, intriguingly in a patient with a childhood adrenocortical carcinoma [29]. There are no reports on a germline codon 158 mutation (see [4]). The high incidence of mutations at these two residues raises the possibility of a hot-spot region for predisposition to adrenocortical tumours. Further studies will be necessary to address this. Although we have been unable to carry out functional studies on cells containing a codon 158 germline mutation, we have examined fibroblasts from a patient with a codon 152 mutation. We have demonstrated that the cells show reduced transient G_1 arrest and an increased cell survival following irradiation, and that the mutant p53 fails to transactivate in a yeast functional assay [25]. These data confirm that the mutant protein is functionally abnormal.

Although the children in the above study were unselected for any family history, particularly one conforming to LFS/LFL, nonetheless a number did have suspicious family histories. Furthermore in a number of cases we were able to demonstrate that the germline mutation was inherited. The types of cancer present in some families are those typical of LFS, notably brain tumours and sarcomas at relatively early ages. In addition a number of other tumours were identified in some of the families which are not normally associated with LFS, including uterine carcinoma and lung cancer. Most strikingly, a number of germline mutation carriers who are unaffected in their 40 s to 60 s were identified. This is extremely unusual in families in which there is a germline *TP53* mutation segregating. However, many of the observations including ages at onset, are based on studies of families conforming to the definitions of LFS or LFL, which necessarily select for highly penetrant mutations. In our own series of families with LFS/LFL, the mean age of onset of the first primary cancers is 28 years [28], and the lifetime risk of developing cancer in a germline mutation carrier exceeds 90% by the age of 60 years [30]. In families of probands with childhood adrenocortical tumours who are unselected for family history of other cancers, the increased incidence of tumours, some of which are typical of LFS, together with the identification of unaffected carriers indicates that the mutations that we have found in these families are not highly penetrant germline *TP53* mutations. This is an extremely important finding, and may indicate that germline *TP53* mutations are more frequent than previously estimated.

The finding of a high incidence of germline *TP53* mutations in childhood adrenocortical tumours also suggests that the probands and their families should be offered genetic counselling, and if appropriate and requested, genetic screening.

Somatic alterations in childhood adrenocortical tumours

We applied the technique of comparative genomic hybridisation (CGH) to a subset of eleven cases of childhood adrenocortical cancer (two adenomas and nine carcinomas, mean age 4.2 years). CGH is a cytogenetic technique that permits a global survey of chromosomal losses and gains. The principle of the technique is the labelling of normal (control) DNA using one fluorochrome, and of the test (tumour) DNA using a second fluorochrome. The two probes are mixed and, after pre-annealling with repetitive DNA, are hybridised to normal human metaphase chromosomes [31]. A comprehensive review of the technique and its applications is given in James et al. [32].

All of the samples, both those classified as adenoma and those as carcinoma, had multiple copy number aberrations (mean 14.5 per tumour, range 7–22). Notably, regions of increased copy number were identified on chromosome 19 in all eleven cases and 1p and 9q in 10/11 cases examined. Other common regions of DNA gain were seen on chromosomes 22q (8/11), 12q (8/11), 11q (5/11), 17q (5/11) and 20q (6/11). The minimal common regions of gain were at 1p35-pter, 9q34, 12q24.1-qter, 17q21 and 11q13. The most frequently observed losses involved chromosomes 18q (8/11), 4 (7/11), X (5/11), 3 (4/11) and 2 (4/11). The minimal common regions of deletion were at 18q21, 4q21-q28, Xp22.2-pcen, Xq21-q25, 3cen-q13.2 and 2q21-q32. There were no significant differences between either the number or the type of copy number changes between adenomas and carcinomas. Furthermore there were no obvious differences in the patterns of chromosomal gains and losses according to germline *TP53* mutation status (see [33]).

The changes observed in the childhood tumours are shown diagramatically in Figure 1. A high proportion of the changes are highly consistent, with three regions showing copy number gains in over 90% of cases (1p36, 9q34 and chromosome 19). There are very few instances of such consistent changes in any solid tumour, either childhood or adult (for review see [34]). A number of the chromosomal regions that we have identified in this study have been reported by others to be implicated in a variety of malignancies. For example gain/amplification of 11q13, 17q21, regions of chromosome 19 and 20q have all been reported in a number of solid tumours (for review see [34]). Similarly losses of chromosome 18 have been described in breast, colorectal, lung, pancreatic and head and neck tumours. Candidate genes have been identified in some of these regions, for example *CCND1* and *INT2* at 11q13, *c-erbB2* at 17q21, and *DCC* at 18q21. However the pattern of copy number aberrations that we have found in childhood adrenocortical tumours is unusual.

A study of CGH analysis of adult adrenocortical tumours has been reported recently [35]. Of great interest is the observation that the changes that we have found in childhood tumours [33] are very different to those reported in adult tumours. A comparison of the most common changes seen in adult tumours compared to those seen in childhood tumours is shown in Figure 1. This observed difference clearly demonstrates that the genetic changes associated with the development and progression of adult and childhood adrenocortical tumours are quite distinct. This idea is further supported by the apparent bimodal age distribution of childhood and adult disease. Finally, childhood adrenocortical tumours are strongly associated with germline *TP53* mutations [28, 29], unlike adult disease.

We have proposed that childhood adrenocortical tumours (both carcinomas and adenomas) are embryonal tumours, derived from foetal adrenal cortex, in contrast to adult-onset tumours that arise from adult cortex [33]. It is possible that inherited defects in *TP53* result in failure of the complete programmed apoptosis of the foetal cortex that occurs around birth [36]. This, coupled with an increase in genomic instability as a consequence of the presence of a mutant p53 protein (and see below), leads to a vastly elevated risk of adrenocortical tumours in children with germline mutations. The copy number aberrations that we have described in childhood tumours would therefore reflect genetic alterations specific to the transformation of foetal, rather than adult, adrenal cortex.

Figure 1. (See color plate XII between pages 256 and 257.) Summary of the most common DNA copy number changes identified in eleven childhood [33] and eight adult adrenocortical tumours [35] by comparative genomic hybridisation (CGH). For the childhood cases, red vertical bars to the left of each chromosome represent DNA copy number losses and green bars to the right of each chromosome show DNA copy number gains. For the adult cases, blue bars represent losses and black bars indicate gains. The number of cases showing each aberration is given next to the bar. Representative hybridisation images of chromosomes 1, 4, 18 and 19 from a case of childhood adrenocortical carcinoma are also shown.

In addition to the types of alteration detected cytogenetically using the technique of CGH, we have also examined childhood adrenocortical tumours for the occurrence of somatic mutations affecting the *TP53* gene, whether loss of heterozygosity (LOH) or further somatic mutation. The simplest model of action for a tumour suppressor gene is that one mutation (usually missense or nonsense) is inherited, and the remaining wild-type allele is lost by some deletion mechanism, or suffers a second mutational hit [37]. We have previously shown, however, that *TP53* does not always obey this general rule [19], and this has subsequently been confirmed by other groups [38]. Under half of the tumours examined from patients with germline *TP53* mutations in our initial study were found to retain the wild-type allele, and indeed some tumours showed loss of the inherited mutant allele [19]. In this latter study, however, we did not look for additional somatic mutations to *TP53* in the tumours. In a panel of childhood adrenocortical tumours we determined not only whether there was loss of the wild-type allele, but also whether there were somatic mutations in the retained allele in tumours showing no LOH. Surprisingly, we found a very complex pattern of loss and mutation in some of the tumours [28]. In tumours from patients with some germline mutations there are few other alterations, limited to loss of the wild-type allele. In contrast, tumours from patients with codon 152 mutations showed an unusually high number of somatic mutations [28]. The types of mutations included loss of the inherited mutant allele and the accumulation of multiple somatic mutations within both exons and introns of *TP53*. In one particularly striking example an adrenocortical tumour from a 10-month female showed a number of separate somatic mutations, together with loss of the mutant allele.

Concluding remarks

Classic Li-Fraumeni and Li-Fraumeni-like families are undoubtedly rare. Adherence to the strict clinical definitions for the LFS and LFL obviously selects for mutant *TP53* alleles that are highly penetrant. Our studies of childhood adrenocortical tumours indicate that germline mutations may be more common than has been previously estimated, and that there may be mutant *TP53* alleles that are of lower penetrance. There have been a number of other studies reported in which patients with a variety of different tumour types have been screened for germline mutations. The tumour types have been selected because of their relatively frequent occurrence within LF families and include sarcomas [39–41], brain tumours [42–45] and breast tumours [46–48]. The proportion of such patients with germline *TP53* mutations is variable and in general low, but further detailed studies are warranted to gain a clear insight into the frequency of mutations within the different tumour types. Only when such studies have been carried out can a clear measure of the frequency and nature of low penetrance *TP53* alleles be determined. In any such studies it will be of importance to screen the entire gene for mutations; there are very few reports in the literature in which this has been the case.

One other important finding to appear from our studies of childhood adrenocortical tumours is the incidence of radiation-induced second primary tumours [28]. Of those children treated with radiotherapy who survived for more than 2 years after diagnosis of their primary tumour, all three developed a second primary tumour within the radiation field. All second primary tumours were of types expected in carriers of a germline *TP53* mutant allele. There are other reports of therapy-induced second primary tumours in germline carriers [49–51]. More detailed studies are needed to determine the clinical implications of radiotherapy and chemotherapy on the development of subsequent primary tumours. It is also not clear whether the survival of patients with comparable primary tumours with and without germline *TP53* mutations is similar. Furthermore, it may be that patients with germline *TP53* mutations are more susceptible to a range of other carcinogenic insults. In our own series of families there are patients who are germline *TP53* mutation carriers with a number of relatively early-onset tumours which are not typical of Li-Fraumeni Syndrome. These tumours include lung cancer in non-smokers, cervical cancer and mesothelioma. Interestingly, there appears to be no increased incidence of tumours associated with exposure to sunlight in germline *TP53* mutation carriers.

Finally, there are still a number of unanswered questions with respect to the risks associated with germline *TP53* mutations. Detailed information regarding age-, site- and morphology-specific risks is still required for accurate management and counselling of patients with germline *TP53* mutations.

References

1 Li FP, Fraumeni JF (1969) Soft-tissue sarcomas, breast cancer and other neoplasms: a familial syndrome? *Ann Intern Med* 71: 747–752
2 Li FP, Fraumeni JF, Mulvihill JJ, Blattner WA, Dreyfus MG, Tucker MA, Miller RW (1988) A cancer family syndrome in twenty-four kindreds. *Cancer Res* 48: 5358–5362
3 Birch JM, Hartley AL, Tricker KJ, Prosser J, Condie A, Kelsey AM, Harris M, Morris Jones PH, Binchy A, Crowther D et al (1994) Prevalence and diversity of constitutional mutations in the *p53* gene among 21 Li-Fraumeni families. *Cancer Res* 54: 1298–1304
4 Varley JM, Evans DGR, Birch JM (1997) Li-Fraumeni syndrome – a molecular and clinical review. *Brit J Cancer* 76: 1–14
5 Malkin D, Li *FP*, Strong LC, Fraumeni JF, Nelson CE, Kim DH, Kassel J, Gryka MA, Bischoff FZ, Tainsky MA et al (1990) Germ line p53 mutations in a familial syndrome of breast cancer, sarcomas, and other neoplasms. *Science* 250: 1233–1238
6 Lavigueur A, Maltby V, Mock D, Rossant J, Pawson T, Bernstein A (1989) High incidence of lung, bone, and lymphoid tumors in transgenic mice overexpressing mutant alleles of the p53 oncogene. *Mol Cell Biol* 9: 3982–3991
7 Hollstein M, Shomer B, Greenblatt M, Soussi T, Hovig E, Montesano R, Harris CC (1996) Somatic point mutations in the *p53* gene of human tumors and cell lines: updated compilation. *Nucl Acid Res* 24: 141–146
8 Soussi T, Caron de Fromentel C, May P (1990) Structural aspects of the p53 protein in relation to

gene evolution. *Oncogene* 5: 945–952
9 Levine AJ, Perry ME, Chang A, Silver A, Ditmer D, Wu M, Welsh D (1994) The 1993 Walter Hubert lecture: The role of the p53 tumour-suppressor gene in tumorigenesis. *Brit J Cancer* 69: 409–416
10 Levine AJ (1997) p53, the cellular gatekeeper for growth and division. *Cell* 88: 323–331
11 Clore GM, Omichinski JG, Sakaguchi K, Zambrano N, Sakamoto H, Appella E, Gronenborn AM (1994) High-resolution structure of the oligomerization domain of p53 by multidimensional NMR. *Science* 265: 386–391
12 Jeffrey PD, Gorina S, Pavletich NP (1995) Crystal structure of the tetramerization domain of the p53 tumor suppressor at 1.7 angstroms. *Science* 267: 1498–1502
13 Davison TS, Yin P, Nie E, Kay C, Arrowsmith CH (1998) Characterization of the oligomerization defects of two p53 mutants found in families with Li-Fraumeni and Li-Fraumeni-like syndrome. *Oncogene* 17: 651–656
14 Milner J, Metcalf EA (1991) Cotranslation of activated mutant p53 with wild-type drives the wild-type p53 protein into the mutant conformation. *Cell* 65: 765–774
15 Eliyahu D, Raz A, Gruss P, Givol D, Oren M (1984) Participation of p53 cellular tumour antigen in transformation of normal embryonic cells. *Nature* 312: 646–649
16 Parada LF, Land H, Weinberg RA, Wolf D, Rotter V (1984) Cooperation between gene encoding p53 tumour antigen and ras in cellular transformation. *Nature* 312: 649–651
17 Varley JM, McGown G, Thorncroft M, Santibanez-Koref MF, Kelsey AM, Tricker KJ, Evans DGR, Birch JM (1997) Germ-line mutations of *TP53* in Li-Fraumeni families: an extended study of 39 families. *Cancer Res* 57: 3245–3252
18 Birch JM, Blair V, Kelsey AM, Evans DGR, Harris M, Tricker KJ, Varley JM (1998) Cancer phenotype correlates with constitutional *TP53* genotype in families with the Li-Fraumeni Syndrome. *Oncogene* 17: 1061–1068
19 Varley JM, Thorncroft M, McGown G, Appleby J, Kelsey AM, Tricker KM, Evans DGR, Birch JM (1997) A detailed study of loss of heterozygosity on chromosome 17 in tumours from Li-Fraumeni patients carrying a mutation to the *TP53* gene. *Oncogene* 14: 865–871
20 Boyle JM, Mitchell ELD, Greaves MJ, Roberts SA, Tricker K, Burt E, Varley JM, Birch JM, Scott D (1997) Chromosome instability is a predominant trait of fibroblasts from Li-Fraumeni families. *Brit J Cancer* 77: 2181–2192
21 Sproston ARM, Boyle JM, Heighway J, Birch JM, Scott D (1996) Fibroblasts from Li-Fraumeni patients are resistant to low dose-rate irradiation. *Int J Radiat Biol* 70: 145–150
22 Mitchell ELD, Scott D (1997) G_2 chromosomal radiosensitivity in fibroblasts of ataxia-telangiectasia heterozygotes and Li-Fraumeni Syndrome patients with radioresistant cells. *Int J Radiat Biol* 72: 435–438
23 Williams KJ, Heighway J, Birch JM, Norton JD, Scott D (1996) No defect in G1/S cell cycle arrest in irradiated Li-Fraumeni lymphoblastoid cell lines. *Brit J Cancer* 74: 698–703
24 Williams KJ, Boyle JM, Birch JM, Norton JD, Scott D (1997) Cell cycle arrest defect in Li-Fraumeni syndrome: a mechanism of cancer predisposition? *Oncogene* 14: 277–282
25 Boyle JM, Greaves MJ, Camplejohn RS, Birch JM, Roberts SA, Varley JM (1999) Radiation-induced G_1 arrest is not defective in fibroblasts from Li-Fraumeni families without *TP53* mutations. *Brit J Cancer* 79: 1657–1664
26 Varley JM, Chapman P, McGown G, Thorncroft M, White GRM, Greaves MJ, Scott D, Spreadborough A, Tricker KJ, Birch JM et al (1998) Genetic and functional studies of a germline *TP53* splicing mutation in a Li-Fraumeni-like family. *Oncogene* 16: 3291–3298
27 Wiedemann H-R (1983) Tumours and hemihypertrophy associated with Wiedemann-Beckwith syndrome. *Eur J Pediat* 141: 129
28 Varley JM, McGown G, Thorncroft M, James LA, Margison GP, Forster G, Evans DGR, Kelsey A, Birch JM (1999) Are there low penetrance *TP53* alleles? Evidence from childhood adrenocortical tumors. *Am J Hum Genet* 65: 995–1006
29 Wagner J, Portwine C, Rabin K, Leclerc J-M, Narod SA, Malkin D (1994) High frequency of germline p53 mutations in childhood adrenocortical cancer. *JNCI* 86: 1707–1710
30 LeBihan C, Moutou C, Brugières L, Feunteun J, Bonaïti-Pellié C (1995) ARCAD: a method for estimating age-dependent disease risk associated with mutation carrier status from family data. *Genet Epidemiol* 12: 13–25
31 Kallioniemi A, Kallioniemi O-P, Sudar D, Rutovitz D, Gray JW, Waldman F, Pinkel D (1992) Comparative genomic hybridization for molecular cytogenetic analysis of solid tumors. *Science*

258: 818–821
32 James LA (1999) Comparative genomic hybridization as a tool in tumour cytogenetics. *J Pathol* 187: 385–395
33 James LA, Kelsey A, Birch JM, Varley JM (1999) Analysis of regions of gain and loss in childhood adrenocortical tumours by comparative genomic hybridisation. *Brit J Cancer* 81: 300–304
34 Knuutila S, Bjorkqvist A-M, Autio K, Tarkkanen M, Wolf M, Monni O, Szymanska J, Larramendy ML, Tapper J, Pere H et al (1998) DNA copy number amplifications in human neoplasms. *Amer J Pathol* 152: 1107–1123
35 Kjellman M, Kallioniemi O-P, Karhu R, Höög A, Farnebo L-O, Auer G, Larsson C, Bäckdal M (1996) Genetic aberrations in adrenocortical tumors detected using comparative genomic hybridization correlate with tumor size and malignancy. *Cancer Res* 56: 4219–4223
36 Mesiano S, Jaffe RB (1997) Developmental and functional biology of the primate fetal adrenal cortex. *Endocrin Rev* 18: 378–403
37 Ponder BA (1988) Gene losses in human tumours. *Nature* 335: 400–402
38 Sedlacek Z, Kodet R, Kriz V, Seemanova E, Vodvarka P, Wilgenbus P, Mares J, Poustka A, Goetz P (1998) Two Li-Fraumeni syndrome families with novel germline p53 mutations: loss of the wild type p53 allele in only 50% of tumours. *Brit J Cancer* 77: 1034–1039
39 Toguchida J, Yamaguchi T, Dayton SH, Beauchamp RL, Herrera GE, Ishizaki K, Yamamuro T, Meyers PA, Little JB, Sasaki MS et al (1992) Prevalence and spectrum of germline mutations of the p53 gene among patients with sarcoma. *N Engl J Med* 326: 1301–1308
40 McIntyre JF, Smith-Sorensen B, Friend SH, Kassell J, Borresen A-L, Yan YX, Russo C, Sato J, Barbier N, Miser J et al (1994) Germline mutations of the p53 tumor suppressor gene in children with osteosarcoma. *J Clin Oncol* 12: 925–930
41 Diller L, Sexsmith E, Gottlieb A, Li *FP*, Malkin D (1995) Germline p53 mutations are frequently detected in young children with rhabdomyosarcoma. *J Clin Invest* 95: 1606–1611
42 Kyritsis AP, Bondy ML, Xiao M, Berman EL, Cunningham JE, Lee PS, Levin VA, Saya H (1994) Germline p53 mutations in subsets of glioma patients. *JNCI* 86: 344–349
43 Chen P, Iavarone A, Fick J, Edwards M, Prados M, Israel MA (1995) Constitutional p53 mutations associated with brain tumors in young adults. *Cancer Genet Cytogenet* 82: 106–115
44 Felix CA, Slavc I, Dunn M, Strauss EA, Phillips PC, Rorke LB, Sutton L, Bunin GR, Biegel JA (1995) p53 gene mutations in pediatric brain tumors. *Med Pediat Oncol* 25: 431–436
45 Li YJ, Sanson M, Hoang-Xuan K, Delattre J-Y, Poisson M, Thomas G, Hamelin R (1995) Incidence of germ-line *p53* mutations in patients with gliomas. *Int J Cancer* 64: 383–387
46 Prosser J, Porter D, Coles C, Condie A, Thompson AM, Chetty U, Steel CM, Evans HJ (1992) Constitutional p53 mutation in a non-Li-Fraumeni cancer family. *Brit J Cancer* 65: 527–528
47 Børresen A-L, Andersen TI, Garber J, Barbier-Piraux N, Thorlacius S, Eyfjörd J, Ottestad L, Smith-Sorensen B, Hovig E, Malkin D et al (1992) Screening for germ line *TP53* mutations in breast cancer patients. *Cancer Res* 52: 3234–3236
48 Sidransky D, Tokino T, Helzlsouer K, Zehnbauer B, Rausch G, Shelton B, Prestigiacomo L, Vogelstein B, Davidson N (1992) Inherited *p53* gene mutations in breast cancer. *Cancer Res* 52: 2984–2986
49 Felix CA, Hosler MR, Provisor D, Salhany K, Sexsmith EA, Slater DJ, Cheung N-KV, Winick NJ, Strauss EA, Heyn R et al (1996) The p53 gene in pediatric therapy-related leukemia and myelodysplasia. *Blood* 87: 4376–4381
50 Felix CA, Megonigal MD, Chervinsky DS, Leonard DGB, Tsuchida N, Kakati S, Block AMW, Fisher J, Grossi M, Salhany KI et al (1998) Association of germline p53 mutation with *MLL* segmental jumping translocation in treatment-related leukemia. *Blood* 91: 4451–4456
51 Hisada M, Garber JE, Fung CY, Fraumeni JF, Li *FP* (1998) Multiple primary cancers in families with Li-Fraumeni Syndrome. *JNCI* 90: 606–611

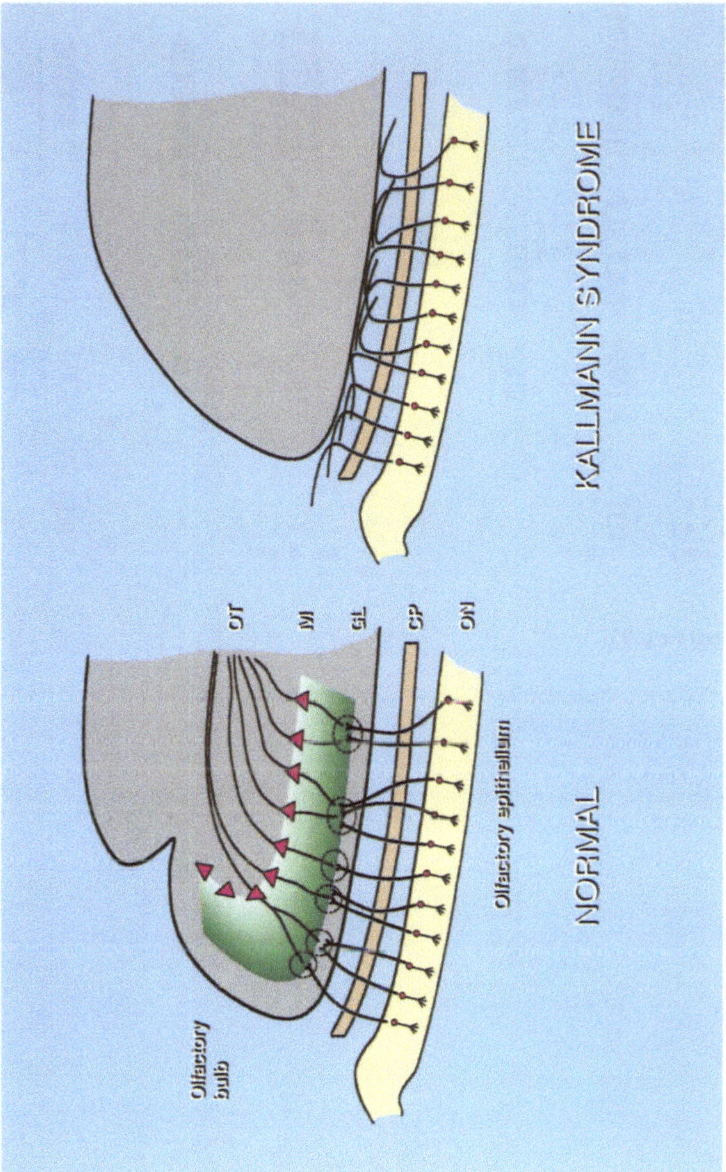

Figure 5. Legend see p. 236

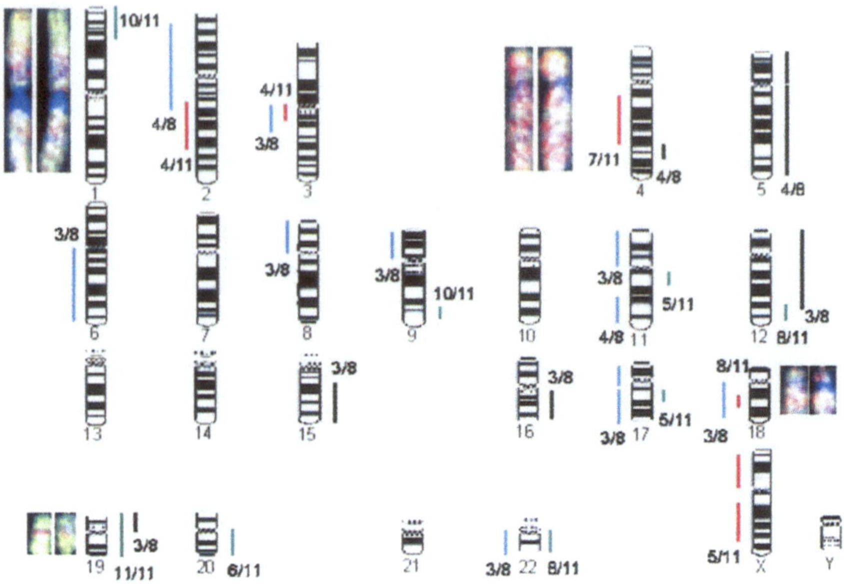

Figure 1. Legend see p. 251

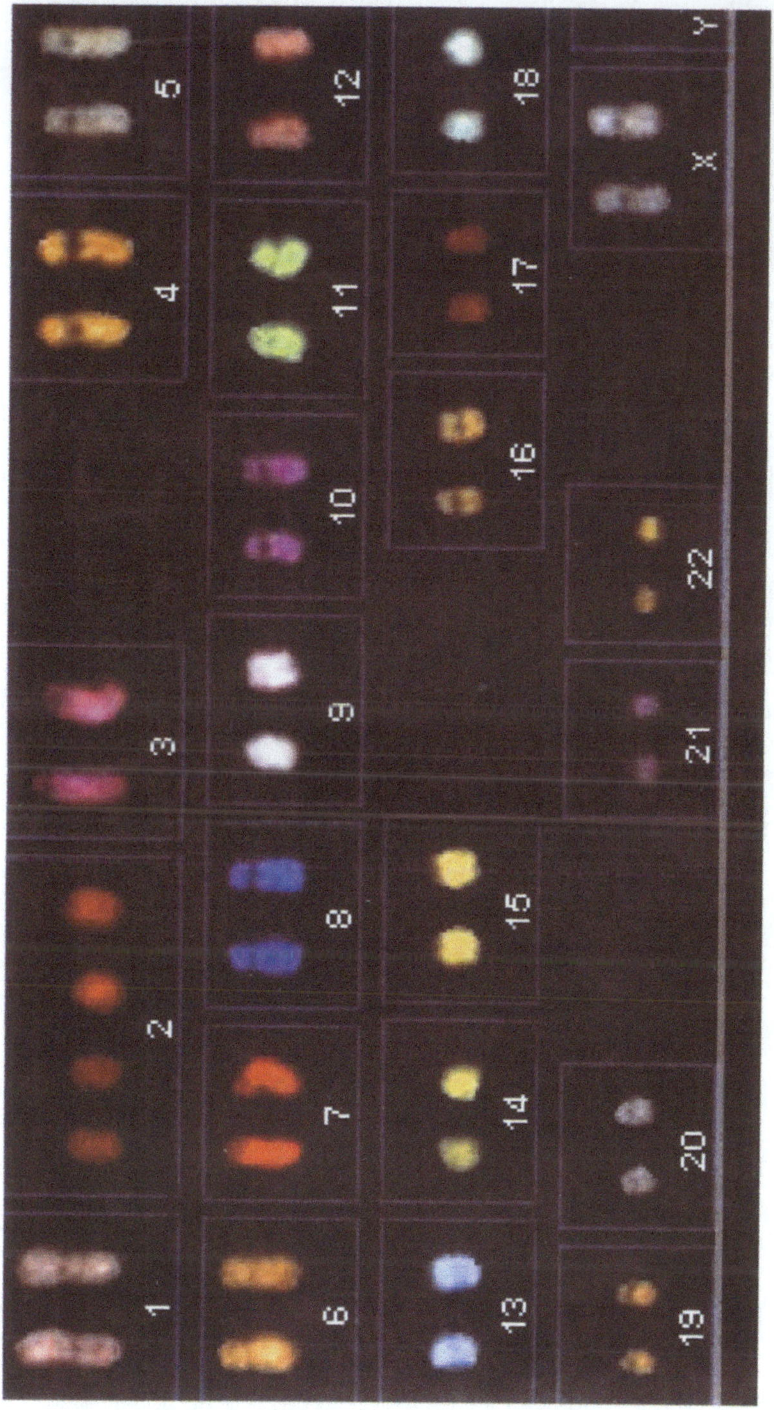

Figure 1. Legend see p. 260

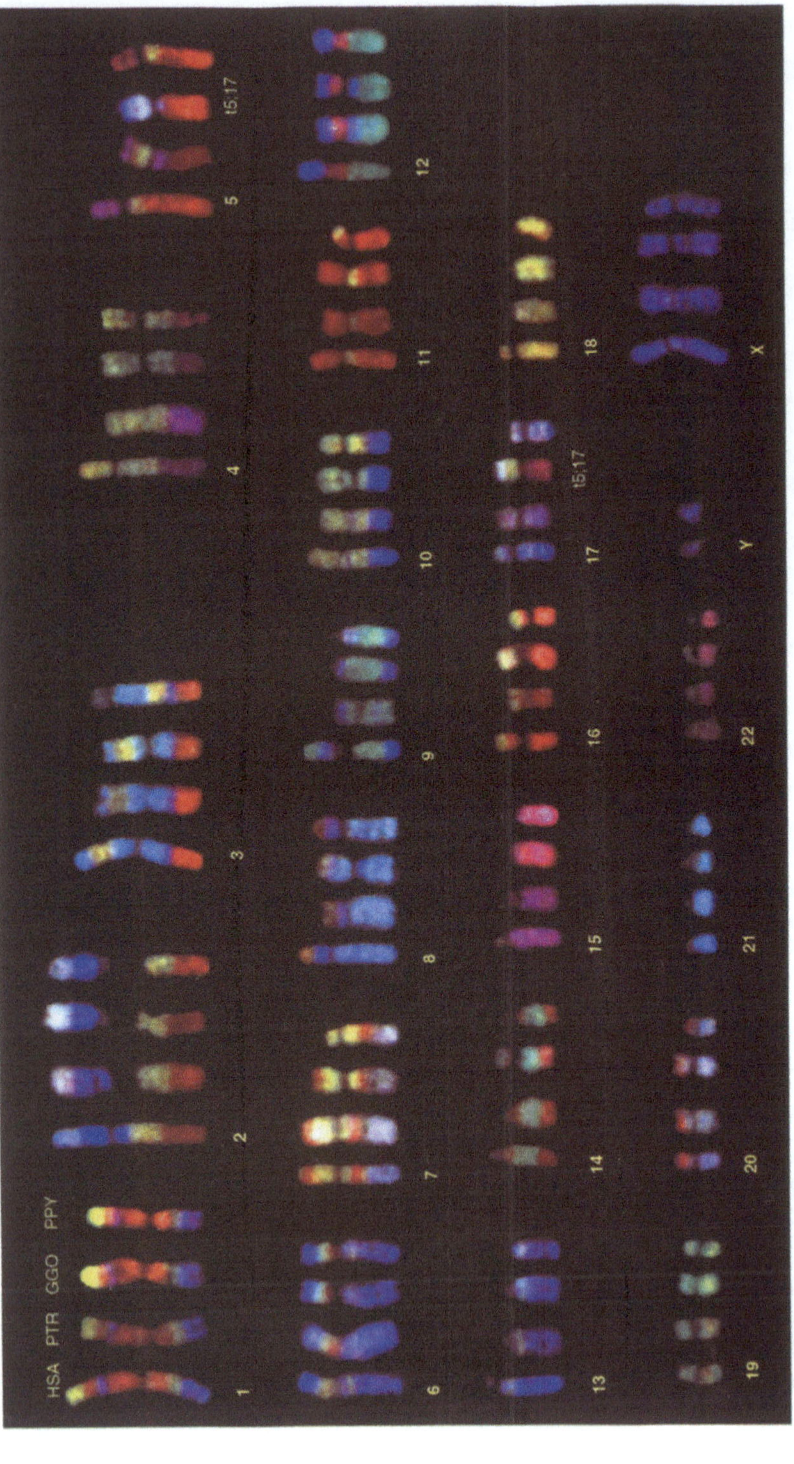

Figure 2. Legend see p. 260

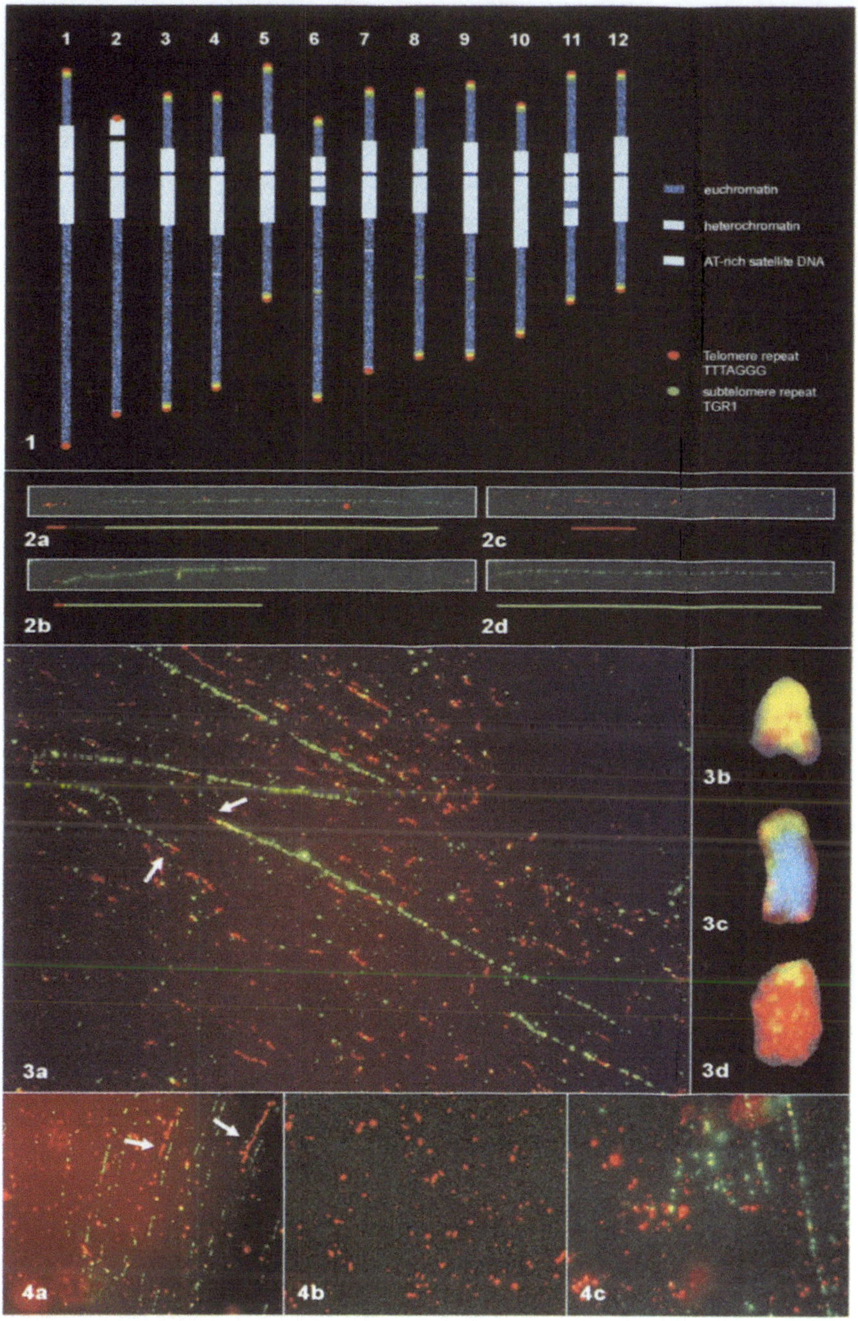

Figures 1–4. Legends see p. 270

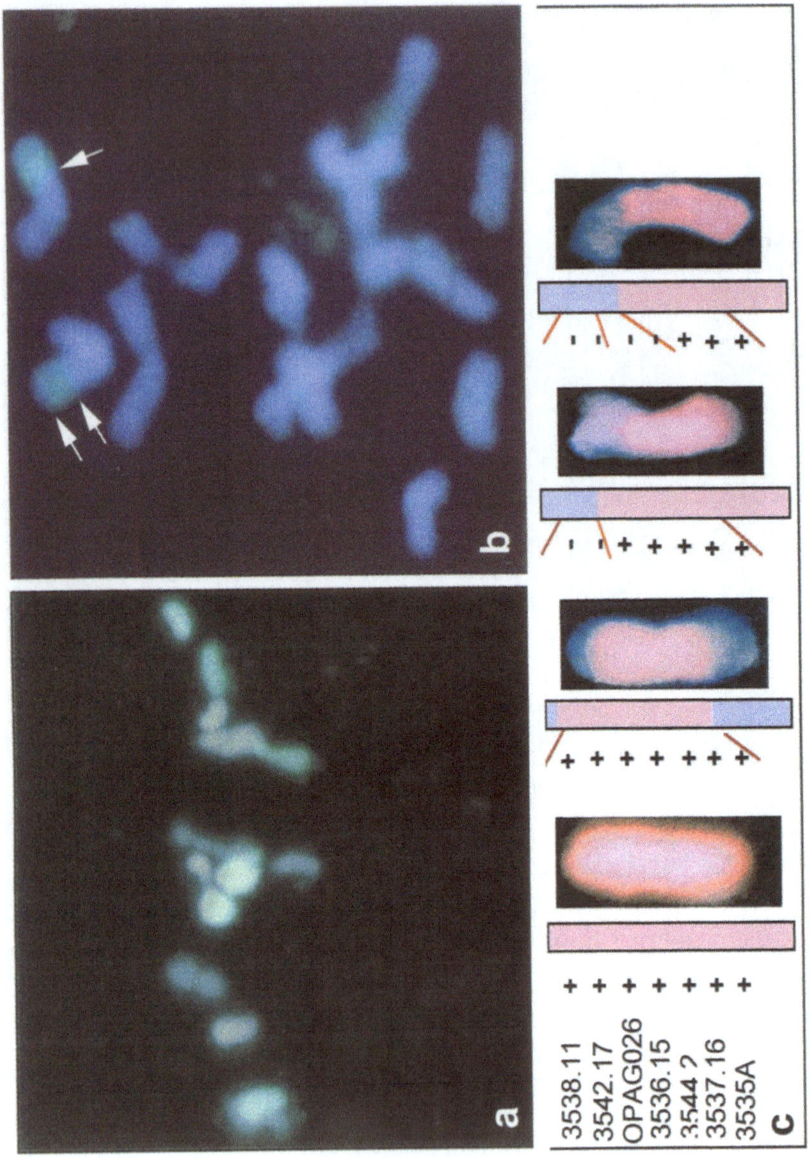

Figure 1. Legend see p. 279

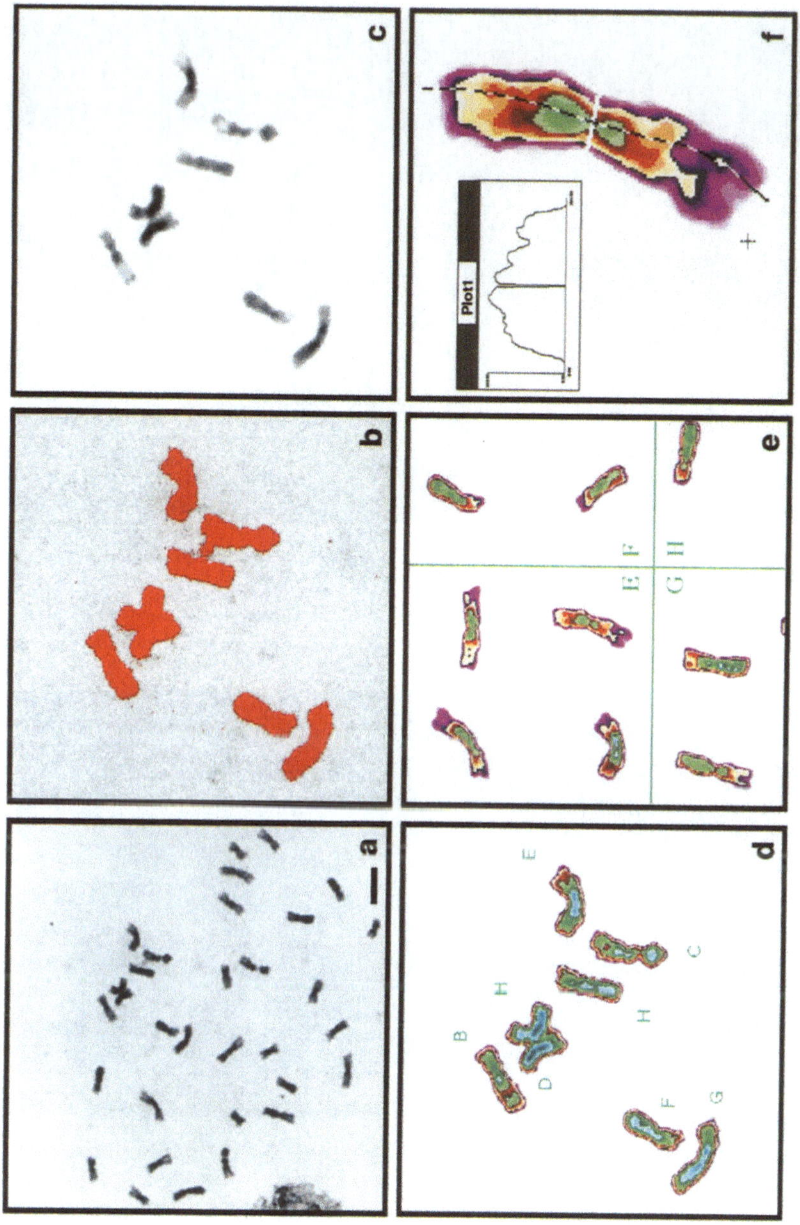

Figure 3. Legend see p. 291

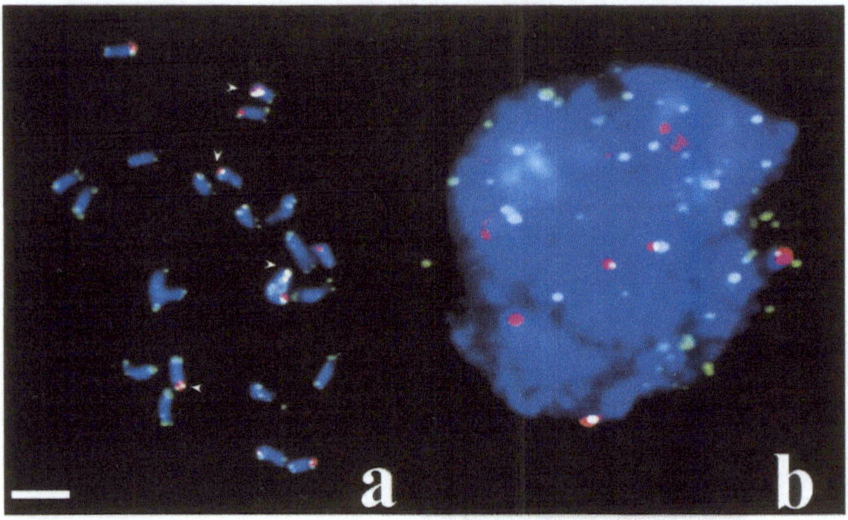

Figure 8. Legend see p. 297

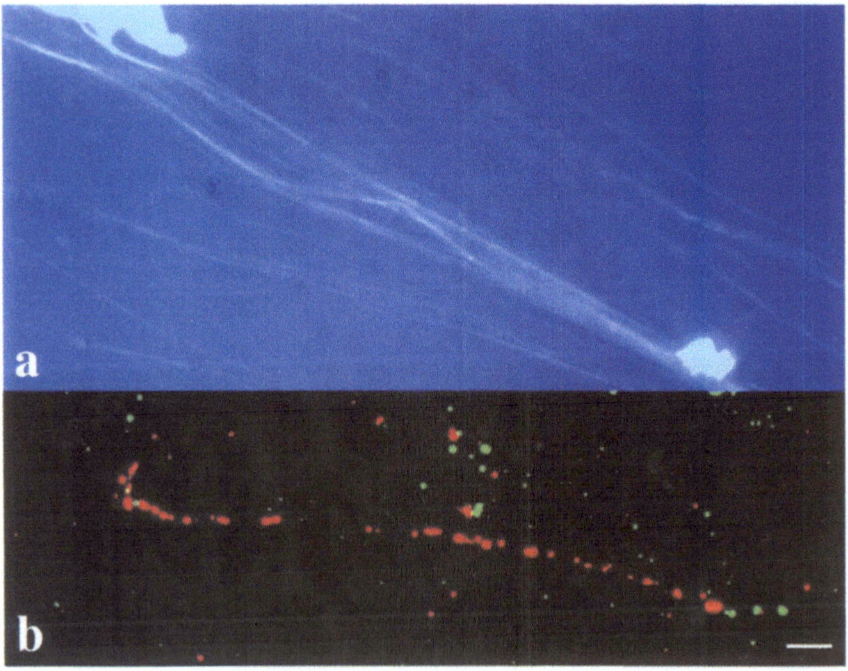

Figure 9. Legend see p. 297

Part six
Mapping and new methods

Malcolm A. Ferguson-Smith

J. Hans de Jong

Ian P. King

Kiichi Fukui

Chromosomes Today, Vol. 13
ed. by E. Olmo and C.A. Redi

Comparative chromosome painting

Malcolm A. Ferguson-Smith, Patricia C.M. O'Brien, Willem Rens, and Fengtang Yang

Molecular Cytogenetics Group, Cambridge University, Centre for Veterinary Science, Madingley Road, Cambridge CB3 0ES, UK

Introduction

Chromosome paint probes are complex genomic probes prepared by random PCR amplification of chromosome samples. Most chromosome samples are obtained by sorting a fluid suspension of chromosomes in a dual laser fluorescence-activated cell sorter [1] although microdissection of chromosomes or chromosome regions may also be used [2]. Flow sorting effectively separates chromosomes by their size and base-pair ratio into a series of groups (the flow karyotype) roughly corresponding to the haploid number of the species. In some species non-identical chromosomes of similar size sort together in one group. For example, human chromosomes 9–12 tend to sort together, and in this case individual cell lines with centric heteromorphisms or intrachromosomal duplications have to be employed to separate them. 300–500 chromosomes sorted into an eppendorf tube are sufficient to make a chromosome-specific amplification product for hybridisation with 50–100 glass slides of air-dried chromosome preparations. DNA amplification of the chromosomal DNA is achieved by two rounds of degenerate-oligonucleotide primed PCR [3]. Nucleotides labelled with biotin or other labels are added to the second round of amplification. Details of labelling, hybridisation and detection of probe signal by fluorescence *in situ* hybridisation (FISH) are described elsewhere [1]. The technique is designed to detect unique sequence DNA and signals due to repetitive sequences are suppressed either by the addition of unlabelled Cot-1 DNA or by a pre-annealing step in which the probe is incubated for one hour before application to the microscope slide [4].

Paint probes prepared in this way are remarkably chromosome specific and cross-contamination with other chromosomes is not a problem. An excellent example of the specificity of the method is provided when human Y-specific chromosome paint is applied to normal human metaphases. The repetitive sequences on Yq heterochromatin are suppressed and the probe anneals exclusively to the euchromatin present in the short arm, proximal long arm and distal tip of the long arm. The probe also anneals to the X-Y homologous regions of the X chromosome, namely the two pairing segments at the ends of Xp and Xq and the block of X-Y homology in Xq21:3.

Cross-species homology

When chromosome-specific paint probes from one species are applied to the chromosome preparations from another species, hybridisation occurs at sites of unique sequence DNA conserved between the two species. As unique sequence DNA is distributed more or less evenly across euchromatin, the chromosome specific paint effectively identifies regions of chromosome homology between the two species. In closely-related species such as the human and great apes, a chromosome paint in one species can be used to identify the equivalent chromosome in the other (Fig. 1). In fact, improved resolution in human chromosome diagnosis is possible if ape chromosome-specific paints are used rather than human chromosome specific, as background signals are substantially reduced owing to the divergence of repetitive DNA between the two species [5].

Flow karyotypes of the great apes are similar to those of humans, reflecting the small number of genomic rearrangements that have occurred during evolution. What variation there is, results mainly from variation in the amount and distribution of repetitive DNA. The exceptions are human chromosome 2, which is represented by two separate acrocentric chromosomes in the great apes (Fig. 1) and the 5;17 rearrangement in the gorilla. G-banded karyotype comparisons between the human and great apes reveal intrachromosomal inversions [6] which are not visible by direct chromosome painting. However, using a paint probe set derived from gibbon chromosomes [7], a colour-banded pattern is produced which can reveal many of these inversions (Fig. 2). This colour banding reflects the extensive interchromosomal rearrangements that have occurred during the divergence of the lesser apes and humans some 20 million years ago compared, for example, to the relatively small number of interchromosomal rearrangements that have occurred within the 70 million years that have elapsed since the divergence of cats and humans.

When human chromosome-specific probes are applied to chromosome preparations from the hoolock gibbon (*Hylobates hoolock*) the gibbon com-

Figure 1. (See color plate XIII between pages 256 and 257.) Orangutan karyotype showing hybridisation of multicoloured human chromosome-specific paint probes by spectral karyotyping [22]. The karyotype is numbered according to the human idiogram. Note that with the exception of chromosome 2, each orangutan chromosome is homologous to one human chromosome. Orangutan chromosomes 11 and 12 are painted by human chromosome 2 (long and short-arms respectively).

Figure 2. (See color plate XIV between pages 256 and 257.) Comparison of colour-banding patterns of human (HSA) chimpanzee (PTR), gorilla (GGO) and orangutan (PPY) chromosomes arranged according to the human idiogram. Note the 5; 17 translocation in gorilla and the numerous examples of inversion, for example involving chromosomes 3, 7, 9, 11, 12 and 20 in PPY.

plement of 24 autosomes is divided into 64 distinct segments of homology. In comparison, the chimpanzee complement of 23 autosomes is divided into 23 segments, the cat complement of 18 autosomes into 33 segments, the sheep complement of 26 autosomes into 42 segments and the cattle complement of 29 autosomes into 56 segments. The mouse autosomal complement of 19 is divided into about 150 segments of human homology and these have been shown to correspond very closely to the number of segments estimated from the distribution of approximately 2000 conserved loci that have been mapped in both species. Mice and humans diverged about 85 million years ago and this to some extent explains the accumulation of such a large number of rearrangements. However, variation in the number of segments of homology among other species indicates that the extent of karyotype rearrangement occurring between two species does not depend solely on the length of time which has elapsed since they diverged from a common ancestor, and that other unknown factors must operate as well.

Chromosome painting has been valuable in identifying ancient syntenies shared by widely divergent species. These syntenies indicate the likely structure of ancestral chromosomes. It can usually be assumed that when large blocks of homologous DNA are retained intact in distantly related species (outgroups), this arrangement represents the ancestral condition. Conversely, when these blocks are divided among different chromosomes in a single species, or in a group of closely-related species, the rearrangements responsible for this division can usually be regarded as recent events in evolution. Occasionally, however, an ancient arrangement may be reconstituted by a more recent rearrangement, i.e. there is a reversion to the earlier condition. The study of karyotype evolution depends on comparisons between closely-related species and outgroup species. Chromosome 2 in humans is the most familiar example of a comparatively recent rearrangement which has occurred as the result of fusion between two acrocentric chromosomes which are separate in all higher primates and so, presumably, in the common ancestor of human and the great apes. In another example, the homologue of the short arm of human chromosome 2 in the gorilla and orangutan has a pericentric inversion which is absent in both human and chimpanzee [8]. As this inversion is absent in Old World monkeys, it must have occurred subsequent to the divergence of humans and chimpanzees from the line leading to the gorilla and orangutan. The human arrangement in this case is the ancestral one.

Hybridisation of human chromosome-specific paints to a wide range of mammalian species has revealed syntenic blocks which are separated on different human chromosomes but contained in one syntenic block on a single chromosome in a number of outgroup species, indicating that the arrangement in the outgroup species is ancestral to them all. Table 1 shows that homologues to human chromosomes 14 and 15 are combined in the same chromosome in all mammals so far studied with the exception of the great and lesser apes, a small group of New World monkeys (the howler monkeys) and the giant

Table 1. Ancient syntenies revealed by hybridisation of human chromosome-specific probes across mammalian species

Human chromosome associations	14/15	3/21	12/22	16/19	1/2	8/18	2/16	10/16
Chimpanzee	–	–	–	–	–	–	–	–
Gorilla	–	–	–	–	–	–	–	–
Orangutan	–	–	–	–	–	–	–	–
Gibbons	–	–	–	–	–	–	–	–
Old World monkeys	+	–	–	–	–	–	–	–
New World monkeys*	(±)	–	–	–	–	(±)	(±)	(±)
Lemurs	+	+	+	–	–	–	–	–
Cattle	+	+	+	+	+	–	–	–
Sheep	+	+	+	+	+	–	–	–
Muntjac	+	+	+	+	+	–	–	+
Pig*	+	+	+	+	–	–	+	–
Horse*	+	–	+	–	–	–	–	–
Cat	+	+	+	+	+	–	–	–
Mink*	+	+	+	+	–	–	–	–
Giant panda	–	+	+	+	+	–	–	–
Dolphin*	+	+	+	+	–	–	–	–
Seal*	+	+	+	+	–	–	–	+
Rabbit	+	+	+	+	–	–	–	–

+, –, (±) = association present, absent, present in some.
* Data referenced in Glas, Wakefield, Toder and Graves (1998) [23]. Remaining data from Cambridge group and their collaborators.

panda. Similar results are shown for homologues for other parts of separate human chromosomes namely, 3/21, 12/22, 16/19 and 1/2. Several groups of New World monkeys show associations for 8/18, 2/16 and 10/16, although these have been lost in some species within the same families. In some cases, particular associations appear to have been lost (e.g. in horse and mink), while in others a particular association found in widely divergent species seems to have been gained (e.g. pig and muntjac). It has not been determined whether similar breakpoints are involved in these apparent reversions.

The construction of comparative maps

The excellent correspondence between the comparative genetic maps of mouse and man, and the regions of homology identified by chromosome specific paint probes [1], not only confirms the value of comparative chromosome painting as a useful tool for the study of karyotype evolution, but also indicates its value as a guide to the construction of simple maps in unmapped species.

Segments of DNA homology identified by chromosome painting are, in effect, genetic syntenic regions containing genetic linkage groups. However, the order of conserved genes within these syntenic regions can be variable due to intrachromosomal rearrangements, notably inversions, undetectable by whole-chromosome-specific paints. Additional resolution is possible using regional paint probes obtained by amplification of chromosome fragments obtained by microdissection. The orientation of such regional paints is determined from FISH analysis of locus specific probes cloned in cosmids which label either end of the region. Ordering of locus specific probes linked as close as 200–500 kb can then be achieved using extended DNA fibres.

In order to determine the genetic content and limits of a particular region of homology, rather than simply its chromosomal origin, it is necessary to undertake reciprocal cross-species painting. This means the painting of chromosome-specific probes from each species to the chromosome preparations of the other. The essential requirements are cell lines from each species and a flow cytometer able to separate the chromosomes from one another. One example should suffice to explain the procedure. Human chromosome 6 paint hybridises to chromosome segments in hoolock gibbon chromosomes 4q and 8q. In order to determine which parts of human chromosome 6 are homologous to these parts of the gibbon karyotype, it is necessary to prepare paint probes of gibbon chromosomes 4 and 8 to paint back onto human chromosome 6. Once this is done the exact origins and breakpoints are clear. Human 6p and 6q share homology with gibbon 8q and 4q respectively, although the orientation of each has then to be determined using locus specific probes.

Comparative maps based on reciprocal cross-species painting are now available for human and pig [9, 10], human and cat [11, 12], human and muntjac [13], mouse and chinese hamster (unpublished observations), mouse and rat [14], human and rabbit (unpublished observations).

Mechanisms of karyotype evolution

In the course of our studies on karyotype evolution in muntjac species [13, 15], we hybridised chromosome-specific paints from a number of species including human, sheep and several deer species to the three pairs of chromosomes of the Indian muntjac. By aligning the hybridisation results of these species against the idiogram of the Indian muntjac the various homologies shared by each species were evident at a glance [16]. In effect, we have used the muntjac idiogram as a universal mapping nomogram. An intriguing observation revealed by the nomogram was the remarkable coincidence of many of the breakpoints across the range of species.

Hybridisation of chromosome specific paints of deer species (brown brocket deer and tufted deer) with large chromosome numbers with the chromosomes of the Chinese muntjac ($2n = 46$) and the Indian muntjac, indicated that a reduction in chromosome numbers from an ancestral species with a chromosome number of 70 to the numbers observed in the muntjacs occurred largely as a result of centromere-telomere fusions of ancestral chromosomes. Using the C5 centromeric repeat sequence common in deer species [17], fragments of ancestral centromeric sequences could be detected at the presumptive fusion sites in the chromosomes of the Indian muntjac. This supports the hypothesis that the sites of rearrangement which occur during evolution contain homologous sequences which render the sites susceptible to illegitimate recombination during meiosis leading to fusions. Robertsonian translocations in humans have been postulated to result from such accidental events [18]. In the case of the origin of human chromosome 2 from two ancestral acrocentric chromosomes (with homologues in the great apes) the fusion site in human chromosome 2 has been shown to contain a centromeric alphoid sequence [19]. In some individuals, the fusion site behaves as an unstable fragile site leading to the formation of acentric fragments and triradials [20], an occurrence highly suggestive of a fission event. The underlying feature of fusion and fission events may well be a particular DNA sequence repeat behaving as a recombinational hot spot. A closer investigation of these sites at the molecular level is clearly indicated and may yield information relevant to these hypotheses.

In muntjacs, the karyotypes of extant species appear to have evolved from ancestral species largely by a series of fusion events. In contrast, the karyotypes of the Cercopithecinae subfamily (guenons) seem to have evolved by a series of fission and translocation events [6]. It is not at all clear what factors drive the direction of karyotype evolution in either of these examples.

Future studies in comparative painting

We have flow sorted chromosomes from over 50 different species of mammals and birds and demonstrated successful cross-species hybridisation between many of these. The most difficult have been the hybridisations between mouse and man and the easiest have been between primate species. We have so far been unsuccessful in our attempts to hybridise human autosomal paints to marsupial chromosomes, but have had some success with a marsupial X chromosome paint on the human X [21]. There is certainly room for considerable technological improvement in both hybridisation and signal detection and it is likely that greater resolution will be achieved in the future. Chromosome painting has the potential to contribute much further to our understanding of genomic evolution and one looks forward to extending the range of species studied.

References

1 Ferguson-Smith MA (1997) Genetic analysis by chromosome sorting and painting: phylogenetic and diagnostic applications. *Eur J Human Genet* 5: 253–265

2 Meltzer PS, Guan XY, Burgess A, Trent JM (1992) Rapid generation of region specific probes by chromosome microdissection and their application. *Nat Genet* 1: 24–28

3 Telenius H, Pelmear AH, Tunnacliffe A, Carter NP, Behmel A, Ferguson-Smith MA, Nordenskjold M, Peragner R, Ponder BAJ (1992) Cytogenetic analysis by chromosome painting using DOP-PCR amplified flow-sorted chromosomes. *Gene Chromosome Cancer* 4: 257–263

4 Wienberg J, Adamski E, Yang F, Müller S, Ferguson-Smith MA (1997) Chromosome painting without competitor DNA. Technical Tips Online (http://www.elsevier.com/locate/tto) T40065

5 Müller S, O'Brien PCM, Ferguson-Smith MA, Wienberg J (1997) A novel source of highly specific chromosome painting probes for human karyotype analysis derived from primate homologues. *Hum Genet* 101: 149–153

6 Rumpler Y, Dutrillaux B (1990) *Chromosome evolution and speciation in primates.* RBC Cell Biology Reviews, Vol. 23, Springer International, Bilbao

7 Müller S, O'Brien PCM, Ferguson-Smith MA, Wienberg J (1998) Cross-species colour segmenting: a novel tool in human karyotype analysis. *Cytometry* 33: 1–8

8 Stanyon R, Wienberg J (1998) Comparative chromosome painting of primate genomes. *ILAR J* 39: 77–91

9 Goreau A, Yerle M, Schmitz A, Riquet J, Milan D, Pinton P, Frelat G, Gellin J (1996) Human and porcine correspondence of chromosome segments using bidirectional chromosome painting. *Genomics* 36: 252–262

10 Milan D, Riquet J, Yerle M, Goureau A, Schmitz A, Cribiu EP, Frelat G, Gelin J (1996) Homologous and heterologous FISH painting with PARM-PCR chromosome-specific probes in mammals. *Mamm Genome* 7: 194–199

11 Rettenberger G, Klett C, Zechner U (1995) ZOO-FISH analysis: cat and human karyotypes closely resemble the putative ancestral mammalian karyotype. *Chromosome Res* 3: 479–486

12 Wienberg J, Stanyon R, Nash WG, O'Brien PCM, Yang F, O'Brien SJ, Ferguson-Smith MA (1997) Conservation of human vs. feline genome organisation revealed by reciprocal chromosome painting. *Cytogenet Cell Genet* 77: 211–217

13 Yang F, Müller S, Just R, Ferguson-Smith MA, Wienberg J (1997) Comparative chromosome painting in mammals: human and the Indian muntjac (*Muntiacus muntjak vaginalis*). *Genomics* 39: 396–401

14 Stanyon R, Yang F, Cavagna P, O'Brien PCM, Bagga M, Ferguson-Smith MA, Wienberg J (1999) Reciprocal chromosome painting shows that genomic rearrangement between rat and mouse proceeds ten times faster than between human and cats. *Cytogenet Cell Genet* 84: 150–155

15 Yang F, O'Brien PCM, Wienberg J, Ferguson-Smith MA (1997) A reappraisal of the tandem fusion theory of karyotype evolution in the Indian muntjac using chromosome painting. *Chromosome Res* 5: 109–117

16 Ferguson-Smith MA, Yang F, O'Brien PCM (1998) Comparative mapping using chromosome sorting and painting. *ILAR J* 39: 68–76

17 Lin CC, Sasi R, Fan Y-S, Chen Z-Q (1991) New evidence for tandem chromosome fusions in the karytype evolution of Asian muntjacs. *Chromosoma* 101: 19–24

18 Ferguson-Smith MA (1973) Human autosomal polymorphism and the non-random involvement of chromosomes in translocations. *In*: K Lewis (ed.): *Chromosome Today.* John Wiley and Sons, New York, 235–246

19 Avarello R, Pedicini A, Caicho A, Zuffardi O, Fraccaro M (1992) Evidence for an ancestral alphoid domain on the long arm of human chromosome 2. *Hum Genet* 89: 247–249

20 Ferguson-Smith MA (1973) Inherited contriction fragility of chromosome 2. *Ann Genet* 16: 29–34

21 Glas R, Marshall Graves JA, Toder R, Ferguson-Smith M, O'Brien PC (1999) Cross-species chromosome painting between human and marsupial directly demonstrates the ancient region of the mammalian X. *Mamm Genome* 10: 1115–1116

22 Schröck E, du Manoir S, Veldman T, Schoell B, Wienberg J, Ferguson-Smith MA, Ning Y, Ledbetter D, Bar-Am J, Soenksen D et al (1996) Multicolor spectral karyotyping of human chromosomes. *Science* 273: 494–497

23 Glas R, Wakefield MJ, Toder R, Graves JAM (1998) Comparative chromosome painting (Poster). *ILAR J* 39: numbers 2–3

Chromosomes Today, Vol. 13
ed. by E. Olmo and C.A. Redi

High resolution FISH reveals the molecular and chromosomal organisation of repetitive sequences of individual tomato chromosomes

J.H. de Jong[1], X.-B. Zhong[4], P.F. Fransz[2], J. Wennekes-van Eden, E. Jacobsen[3] and P. Zabel[2]

[1] *Laboratory of Genetics,* [2] *Laboratory of Molecular Biology and* [3] *Laboratory of Plant Breeding, Wageningen Agricultural University, Graduate School of Experimental Plant Sciences, Dreijenlaan 2, 6703 HA Wageningen, The Netherlands*
[4] *Yale University, School Of Medicine I-310 SHM, New Haven, CT 06520-8005, USA*

Introduction

High-resolution fluorescence *in situ* hybridisation (FISH) to pachytene complements and extended DNA fibres, along with pulse field gel electrophoresis, were used to study the chromosomal distribution and molecular organisation of the telomere repeat (TR) and the tomato-specific subtelomeric repeat (TGR1) in tomato. Pachytene chromosomes were chosen because of their superior resolution as compared to metaphase chromosomes. In addition, chromosomes at this meiotic prophase stage display a clear pattern of distinctive chromomeres and diagnostic heterochromatin blocks along the chromosome arms which enables identification of all bivalents. FISH to extended DNA fibres from lysed leaf nuclei provided a supplementary tool in allowing the reconstruction of physical maps showing the relative positions of both unique and repeated DNA sequences. The most striking result of the FISH study was the detection of a specific molecular organisation of TR with or without adjacent TGR1 sequences. A monosomic addition with tomato chromosome 6 added to tetraploid potato, which was found in the BC2 family of a hexaploid somatic potato (+) tomato hybrid, allowed the assessment of the telomere ends of the tomato chromosome. The integrity of the alien chromosome was tested with fourteen restriction fragments length polymorphism (RFLP) markers covering the entire linkage map of the alien chromosome. Although the presence of all markers could be established, indicating that chromosome 6 was intact, supplementary FISH using total genomic tomato DNA and the tomato-specific TGR1 and TGR2 satellite repeats revealed that the distal 7% of the long arm was replaced by potato chromatin. As expected, FISH to extended DNA fibres revealed red-green tracks for the remaining short arm TR/TGR1 combination and single green tracks for the long arm TGR1 site under the fluorescence microscope. Quantitative analysis of the fluorescing tracks revealed molecular size estimates of 424 kb for the short arm TR/TGR1 telomere and 163 kb for the long arm TGR1 sites, respectively. Southern analysis of *Bgl*II- and *Eco*RV-digested high molecular weight DNA, separated by Pulse Field Gel Electrophoresis (PFGE) and using TR and TRG1 as probes, demonstrated a dramatic decrease in TR size and the disappearance of TGR1 sequences in the intergeneric hybrid and the BC1 plant. This unexpected observation could be confirmed by microscopic analysis of TGR1 on mitotic metaphase complements showing a clear decrease in TGR1 sites. The results of high resolution FISH to extended DNA fibres were also in agreement with the PFGE data showing that telomere repeats in the somatic hybrid and backcross plants are significantly smaller than in the tomato and potato parent. In addition, the combination of long TGR1 tracks with adjacent TR tracks, which is typical for tomato telomeres, could not be found in the hybrid and BC1 plants. The results indicate that telomeres in the somatic hybrids become particularly unstable giving rise to breakage of TRs and TR/TGR1 combinations.

Complex genomes of higher plants are characterised by numerous types of repetitive DNA sequences. On the basis of their molecular organisation and chromosomal distributions two major classes can be distinguished, namely the group of highly repetitive (tandem) repeats and the group of middle- and low-copy repeats. The former is mainly confined to pericentromeric and (sub)telomeric heterochromatin blocks, whereas the latter are dispersed along large parts of the chromosomes that occur in both euchromatin and heterochromatin areas.

Knowledge on the organisation of repeats was originally based on molecular studies (reviews in [1–3]), but has since been supplemented with a vast body of studies on the DNA sequence, evolutionary relationships and chromosomal locations in a variety of plant species. It thus became obvious that, although most of them are non-coding, repeats play a unique role in chromosome house-holding and genome stability. Among the most conserved tandem repeats are the 45S and 5S rDNA families, and the telomere repeat. A second group includes the hypervariable and heterogenous group of satellite repeats, which occurs in long arrays and are generally species specific. In tomato, three such repeats have been described [4, 5] including (*i*) TGR1, which occurs in the subtelomeric regions of most chromosome arms and on some specific interstitial long arm chromomeres, and amounts to 25–1000 kb; (*ii*) TGR2, which is scattered along all chromosomes except chromosome 2, and amounts to c. 4200 copies per genome, and (*iii*) TGR3, which is located in the centromere regions and at interstitial regions of a few chromosomes, and amounts to c. 2100 copies per genome.

The chromosomal distribution of repetitive sequences has initially been studied using fluorescence *in situ* hybridisation (FISH) to mitotic metaphase cells. The introduction of multicolour FISH with different probes has given the additional advantage of simultaneously mapping the relative position of the different repeats along the chromosomes [6]. The number of probes though that can be mapped simultaneously depends on the fluorochromes distinguishable by spectral means. With basic microscopic equipment that does not involve the application of technically advanced CCD image capturing and multicolour image processing techniques, FISH usually employs two fluorescence systems, such as the green fluorescing isothiocyanate FITC and red fluorescing Texas Red or similar, in combination with the blue fluorescing DNA specific 4',6-diamindino-2-phenylindole (DAPI). Such a system with green fluorescing FITC for detecting TGR1 and red fluorescing Texas Red for TR has been applied in the present FISH experiments.

High resolution FISH

Since its introduction in the mid-eighties, FISH has shown great progress in resolution enhancement and detection sensitivity. The use of FISH was initially limited to metaphase chromosomes with target sequences spatially separated by merely 1–2 Mb, the development of interphase nuclei as targets provided further improvement in mapping resolution by as little as 50 kbp [7], but with a loss of information on recognisable chromosome structure. Similar advances were obtained for plants using both interphase [8] and prophase chromosomes for the *in situ* hybridizations. Meiotic prophase cells at pachytene, in which the chromosomes are 7–40 times longer than their metaphase counterparts, proved highly appropriate for high resolution FISH. Pachytene chromosomes are also superior in displaying well-differentiated patterns of chromomeres and heterochromatin blocks. Such chromosomal markers, together with chromosome length and centromere position, allow identification of individual chromosomes in the karyotype. Tomato, with its splendid pachytene morphology [9], has therefore become a cytogenetic model species par excellence for high resolution FISH [10].

A dramatic jump towards enhanced spatial resolution of neighbouring targets on the chromosomal DNA was aided by the development of protocols for preparing free chromatin and extended genomic DNA fibres from interphase nuclei that are amenable to hybridisation. The technology was first developed for human applications [11, 12], but soon became accommodated to plants [13], using a procedure for releasing free DNA from lysed interphase leaf nuclei and spreading the released DNA fibres along the receding drop across the surface of a tilted microscopic slide. Extended fibres were thus produced with a stretching degree of 3.0–3.5 kb/μm [13]. FISH of probes from yeast artificial chromosomes (YACs), bacterial artificial chromosomes (BACs), cosmids and λ-clones to DNA fibres allowed mapping targets at an accuracy of less than a few kilobases and a detection sensitivity of only a few hundred basepairs [13, 14].

Tomato telomeres

Concurrent FISH to pachytene complements and extended DNA fibres was found very effective in elucidating the chromosomal distribution of specific sequences relative to heterochromatin, centromeres and telomeres, as well as their molecular organisation on linearised DNA molecules. In this report, we describe the potential of this technology for revealing the organisation of two components of the chromosomal ends in tomato: the telomere repeat (TR) with the TTTAGGG motif, typic of most plant species [15]; and the telomere-specific tandem repeat TGR1 [5, 16, 17]. In a previous study [18] we utilised FISH on pachytene chromosomes and extended DNA fibres to reveal the molecular organisation of these repeats in tomato. The TR and TGR1 signals were mapped to individual pachytene chromosomes on the basis of chromosome length, centromere position and heterochromatin pattern. While TRs occur on all chromosome ends, TGR1 was not found on the chromosome termini of the long arm of chromosome 1, both long arm and short arm telomeres of chromosome 2, and the long arm of chromosome 7. On the other hand, three additional TGR1 sites were detected at the interstitial chromomeres of the long arm of chromosomes 6, 8 and 9. The ideogram with DAPI-counterstained pachytene chromosomes depicting their distributions of TR and TGR1 repeats is shown in Figure 1. Three classes of signals could be discerned: (*i*) the combination of partly overlapping green fluorescing TGR1 and red fluorescing TR signals, producing the typical green – yellow – red spots, was observed on twenty chromosome ends; (*ii*) four termini with only the red TR spot and (*iii*) three more interstitial sites with a green TGR1 spot.

Figures 1–4 see color plate XV between pages 256 and 257.

Figure 1. Ideogram of tomato chromosomes at pachytene, depicting the regions of heterochromatin, euchromatin and AT-rich microsatellite (nucleolar organiser region of chromosome 2) DNA. The constrictions in the heterochromatin areas represent the position of the centromeres. Chromosomes are ordered in sequence of decreasing length and classified according to Ramanna and Prakken [9]. The positions of the telomere repeat TTTAGGG and the subtelomere repeat TGR1 are indicated.

Figure 2. Two-colour FISH of the tomato telomeric repeat TR (red) and the subtelomeric repeat TGR1 (green) on extended DNA fibres demonstrates four classes according to Zhong et al. [18, 21]: (a) TR – spacer – TGR1; (b) TR – TGR1; (c) only TR and (d) only TGR1.

Figure 3. FISH to extended DNA fibres and to the alien chromosome at mitotic metaphase of the monosomic addition BC2 plant. The plant was found to contain the tomato chromosome 6, except for the distal end of the long arm that is of potato origin. (a) The arrows indicate the positions of the short arm TR and adjacent TGR1 segment of the DNA fibres from two chromosomes. (b) GISH using total genomic DNA from tomato. Only the distal end of the long arm shows faintly the red fluorescing Propidium Iodide counterstain of the potato chromosome segment. (b) Alien chromosome, counterstained with DAPI and painted with the TGR1 probe, only shows the distal target on the short arm of the chromosome. (c) Same chromosome, now painted with the green TGR1 signal and red TGR2 signal.

Figure 4. Comparison of two-colour FISH signals on extended DNA fibres. (a) Tomato parent. Arrows indicate examples of long tomato TR tracks; (b) Potato parent with only very short TR tracks; (c) Somatic hybrid: note the absence of long tomato TRs.

Additional information on the molecular size of the TR and TGR1 signals was obtained from FISH studies of extended DNA fibres. Molecular size estimates of the TR and TGR1 sites were based on the 3.27 kb/μm extension degree [13]. As expected, the fluorescence tracks of the two-colour FISH to extended fibres were in agreement with our data on the pachytene FISH map. Typically, red fluorescing tracks of the TR were found in the range of 25–223 kb, and green fluorescing tracks of the TGR1 varying from 88–1330 kb. In most cases, TR and TGR1 were in line, sometimes being separated by a spacer that measured 98 kb at most [18]. We were able to discriminate 27 different TR plus TGR1 combinations, subdivided into four classes as shown in Figures 2 a–d: a) TR-spacer-TGR1 (thirteen cases); b) TR-TGR1 (7); only TR (4) and d) only TGR1 (3). These results suggest that each chromosome end possesses a unique telomere organisation, with only TR, or a combination of TR and TGR1 partly separated by a spacer. The four TR sites without TGR1 represent the chromosome termini of 1L, 2S, 2L and 7L arms, whereas the three remaining TGR1 tracks can be mapped to the three interstitial long arm chromomeres on chromosomes 6, 8 and 9, known to contain the subtelomeric repeat (cf. Fig. 1).

Chromosome 6 of tomato in the monosomic addition

The next question we addressed was how to assign individual TR-TGR1 combinations in extended DNA fibre preparations to specific chromosome ends. In principle, multicolour FISH with probes of unique sequences diagnostic for each distal chromosome segment, and adjacent to the (sub)telomeric repeats, should be able to resolve this question. However, the abundance of many other repeats, which are common for most chromosome regions, profoundly impedes proper interpretation. In an alternative approach, the use of monosomic additions, containing single tomato chromosomes added to the cell complement of potato or other related species, reduces the number of TR/TGR1 sites to the two chromosome arms of the alien chromosome. Such aneuploids have only recently been isolated in back-crossed families of somatic potato (+) tomato hybrids [19, 20]. The first monosomic addition analysed for this purpose possesses the tomato chromosome 6, as found in the second backcross family of a allohexaploid potato (+)-tomato hybrid, pollinated with 4x potato. RFLP analysis using fourteen tomato chromosome 6 markers demonstrated full integrity of the entire linkage map [21]. However, additional FISH to metaphase and pachytene complements, using total genomic tomato DNA and the tomato-specific TGR1 and TGR2 repeats as probes, revealed that the distal 7% of the long arm of alien tomato chromosome 6, inclusive the distal TGR1 site, had been replaced by potato DNA. This chimeric chromosome, which is the apparent result of a translocation or homeologous recombination between tomato chromosome 6 and one of the potato chromosomes, had lost the distal long arm TGR1 site and adjacent chromatin, but not the distal short arm and interstitial long arm TGR1 sites (cf. Fig. 1, Figs 3b–d).

FISH to extended DNA fibres using TR and TGR1 as probes produced preparations with fluorescent tracks that were in agreement with the results of previous FISH experiments to metaphase and pachytene spread preparations. We detected bicolour tracks consisting of short red fluorescing TR tracks, measuring 5 ± 1 μm in length, and a long green fluorescing TGR1 track of 125 ± 10 μm, without spacer (Fig. 4a). The second group of TGR1 tracks, which were not connected to any red signal and likely to represent the interstitial TGR1 site, measured 50 ± 3 μm. Converting these values into molecular sizes using the previously described extension degree of 3.27 kb/μm gave a size estimate of 16 ± 3 kb for the short arm TR, 408 ± 33 kb for the adjacent TGR1 site, and 163 ± 10 kb for the interstitial long arm TGR1 site, respectively.

Telomeres of tomato change in the somatic hybrid

We next applied Southern analysis with TR and TGR1 as probes to digested high molecular weight DNA separated by pulse field gel electrophoresis (PFGE). To this end DNA samples from the tomato and potato parents, the somatic hybrid, and the BC1 and monosomic addition BC2 plants were digested with the restriction enzymes *Bgl*II and *Eco*RV, which were known to cut outside the TR and TGR1 repeats [16]. The Southern plots of the TR demonstrated smears for both parents, with a fragment size of less than 150 kb for potato and 50–250 kb for tomato. Tomato also depicted a few discrete bands between 600 and 1900 kb. In the somatic hybrid, the smear of TR fragments was positioned slightly lower than that of potato, and most of the tomato-specific bands were missing. In the BC1 and BC2 plants the smear of small fragments was close to that of the potato parent, with only a few tomato bands remaining in the BC1 plant and a single band for the BC2 plant. The latter fragment amounted to 560 kb in the *Bgl*II digest and 470 kb in the *Eco*RV digest, values that are slightly larger than the 424 kb as established by FISH to extended DNA fibres.

The same blot, hybridised with TGR1, confirmed the species specificity of this subtelomeric repeat for tomato whereas the fragment sizes were in the range observed in the FISH to extended DNA fibres. In the somatic hybrid most of the TGR1 fragments were missing, the absence of the larger TGR1 fragments in the *Bgl*II digested DNA being especially obvious. In the BC1 and BC2 plants, the number of TGR1 fragments diminished even further, leaving only one 560 kb band in the *Bgl*II lane and 470 and 580 kb bands in the *Eco*RV lane.

In order to better understand the unexpected behaviour of the telomere repeats in the somatic hybrid, BC1 and BC2 plants, we performed additional

FISH of the TR and TGR1 sequences to mitotic metaphase and extended DNA fibres preparations. The tomato-specific TGR2 repeat [4] was included to paint the tomato chromosomes in the chromosome spreads of the somatic hybrid. Analysis of the mitotic cells of the somatic hybrid revealed that at least six telomere TGR1 sites were lost. In the FISH preparations of TR and TGR1 probed to extended DNA fibres of the somatic hybrids, no long tracks of tomato TRs were detected, whereas almost any of the green fluorescing TGR1 tracks showed a clear adjacent TR signal. The BC1 plant displayed a TR and TGR1 track that was comparable to that of the somatic hybrid. The BC2 plant, the monosomic addition with the alien tomato chromosome 6 containing a small distal potato segment, exhibited only the previously described TR and TGR1 combination of the short arm and the interstitial long arm TGR1 tracks.

The results of both Southern analysis and FISH to metaphase complements and extended DNA fibres point at some kind of chromosomal instability and/or breakdown in the somatic hybrids occurring at telomeres of potato and tomato chromosomes. It was to be expected that the *Bgl*II and *EcoR*V hybridisation profiles of the Southerns in the somatic hybrid would be a combination of those of the tomato and the potato parent. Instead, most bands of the large tomato fragments disappeared, while the smear of smaller TR fragments in the somatic hybrid was even smaller than that of the potato parent. This band shift and/or complete loss of telomere DNA was in agreement with the disappearance of TGR1 sites in mitotic metaphase complements and the absence of large TR tracks and TR/TRG1 combinations in extended DNA fibre preparations.

The phenomenon of genome instability in newly synthesised polyploid hybrids between related species or genera resulting in loss or gain of DNA sequences has been reported in only a few cases. Pestsova et al. [22] described a dramatic reduction in copy number on slot blots and a decrease in chromosomal sites from 13 to 2 per haploid genome of the *Aegilops speltoides*-specific subtelomeric Spelt1 repeat in related allopolyploid *Triticum* species. This suggests that the Spelt1 repeat was eliminated in the course of wheat allopolyploidisation. A similar process may have occurred during the production of newly synthesised amphiploids with the genomic constitution of hexaploid wheat, revealing rapid, non-random elimination of specific, low-copy, probably non-coding sequences [23]. Such sudden modifications of, possibly, specific sequences were considered to be brought about by DNA methylation, perhaps resulting from ploidy changes [24].

The important question as to whether the TR and TGR1 repeats in chromosome 6 of the monosomic addition BC2 plant remain unaffected during protoplast fusion, regeneration of the somatic hybrid or backcrossings is still to be answered. Although the lengths of the TR and TGR1 tracks of the tomato chromosome in the monosomic addition precisely match the class IIc TR/TGR1 combination in wild type tomato [18], it cannot be excluded that the short arm telomere end of this chromosome becomes affected by the previously mentioned process of telomere degradation. Our data from the Southern analysis suggest that especially the large TR/TGR1 repeats become eliminated or affected during the formation of the somatic hybrid, thereby leaving the relatively small TR/TGR1 combination of the short arm chromosome 6 unchanged.

Acknowledgements

The research was supported financially by a grant from the Dutch organisation for Scientific Research – NWO to XBZ. PF is a postdoctoral fellow supported by a grant (No. PL960443) of the Commission of the European Community. The authors thank Drs. E.J. Richards and S.D. Tanksley for providing DNA clones.

References

1 Flavell RB (1980) The molecular characterisation and organisation of plant chromosomal DNA sequences. *Annu Rev Plant Physiol* 31: 569–596
2 Flavell RB (1986) Repetitive DNA and chromosome evolution in plants. *Phil Trans R Soc Lond B* 312: 227–242
3 Dean C, Schmidt R (1995) Plant genomes: a current molecular description. *Annu Rev Plant Physiol Plant Mol Biol* 46: 395–418
4 Ganal MW, Lapitan NLV, Tanksley SD (1988) A molecular and cytogenetic survey of major repeated DNA sequences in tomato. *Mol Gen Genet* 213: 262–268
5 Lapitan NLV, Ganal MW, Tanksley SD (1989) Somatic chromosome karyotype of tomato based on *in situ* hybridisation of the TGRI satellite repeat. *Genome* 32: 992–998
6 Zhong XB, de Jong JH, Zabel P (1996) Preparation of tomato meiotic pachytene and mitotic metaphase chromosomes suitable for fluorescence *in situ* hybridisation (FISH). *Chromosome Res* 4: 24–28
7 Trask BJ (1991) Fluorescence *in situ* hybridization: application in cytogenetics and gene mapping. *Trends Genet* 7: 149–154
8 Jiang J, Hulbert SH, Gill KS, Ward DC (1996) Interphase fluorescence *in situ* hyridization mapping: a physical mapping strategy for plant species with large complex genomes. *Mol Gen Genet* 252: 497–502
9 Ramanna MS, Prakken R (1967) Structure of and homology between pachytene and somatic metaphase chromosomes of the tomato. *Genetica* 38: 115–133
10 Zhong XB, Fransz PF, Wennekes-van Eden J, Zabel P, van Kammen A, de Jong JH (1996) High resolution mapping by fluorescence *in situ* hybridisation to pachytene chromosomes and extended DNA fibres. *Plant Molec Biol Reporter* 14: 232–242
11 Wiegant J, Kalle W, Mullenders L, Brookes S, Hoovers JNM, Dauwerse JG, Van Ommen JGB, Raap AK (1992) High-resolution *in situ* hybridization using DNA halo preparations. *Hum Mol Genet* 1: 587–591
12 Parra I, Windle B (1993) High resolution visual mapping of stretched DNA by fluorescent hybridisation. *Nat Genet* 5: 17–21
13 Fransz PF, Alonso-Blanco C, Liharska TB, Peeters AJM, Zabel P, de Jong JH (1996) High resolution physical mapping in *Arabidopsis thaliana* and tomato by fluorescence *in situ* hybridisation to extended DNA fibre. *Plant J* 9: 421–430
14 Jackson SA, Wang ML, Goodman HM, Jiang J (1998) Application of fibre-FISH in physical mapping of *Arabidopsis thaliana*. *Genome* 41: 566–572
15 Richards EJ, Ausubel FM (1988) Isolation of a higher eukaryotic telomere from *Arabidopsis thaliana*. *Cell* 53: 127–136
16 Ganal MW, Lapitan NLVTanksley SD (1991) Macrostructure of the tomato telomeres. *Plant Cell* 3: 87–94

17 Ganal MW, Broun P, Tanksley SD (1992) Genetic mapping of tandemly repeated telomeric DNA sequences in tomato (*Lycopersicum esculentum*). *Genomics* 14: 444–448
18 Zhong XB, Fransz PF, Wennekes-van Eden J, Ramanna MS, van Kammen A, Zabel P, de Jong JH (1998) FISH studies reveal the molecular and chromosomal organisation of individual telomere domains in tomato. *Plant J* 13: 507–517
19 Jacobsen E, de Jong JH, Kamstra SA, van den Berg PMMM, Ramanna MS (1995) Genomic *in situ* hybridisation (GISH) and RFLP analysis for the identification of alien chromosomes in the back-cross progeny of potato (+) tomato fusion hybrids. *Heredity* 74: 250–257
20 Carriga-Calderé F, Huigen DJ, Filotico F, Jacobsen E, Ramanna MS (1997) Identification of alien chromosomes through GISH and RFLP analysis and the potential for establishing potato lines with monosomic additions of tomato chromosomes. *Genome* 40: 666–673
21 Zhong XB, Fransz PF, Wennekes-van Eden J, Jacobsen E, de Jong JH, Zabel P (1998) A novel strategy for establishing the molecular organisation of specific repeat sites using high resolution FISH to alien chromosomes of tomato-potato monosomic additions. *In*: XB Zhong – PhD thesis: *Tomato genome mapping by fluorescence* in situ *hybridisation.* Wageningen Agricultural University, 71–92
22 Pestsova EG, Goncharov NP, Salina EA (1998) Elimination of a tandem repeat of telomeric heterochromatin during the evolution of wheat. *Theor Appl Genet* 97: 1380–1386
23 Feldman M, Liu B, Segal G, Abbo S, Levy AA, Vega JM (1997) Rapid elimination of low-copy DNA sequences in polyploid wheat: a possible mechanism for differentiation of homoeologous chromosomes. *Genetics* 147: 1381–1387
24 Scheid OM, Jakovleva L, Afsar K, Maluszynska J, Paszkowksi J (1996) A change of ploidy can modify epigenetic silencing. *Proc Natl Acad Sci USA* 93: 7114–7119

Chromosomes Today, Vol. 13
ed. by E. Olmo and C.A. Redi

Species relationships, introgression and physical mapping in *Lolium perenne* × *Festuca pratensis* hybrids

King I.P.[1], Morgan W.G.[1], Armstead I.P.[1], King J.[1], Jones T.A.H.[1], Nash J.V.[1], Jones R.N[2]., Harper J.A.[1], Roberts L.A.[1], Steve Taylor[1] and Thomas H.M.[1]

[1] *Institute of Grassland and Environmental Research, Plas Gogerddan, Aberystwyth, Ceredigion, SY23 3EB, Wales, UK*
[2] *Institute of Biological Sciences, University of Wales Aberystwyth, Aberystwyth, Ceredigion, SY23 3DA, Wales, UK*

Introduction

The introgression of genes from alien species provides a means of extending the gene pool available to plant breeders for the development of new varieties. However, the process of alien introgression has proved to be difficult in many crop species. For example, the introgression of desirable target genes from an alien species into wheat via recombination at meiosis is not straight forward (for review see [1]). The chromosomes of many alien species will not recom bine with those of wheat due to the presence of a pairing control gene, Ph1, located on the long arm of chromosome 5B. This gene, which inhibits homologous recombination, has to be removed before the chromosomes of wheat and the chromosomes of alien species such as rye can recombine. However, even when the Ph1 gene is removed the frequency of alien recombination is still very low. Another problem with wheat/alien introgression is that when a chromosome segment carrying a desirable gene is transferred to wheat the alien segment also frequently carries deleterious genes. Because of the pairing control gene and the low frequency of wheat/alien recombination it is normally very difficult to break the linkage between the target gene and the deleterious genes. Finally, even though the frequency of recombination between wheat and alien species is low, in many cases it is not possible to routinely screen for the presence or absence of introgressed chromosome segments using genomic *in situ* hybridization (GISH).

The transfer of genes from *F. pratensis* (*Fp*) into *Lolium perenne* (*Lp*), two of the most important forage grasses worldwide, has considerable advantages over other alien introgression systems [2–4]. The chromosomes of *Lp* and *Fp* in diploid *LpFp* and triploid *LpLpFp* hybrids recombine at high frequency, i.e. there are no problems with pairing control genes such as Ph1. Chromosome segments of *Fp* introgressed into *Lp* do carry deleterious genes along with desirable genes. However, because of the absence of pairing control genes which inhibit homologous recombination and the high frequency of *Lp*/*Fp* recombination it is relatively easy to break the linkage between desirable and deleterious genes. Another positive feature, of even greater importance, is the fact that although *Lp* and *Fp* chromosomes recombine at a high frequency it is very easy to distinguish between the genomes of the two species using genomic *in situ* hybridization (GISH). This unique combination of features exhibited by the *Lp* and *Fp* hybrids i.e. the high recombination frequency and the ability to differentiate the chromosomes of the two species with GISH, makes the hybrids ideally suited for alien introgression.

Species relationships

The phylogenetic relationship between *Lp* and *Fp* is unclear. The fact that *Lp* and *Fp* hybridize readily and the hybrids show full chromosome pairing indicates that *Lp* and *Fp* are closely related [5]. Further evidence of this closeness comes from a preliminary molecular analysis based on the use of random amplified polymorphic DNAs (RAPDs) [6]. These kinds of observations have led various researchers to suggest that *Fp* should be reclassified as a *Lolium*. In contrast, there is other evidence which indicates that *Lp* and *Fp* are distantly related. Firstly, *Lp*/*Fp* hybrids are completely male sterile [7, 5]. Secondly, the fact that *Lp* and *Fp* chromosomes can be distinguished using GISH [4] indicates that considerable divergence of repetitive sequences has occurred. In addition, the position of rDNA sites also indicates that the two species have diverged [8, 9]. In *Lp* and its close relative *Lolium multiflorum*, the 18S-5.8S-26S rDNA and 5S rDNA sites are syntenic. However, in *Fp* the 18S-5.8S-26S rDNA sites are located on different chromosomes as is the case in other closely related *Festuca* species, i.e. *F. glaucescens, F. arundinacea, F. marei, F. gigantea.*

In order to get a better insight into the phylogenetic relationship between *Lp* and *Fp*, and also to quantify the frequency of *Lp*/*Fp* recombination, we analysed metaphase I of meiosis in the *LpLpFp* triploid hybrid using GISH. Analysis of meiosis in hybrids has been a long established method of determining the relationships of genomes from different species [10–15].

Using the GISH technique it was possible to distinguish the *Lp* and *Fp* chromosomes in each pollen mother cell observed (Fig. 1a) [4]. Thus it was possible to record the presence or absence and position of the *Lp* and *Fp* chromosomes in each meiotic configuration (Tab. 1).

Each of the 14 *Lp* and seven *Fp* chromosomes can be placed in one of seven homologous/homoeologous groups such that each group contains two homologous *Lp* chromosomes and one homoeologous *Fp* chromosome. In the absence of preferential pairing the three chromosomes in each group will pair at random with each other. Thus, at any single position on the chromosome,

Figure 1. (See color plate XVI between pages 256 and 257.) (a) Metaphase I (meiosis) of the *LpLpFp* triploid with the *Fp* chromosomes in green following GISH using *Fp* DNA as a probe, showing five frying pan trivalents, four with the *Fp* chromosome in the ring and one with the *Fp* chromosome as handle, one *Lp* bivalent, two *Lp* univalents and two *Fp* univalents; (b) metaphase I (mitosis) of the *LpLpFp* triploid with the *Fp* chromosomes in green following GISH using *Fp* DNA as a probe, showing an introgression in two different chromosomes, one of which is the result of a double recombinant event; (c) (mitosis) shows a series of different lines either carrying a complete *Fp* chromosome or a single *Fp* chromosome segment (in red) derived from the same chromosome. The physical position of the genetic markers can be determined by the presence or absence of *Fp*-specific polymorphisms (determination of the exact order of markers 3538.11 and 3542.17; OPAGO26 and 3536.15; and 3544.2 and 3535A would require analysis of further recombination or the production of a *Lp*/*Fp* genetic map).

Table 1. Frequency of meiotic configurations in the *LpLpFp* triploid hybrid

Univalents		*Lp* bivalents		*Lp/Fp* bivalents		V-shaped or linear trivalents		Frying pan trivalents (handle in brackets)	
Lp	*Fp*	Rods	Rings	Rods	Rings	*LpLpFp*	*LpFpLp*	*LpFp(Lp)*	*LpLp(Fp)*
39	239	31	204	6	22	85	14	65	72

Four other configurations, i.e. two *Fp*/*Fp* bivalents and two triradial bivalents are not included in the above table.

each *Lp* chromosome can pair homologously with the other *Lp* chromosome or homoeologously with the *Fp* while the *Fp* chromosome can only pair homoeologously with either of the *Lp* chromosomes. Consequently, in the absence of preferential pairing, the ratio of homoeologous chromosome to homologous chromosome pairing is 2:1. Therefore, in the absence of trivalent formation, *Lp*/*Fp* bivalents should outnumber *Lp*/*Lp* bivalents by 2:1. However, in the 72 pollen mother cells analysed, only 28 *Lp*/*Fp* bivalents compared to 235 *Lp*/*Lp* bivalents were observed (Tab. 1). Clearly this does not fit a 2 *Lp*/*Fp*:1*Lp*/*Lp* ratio $\chi^2_1 = 368.93$, $P < 0.001$). Again, with random pairing, there should be twice as many *Lp* univalents as *Fp* univalents; but there were significantly more *Fp* univalents than *Lp* univalents observed $\chi^2_1 = 142.46$, $P < 0.001$) (Tab. 1).

Ninety-nine chain trivalents were observed (Tab. 1). There are two opportunities for the *Fp* chromosome to be at the end of a linear trivalent and only one at the middle position. Thus, if there is no preferential pairing, the ratio at which the *Fp* chromosome should occur at the end of a trivalent compared to the middle of the trivalent is 2:1. Thus, in the 99 chain trivalents observed the *Fp* chromosome should be at the end of the configuration on approximately 66 occasions and in the middle on 33 occasions. However, the *Fp* chromosome was at the middle position in only 14 trivalents and was at the end position in 85 trivalents, which was significantly different from the expected 2:1 ratio $\chi^2_1 = 15.56$, $P < 0.001$). The two chromosomes at the ends of the configuration each only need to pair once to the chromosome in the centre of the configuration. In contrast, the chromosome at the centre of the configuration has to pair twice. The *Fp* chromosome therefore occupied a position that required the least number of chiasmata between homoeologous chromosomes.

Analysis of the 137 "frying pan" trivalents revealed a similar pattern (Tab. 1). In the absence of preferential pairing, the ratio at which the *Fp* chromosome should appear in the ring relative to the handle is 2:1. However, the *Fp* chromosome was located within the ring significantly less frequently than expected ($\chi^2_1 = 21.93$, $P < 0.001$). The *Fp* chromosome is not therefore randomly located within the configuration. Chromosomes within the ring have to form chiasmata either twice with one chromosome or three times with two chromosomes whereas the chromosome forming the handle only needs to form one chiasma with one chromosome. As in the chain trivalents, the *Fp* chromosome is preferentially located to form the least number of chiasmata with *Lp* chromosomes.

In the absence of preferential pairing in the *LpLpFp* hybrid, it is expected that the ratio of homoeologous to homologous paired arms would be 2:1. Thus, out of the 1098 paired arms observed, approximately 732 would be expected to be between homoeologous *Lp*/*Fp* chromosomes and 366 between homologous *Lp* chromosomes. However, only 364 of the paired arms were between homoeologous chromosomes. Not surprisingly, a chi-squared analysis rejected a 2*Lp*/*Fp*: 1*Lp*/*Lp* ratio ($\chi^2_1 = 553.5$, $P < 0.001$).

In summary, the GISH analysis shows overwhelmingly that preferential pairing occurs between the *Lp* homologues in the *LpLpFp* triploid and demonstrates that the *Fp* genome is distinct from the *Lp* genome.

The result of the meiotic analysis does not support the reclassification of *Fp* as a *Lolium*. However, with regard to alien introgression the phylogenetic relationship of *Lp* and *Fp* is of no consequence. The important factors are that, 1) the two species can be distinguished using GISH thus allowing the presence and size of *Fp* chromosome segments introduced into *Lp* to be monitored from one generation to the next, 2) there is a very high frequency of homoeologous recombination between the chromosomes of the two species. The frequency of homoeologous recombination between the two species was not 66% as would have been expected in the absence of preferential pairing but, at 33%, was still very high.

Introgression

In order to ascertain the frequency at which *Fp* chromosomes and *Fp* chromosome segments were transferred to the next generation the *LpLpFp* triploid hybrid described above was backcrossed to diploid *Lp* [3]. Seventy eight BC_1 individuals were analysed by GISH, and of these 75 had 14 chromosomes and three had 15. Fifty-eight plants, including the three with 15 chromosomes, carried one or more *Fp* chromosomes or chromosome segments (Fig. 1b). Three lines carried *Lp*/*Fp* chromosome substitutions and these were fully fertile, near-normal plants, indicating the substitution of a homoeologous chromosome. Fourteen plants carried a single *Fp* chromosome segment which ranged in size from a small segment to more than a whole chromosome arm. Thirty-one plants had two *Fp* chromosome segments, nine had three and one had six. The remaining 20 plants analysed had no *Fp* chromosome segment detectable by GISH.

Analysis of the *Lp*/*Fp* recombinant chromosomes revealed that homoeologous recombination in the parental *LpLpFp* triploid plant appeared to occur along the entire length of the *Lp* and *Fp* chromosomes. Some *Lp*/*Fp* recombinant chromosomes demonstrated that double-recombinant events occurred between *Lp* and *Fp* homoeologous arms.

Amplification fragments length polymorphisms (AFLPs), restriction fragments length polymorphisms (RFLPs) and RAPDs were used to determine which linkage group each BC_1 individual represented. From these data three linkage groups were identified as having whole *Fp* chromosome substitutions and two linkage groups were identified as including individuals with single chromosome arm substitutions. The size of the *Fp* chromosome segments, determined by GISH, was related to the presence or absence of *Fp*-specific polymorphisms (Fig. 1c). This determined the approximate physical position of marker loci [3].

Physical mapping

In the *Lp*/*Fp* system not only is there a high frequency of homoeologous recombination, and a means to distinguish the chromosomes of the two species using GISH, but also recombination occurs throughout the length of the chromosomes. These three features are being exploited for physical mapping, as follows: (i) Fourteen chromosome arm substitution lines are being isolated. Each chromosome arm substitution line having a different chromosome arm of *Lp* replaced by the homoeologous *Fp* chromosome arm (Fig. 2). To date eight chromosome arm substitution lines have been isolated [3]. (ii) Once a chromosome arm substitution line has been characterized it is then backcrossed to *Lp* to produce a BC_2 population. The BC_2 population will be comprised of individuals carrying different size *Fp* chromosome segments. From the BC_2 population a series of plants can be selected which carry smaller and smaller *Fp* chromosome segments i.e. a recombinant chromosome series (Fig. 3). (iii) Once the recombinant chromosome series have been isolated they can be used to determine the genetic and physical position of markers (Fig. 3). In Figure 3 we have only shown a recombinant chromosome series comprised of six plants. However, recombinant chromosome series comprised of thousands of plants can be generated by simply backcrossing chromosome arm substitution lines with *Lp*. Thus, the generation of a high resolution physical map is a relatively simple process. In addition, by screening just for the presence or absence of each of the genetic markers in the BC_2 population a genetic map of

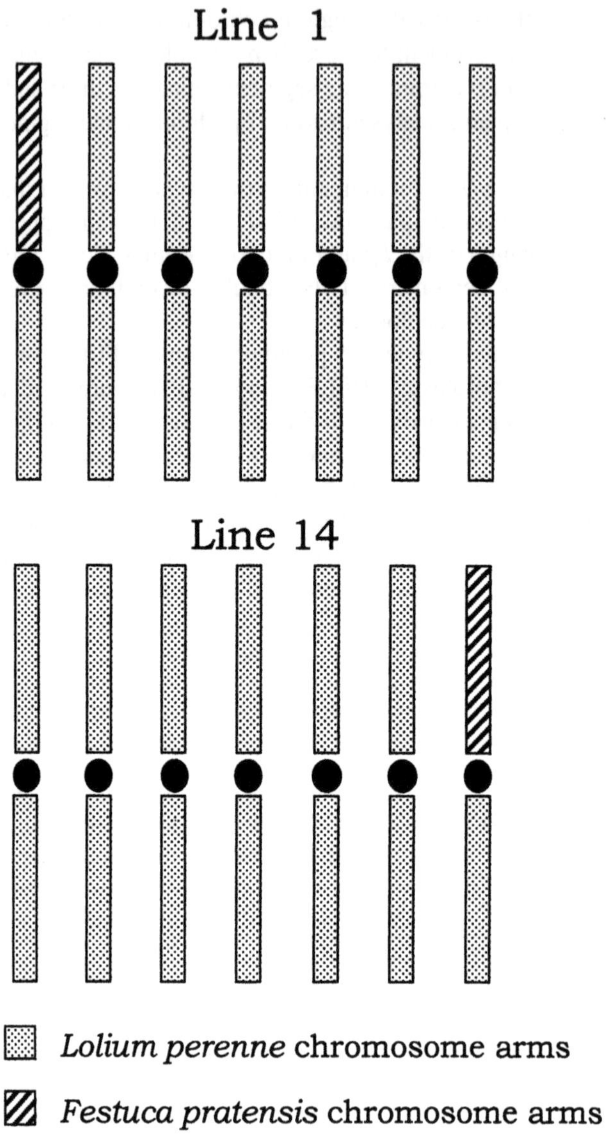

Figure 2. Chromosome arm substitutions lines 1 and 14 (haploid chromosome compliment).

the chromosome arm can also be generated. By comparing the two types of maps we are determining how the genetic distance in centimorgans relates to the physical distance from one part of the genome to another [3].

In a preliminary experiment we have isolated two BC_1 individuals which carry the same introgressed *Fp* chromosome segment. Fourteen genetic markers, AFLPs, RFLPs, and RAPDs, located on the *Fp* chromosome segment have

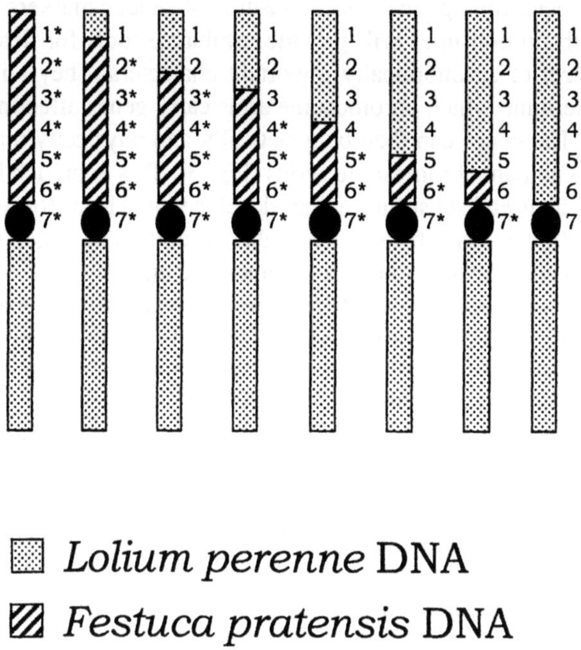

Figure 3. Recombinant chromosome series 1. Numbers 1 to 7 indicate genes or markers (showing *Lp*/*Fp* polymorphism) located on a single chromosome arm. Numbers with an asterisk indicate *Fp* alleles/polymorphisms, numbers without an asterisk indicate *Lp* alleles/polymorphisms.

been used to screen the two BC_2 populations derived from them. Analysis of the segregation of these markers in 94 individuals has revealed that the *Fp* chromosome segment is 54 cM long. Twenty-two different types of recombinant chromosome were detected most of which were derived from a single *Lp*/*Fp* recombinant event. However, some of the recombinant chromosomes resulted from up to five *Lp*/*Fp* recombinant events. A comparison of genetic distance and physical distance revealed that, at the centromere 7%, of the genetic distance related to 21% of the physical distance of the *Fp* chromosome segment. In contrast, interstitially 21% of the genetic distance related to11% of the physical distance of the *Fp* chromosome segment. Near the ends of the chromosome 15% of the genetic distance equated to 17% of the genetic distance. Thus, from this data it appears that the lowest frequency of recombination occurs at the centromeric and telomeric regions of the *Fp* chromosome segment and that the highest frequency of recombination occurs interstitially (Armstead et al., in preparation).

At present a much larger experiment to determine how physical distance relates to genetic distance is being undertaken. This involves the mapping of a whole *Fp* chromosome from a *Lp*/*Fp* monosomic substitution line. The *Fp* chromosome in question is presently being mapped with over 100 AFLP markers.

In addition to determining the physical position of genetic markers the chromosome arm substitution lines will provide invaluable tools for determining the genetic control of agronomically important characters. They will enable screening to determine which chromosome arms carry genes affecting particular characters. The genes can then be genetically and physically mapped by analysing a recombinant chromosome series (Fig. 3). This approach has similarities to the intervarietal substitution lines developed in wheat (for review see [16]) but the *Lp*/*Fp* system offers much higher resolution as it is possible to assign genes to a specific location on a chromosome arm.

A consequence of developing the *Lp*/*Fp* recombinant chromosome series is that many lines of *Lp* will be produced with a single small *Fp* chromosome segment carrying an important gene. We have already achieved the transfer of genes for delayed senescence [17], drought [18] and cold tolerance, crown rust resistance, nutritive quality, late heading, perenniality, various morphological characters and some isozymes from the Festuceae into *Lolium* (unpublished observations). The commercial value to the UK market alone of the gene transfer for delayed senescence and drought tolerance, respectively, has been estimated at £15 and £45 millions per annum.

As previously described, a consequence of developing the *Lp*/*Fp* recombinant chromosome series is that many lines of *Lp* will be produced with a single small *Fp* chromosome segment carrying an important gene. The isolation of genetic markers linked to important *Fp* genes will be valuble for marker assisted selection in breeding programmes. We have already had success in isolating AFLP markers located on introgressed *Fp* chromosome segments. In addition, once an RFLP marker is found to be located on a *Fp* chromosome segment, other tightly linked RFLPs can be used to further characterize the segment. We are also exploring the potential of another strategy to generate markers from introgressed *Fp* chromosome segments. We are presently attempting to isolate *Fp*-specific repetitive DNA sequences. These will then be used to screen bacterial artificial chromosomes and Lambda libraries generated from lines of *Lp* carrying a single *Fp* chromosome segment. The rationale is that by screening bacterial artificial chromosome (BAC) and Lambda clones with the *Fp*-specific repetitive sequence we will be able to identify those clones which are derived specifically from the *Fp* segment.

References

1 Gale MD, Miller TE (1987) The introduction of alien genetic variation into wheat. *In*: FGH Lupton (ed.): *Wheat Breeding. Its Scientific Basis*. Chapman and Hall, London, 173–210

2 King IP, Morgan WG, Harper JA, Meredith MR, Jones RN, Armstead IP, Hayward, MD, Forster, JW, Evans, GM, Bollard et al (1997) Introgression mapping: A tool to integrate physical and genetic maps. *In*: *Plant and Animal genome V*. San Diego, California, 77

3 King IP, Morgan WG, Armstead IP, Harper JA, Hayward MD, Bollard A, Nash JV. Forster JW, Thomas HM (1998) Introgression mapping in the grasses. I. Introgression of *Festuca pratensis*

chromosomes and chromosome segements into *Lolium perenne*. *Heredity* 81: 462–467
4 King IP, Morgan WG, Harper JA, Thomas HM (1999) Introgression mapping in the grasses. II. Meiotic analysis of the *Lolium perenne/Festuca pratensis* triploid hybrid. *Heredity* 82: 107–112
5 Jauhar PP (1975) Chromosome relationships between *Lolium* and *Festuca* (Gramineae). *Chromosoma* 52: 103–121
6 Stammers M, Harris J, Evans GM, Hayward MD, Forster JW (1995) Use of random PCR (RAPD) technology to analyse phylogenetic relationships in the *Lolium/Festuca* complex. *Heredity* 74: 19–27
7 Lewis EJ (1966) The production and manipulation of new breeding material in *Lolium-Festuca*. *In*: AGG Hill (ed.): *Proc X Intern Grassland Congr Helsinki*. Valtioneuvoston Kirjapaino, Helsinki, 688–693
8 Thomas HM, Harper JA, Meredith MR, Morgan WG, Thomas ID, Timms E, King IP (1996) Comparison of ribosomal DNA sites in *Lolium* species by fluorescence *in situ* hybridization. *Chromosome Res* 4: 486–490
9 Thomas HM, Harper JA, Meredith MR, Morgan WG, King IP (1997) Physical mapping of ribosomal DNA sites in *Festuca arundinacea* and related species by *in situ* hybridization. *Genome* 40: 406–410
10 Kihara H (1937) Genomanalyse bei *Triticum* und *Aegilops*. VII Kurzübersicht über die Ergebnisse der Jahre 1934–36. *Mem Coll Agric Kyoto Univ* 41
11 Alonso LC, Kimber G (1981) The analysis of meiosis in hybrids. II. Triploid hybrids. *Can J Genet Cytol* 23: 221–234
12 Chapman CGD, Kimber G (1992) Developments in the meiotic analysis of hybrids. I. Review of theory and optimization in triploids. *Heredity* 68: 97–103
13 King IP, Purdie KA, Orford SE, Reader SM, Miller TE (1993) Detection of homoeologous chiasma formation in *Triticum aestivum* × *Thinopyrum bessarabicum* hybrids using genomic *in situ* hybridization. *Heredity* 71: 369–372
14 King IP, Reader SM, Purdie KA, Orford SE, Miller TE (1994) A study of the effect of a homoeologous promoter on chromosome pairing in wheat/rye hybrids using genomic *in situ* hybridization. *Heredity* 72: 318–321
15 Sybenga J (1996) Chromosome pairing affinity and quadrivalent formation in polyploids: do segmental allopolyploids exist. *Genome* 39: 1176–1184
16 Law CN, Snape JW, Worland AJ (1987) Aneupolyploidy in wheat and its use in genetic analysis. *In*: FGH Lupton (ed.): *Wheat breeding. Its scientific basis*. Chapman and Hall, London and New York, 71–127
17 Thomas HM, Morgan WG, Meredith MR, Humphreys MW, Thomas H, Leggett JM, (1994) Identification of parental and recombined chromosomes in hybrid derivatives of *Lolium multiflorum* × *Festuca pratensis* by genomic *in situ* hybridization. *Theor Appl Genet* 88: 909–913
18 Humphreys MW, Pasakinskiene I (1996) Chromosome painting to locate genes for drought resistance transferred from *Festuca arundinacea* into *Lolium multiflorum*. *Heredity* 77: 530–534.

Chromosomes Today, Vol. 13
ed. by E. Olmo and C.A. Redi

Smallness: gain and loss in plant chromosome research

Kiichi Fukui[1], Nobuko Ohmido[2] and Toshiyuki Wako[3]

[1] *Department of Biotechnology, Graduate School of Engineering, Osaka University, Suita 565-0871, Osaka, Japan*
[2] *Laboratory of Rice Genetic Engineering, Hokuriku National Agricultural Experiment Station, Joetsu 943-0193, Niigata, Japan*
[3] *Genetic Resources II Department, National Institute of Agrobiological Resources, Tsukuba, 305-8602, Japan*

Introduction

Once, a good microscope was the pride of a laboratory in the field of cytology and cytogenetics. Now the situation seems to have completely changed. Biochemical and molecular biological equipment replace the position of microscopes even in the laboratories of cytology and cytogenetics. In this paper, the importance of the visual sense to the study of small plant chromosomes is again demonstrated through typical examples. New methodologies related to visual approaches in chromosome research are also reviewed based on the results obtained mainly in our research group. Smallness is, thus, no longer disadvantageous for chromosome research.

Large and small plant chromosomes

Chromosome size as invaluable information

Looking at plant chromosomes under a microscope, it is easily perceived that the size of chromosomes varies from plant species to species. Figure 1 shows the typical large (barley) and small chromosomes (rice) observed in the same optical field [1]. Now the genome size of barley, 5,360 Mbp and that of rice, 487 Mbp [2] are known. Significant correlation ($r = 0.939$) exists between genome sizes and chromosome lengths as shown by comparison of nuclear sizes and chromosome lengths using several rice species [3]. Thus, chromosome length would be a good estimate of a genome size.

Because large chromosomes demonstrate various visual features, such as positions of primary, secondary, and tertiary constrictions, number and positions of bands, sizes and positions of condensed regions, etc., it is reasonable that the studies on plant chromosomes have mainly been carried out using the

Figure 1. Barley and rice chromosomes in the same optical field. Barley chromosomes were pretreated with 0 °C cold water overnight and rice chromosomes were not pretreated. The root tips of barley and rice were simultaneously macerated and prepared on the same glass slide. Bar shows 10 μm [1].

large sized chromosomes to date. In contrast to the several prominent morphological characteristics occurring in the large chromosomes, small chromosomes often show a completely condensed dot- or rod-like shape without any morphological characteristic at the mitotic metaphase stage. Pretreatment with chemicals or low temperature, which is quite effective for collecting chromosomes at metaphase, effectively removes most of the authentic characteristics of the chromosomes at the same time.

Recent studies on plant chromosomes revealed that the volume of nucleotide sequences in a plant chromosome affects not only the size of the chromosome, but also its structural differentiation. For example, large chromosomes represented by barley [4], *Paris* [5], *Crepis* [6], etc., usually have typical banding patterns after C- and/or N-banding treatments. On the other hand, small chromosomes show only simple centromeric bands (e.g. soybean [7], *Brassica* spp. [8, 9]) or even no bands (e.g., rice).

Condensation pattern as a common feature in small plant chromosomes

A prominent feature of small plant chromosomes is uneven staining appearing at the mitotic prometaphase stage. Figure 2 shows rice chromosomes at the metaphase and prometaphase stages. Based on the experience of looking at the numerous rice chromosomes without pretreatment, most of the spreads at mitosis show prometaphase-like morphology. In other words, only a small number of typical metaphase spreads are observed. This fact means that the prometaphase stage is the longest among mitotic stages and most of the dot-like condensed chromosomes that are assumed to be the metaphase chromosomes may even be artifacts from the chemical pretreatments, at least, in rice.

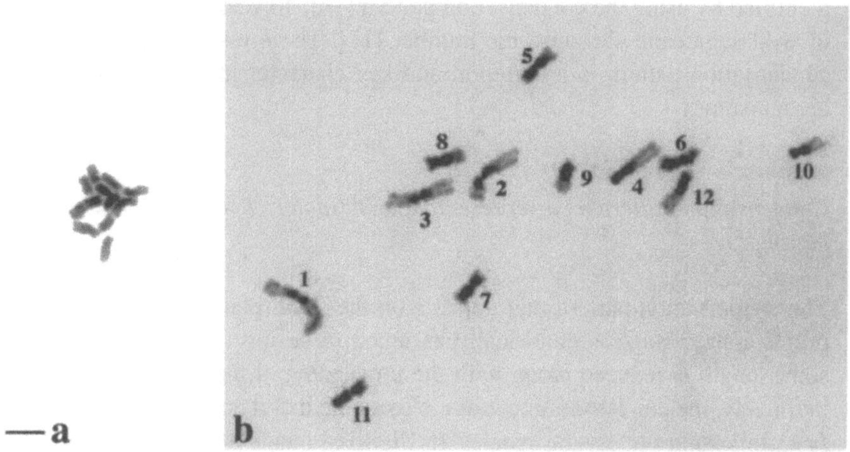

Figure 2. Rice haploid chromosomes Giemsa stained at the metaphase and prometaphase stages. a; Metaphase chromosomes. All the chromosomes are more or less evenly stained with a Giemsa solution. Bar shows 5 µm. b; Prominent condensation patterns are observed in the prometaphase chromosomes [55].

It is obvious that the rice chromosomes at the prometaphase stage depict a prominent uneven staining pattern with Giemsa staining. The same pattern is observed by phase-contrast microscopy and staining with several fluorescent dyes. Sequential staining with a DNA-binding fluorescent dye, ethidium bromide, and with a Giemsa solution revealed that the uneven staining pattern is not due to preferential affinity of the dye, but to the uneven condensation of chromatin fibers along the chromosome axis [10]. This fact is quite different from the cases in the G-banded vertebrate chromosomes. As a result, the uneven staining pattern was designated as the condensation pattern [10, 11]

The condensation pattern facilitates objective identification of individual chromosomes. Small plant chromosomes are, at least, categorized into several groups by visual inspection under a microscope as in the case of soybean ($2n = 40$[7]). No specific treatment is necessary to obtain the chromosomes with the condensation pattern. Prometaphase chromosomes can simply be obtained by omission of pretreatment. Chemical pretreatment may also change the pattern of condensation itself [12]. On the other hand, no distortion of the condensation pattern has been reported by the low temperature treatment and it is effective for collecting prometaphase chromosomes in some plant species.

By visual inspection of the condensation pattern, all the rice chromosomes were completely identified by 1989 [10], 80 years after the first accurate report on rice chromosome number [13]. All the chromosomes of three *Brassica* species with a basic genome of A, B or C have been completely identified by their condensation patterns [14], 75 years after the first report on *Brassica* chromosome number [15]. All the chromosomes of a wild sugarcane have also been identified by using the condensation pattern [16], 40 years after the first report of wild sugarcane chromosome number [17]. These results indicate that the condensation pattern is a common and key character to identify small plant chromosomes.

Condensation pattern as a representative character of chromosomal morphology

The condensation pattern that appears on the small plant chromosome at the prometaphase stage is empirically assumed to be unstable because chromosome length is reduced along with the proceeding of the cell cycle. Counter-intuitively, the condensation pattern is so stable that it could be used to identify chromosomes in several species. Its digitized image parameter, CP, which is defined as a density profile at the mid-rib of a chromatid, suitably represents most of the visual characteristics that are employed in identification of chromosomes.

The chromosome number determined both by visual inspection and by automatic identification based on CPs were compared using 360 rice chromosomes from 30 prometaphase spreads. As a result, 92.2% of the rice chromosomes agreed in their number between the visual and computational methods [18]. The results demonstrate the effectiveness of CP in identification of chromosomes and justify the definition of the CP as a new image parameter for small plant chromosomes alongside the conventional parameters of total length and arm ratio.

Imaging methods in plant chromosome research

Brief history of imaging methods in plant chromosome research

A comprehensive image analysis system for plant chromosome research, CHIAS, was developed in 1985 [19, 20]. CHIAS was applied in various fields of chromosome research, such as karyotyping of barley chromosomes using 250 haploid spreads [21], simulation of barley chromosome images generated by several optical equipment and staining solutions [22], simulation of visual sense to quantify C-banding patterns of *Crepis* chromosomes [6], etc. The third generation of the CHIAS, CHIAS III, has now been released after development of a personal system, CHIAS-mini [23] and the second generation, CHIAS II [24]. The newest chromosome image analyzing system, CHIAS III features the Internet-based public domain software, NIH Image as the main software and a Macintosh personal computer as the main hardware [25]. All the necessary

Figure 3. (See color plate XVII between pages 256 and 257.) Representative steps of the image analysis of CP in wild sugarcane (*Saccharum spontaneum* L.) chromosomes. a; Giemsa stained original chromosome image. Bar shows 5 µm. b to h; Representative steps for the CP analysis using CHIAS III [25]. The allocation of grey values to the density levels was adjusted to that of the CHIAS and CHIAS II in ref [25].

steps in the CP analysis are summarized in a written form [26] and obtained via the Internet (http://133.1.131.81/Evolejas/chias3/chias3.html).

Image analysis of the condensation pattern

The representative steps of image analysis of the condensation pattern by CHIAS III is exemplified using haploid sugarcane (*S. spontaneum* L., $n = 4x = 32$) chromosomes (Fig. 3) [25]. The sugarcane chromosomes ($2n = 32$) show the condensation pattern at the prometaphase stage as in the small plant chromosomes (Fig. 3a). A part of the total image frame with 8 chromosomes is captured by a CCD camera and digitized as a 781-kbyte image, an 800×1000 pixel matrix (Fig. 3b). Then the chromosomal areas are discriminated from the background field by setting a threshold in the grey levels. The grey values 102 and over are recognized as the chromosomal areas and colored red (Fig. 3c).

Then the background field other than the chromosomal areas is digitally eliminated and grey chromosome images are pseudo-colored by using a look-up-table. A different color is allocated to each pixel based on its original grey value to enhance the contrast of the condensation pattern. All the 32 chromosomes are categorized into eight homologous groups and one of the letters from A to H is designated according to the enhanced or pseudo-colored chromosome images with condensation pattern. A to H represent each of eight homologous groups (Fig. 3d). The condensation pattern presented in pseudo-color is much more tangible for human vision than that in the grey image. Then each of four homologous chromosomes was selected and placed into one of eight sub-panels (a–h) (Fig. 3e) of a chromosome gallery (Fig. 4). After cutting and pasting of all chromosomes into one of eight sub-panels, all the chromosomes are again checked in detail by comparing the condensation pattern of each chromosome within a sub-panel and among the sub-panels. Finally, the CP of each identified chromosome is measured along the mid-rib of a chromatid and the numerical data are stored in the memory of the CHIAS III (Fig. 3f).

Development of an idiogram based on the CP

Based on the CP, idiograms or chromosome maps were developed in rice [10], three diploid *Brassica* species [14], Oilseed rape [27], wild sugarcane [16] and *Arabidopsis* [28].

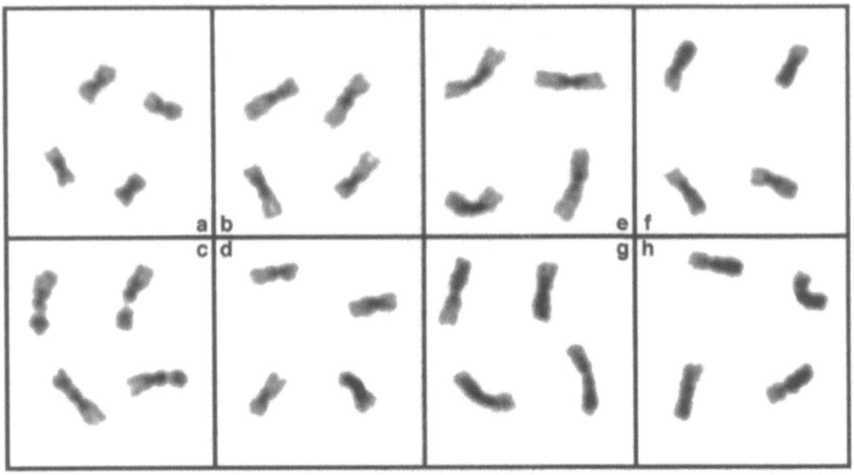

Figure 4. Chromosome gallery of wild sugarcane (*S. spontaneum* L.). Four homologous chromosomes are assigned in each of eight sub-panels [25].

The typical method for development of a quantitative idiogram based on the CP is as follows. Setting of two thresholds at adequate grey values delimits a density profile into three parts, upper, middle and lower grey levels, which correspond to dispersed, condensed and heavily condensed chromosomal regions in the case of CHIAS and CHIAS II, and heavily condensed, condensed and dispersed regions in the CHIAS III (The allocation of grey values to actual grey level is reversed between CHIAS III and the others). The border between condensed and dispersed regions of the density profile should correspond to the actual border between condensed and dispersed regions of a prometaphase chromosome. The other border between condensed and heavily condensed regions could be determined in order to enhance the visual characteristics within the condensed regions. For example, a gap between two heavily condensed regions within a long arm of the sugarcane chromosome was adequately depicted in the idiograms by setting this border [25].

Then development of an idiogram for plant chromosomes is practically divided into three steps [25]. First, standard CPs (stCPs) are obtained by averaging the CPs from homologous chromosomes. The stCP is an averaged CP of each homologous chromosome from an adequate number of chromosomal spreads at a similar mitotic stage. The mitotic stage of a spread is determined according to the total length of the chromosomes in the spread. Second, the stCPs are thresholded at two grey values dividing the profile or the chromosomal area into three regions as mentioned above. Finally, addresses are given to each region of the idiogram developed. In the case of sugarcane, the lower border (grey value = 125) was fixed for all chromosomes and corresponds to the border between grey and black bars in the idiogram as shown in Figure 5. The

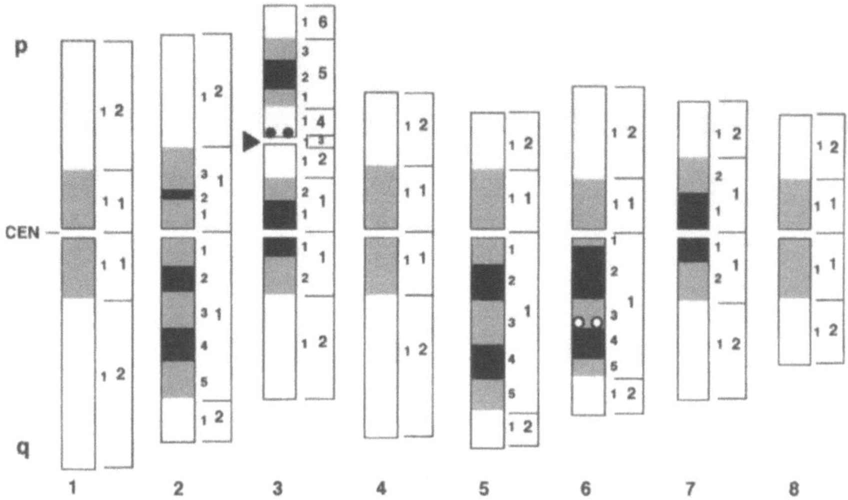

Figure 5. Idiogram of wild sugarcane (*S. spontaneum* L.) based on the CP. Black, grey and white blocks correspond to heavily condensed, condensed and dispersed regions, respectively. CEN, p, q, solid triangle, solid and open circles indicate centromere, short and long arms, NOR, 45S rDNA and 5S rDNA loci, respectively [16].

lower border (grey value = 86) between grey and black bars of the idiogram was set in order to depict the characteristics on the individual chromosomes [16].

Visual inspection and observation: the resurrection of the visual sense

Centromeric condensation visually observed

Visual inspection enables not only identification of individual chromosomes but also detection of specific features of small plant chromosomes. Specific condensations observed at the centromere and telomere regions, sometimes present crucial information on the chromosomal constitution. For example, chromosomes are similar to each other even at prometaphase in *Brassica* species. There, however, are some *B. rapa* chromosomes with a conspicuous different character from *B. oleracea* chromosomes at the centromeric regions [14]. Comparison of the sizes of both the proximal condensed regions on short and long arms reveals that the proximal condensation on the long arm in a half of *B. rapa* chromosomes is much larger than that of the short arms. No *B. oleracea* or *B. nigra* chromosome demonstrates such imbalanced condensation at the proximal region. Figure 6 demonstrates five *B. rapa* chromosomes with a greater condensation at the long arms. This fact especially facilitates the dis-

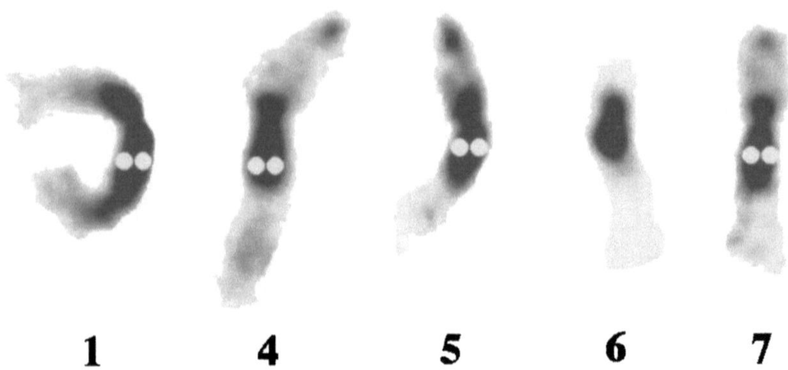

Figure 6. Heavy proximal condensation on long arms observed in a half of *B. rapa* ($2n = 10$) chromosomes. Four of the five heavy condensations are due to the localization of 45S rRNA genes as indicated by solid white circles.

crimination of *B. rapa*-origin chromosomes from *B. oleracea*-origin chromosomes among *B. napus* chromosomes, which is an amphidiploid of *B. rapa* and *B. oleracea* [27]. FISH using 45S rDNA as a probe revealed that the larger condensation is due to the localization of 45S rDNAs exclusively at that region of the *B. rapa* chromosomes. Among five disproportional chromosomes with the larger proximal condensation, four harbor the 45S rDNA as shown in Figure 6.

Telomeric condensation visually observed

In rice chromosomes, faint, unstable and small condensations (FUSCs) are often observed at the dispersed and telomeric regions [29, 30]. FISH revealed that a tandem repeated sequence, TrsA (Tandem repeat sequence A), is concentrated at the terminal FUSCs. TrsA is specific to rice A genome species with a unit length of 355 bp [31]. Although the reason and mechanisms are unknown to date, they occur at the end of long arms only in the cultivated rice species. Thus, the fact facilitates the identification of chromosomal orientation of 6, 2, 3 and 11 chromosomes in *O. sativa* ssp. *indica* and *japonica* [32], *O. glaberrima*, and *O. meridionalis* [32, 33], respectively (Fig. 7). An exceptional case of long arm specific localization of TrsA is found in Australian endemic species of *O. meridionalis*, where TrsA occurs at the end of short arms in the four largest chromosomes and an intercalary region of the long arm of the fourth chromosomes.

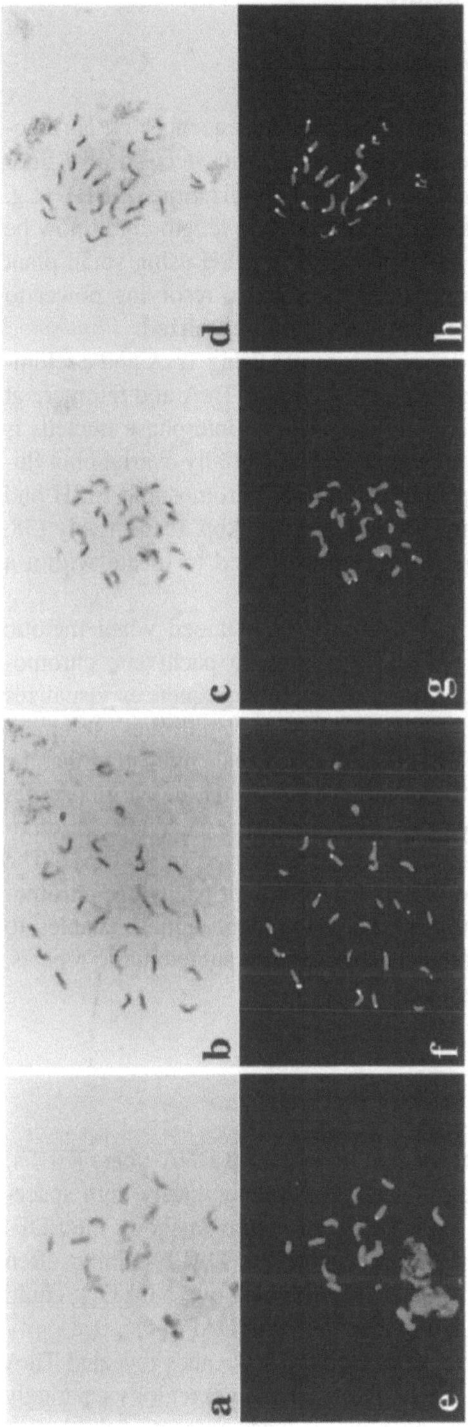

Figure 7. Terminal FUSCs (faint, unstable, small condensations) visualized with Giemsa stain (Upper panels) and FISH with the TrsA probe (Lower panels). a and e; *O. sativa* ssp. *japonica*, b and f; *O. sativa* ssp. *indica*, c and g; *O. glaberrima*, d and h; *O. meridionalis* [33].

Visual verification of telomere structure

Multi-color FISH visualizes the telomere structure

Visual perception to enable the location of nucleotide sequences on chromosomes is much strengthened by the development and introduction of a fluorescent detection method (fluorescence *in situ* hybridization, FISH) (e.g. [34–36]). Because a single RFLP marker with 1.2 kb in length could now be visualized by FISH [37], the main technical target in FISH using small plant chromosomes moves towards the improvement of space resolving power to discriminate two different nucleotide sequences closely localized.

Multi-color FISH (McFISH) on rice chromosomes using TrsA and a telomere sequence as the probes visualizes co-localization of TrsA and telomere at the telomeric regions as shown in Figure 8a. When an interphase nucleus is used, TrsA and telomere sequences are recognized as partially overlapping fluorescence signals (Fig. 8b). The space resolution of chromosome FISH and interphase mapping are estimated at 2–5 Mbp and 100 kbp, respectively [38, 39]. Thus, TrsA and telomere sequences were estimated to locate within a 100 kbp area [32].

The space resolution of 1.2 Mbp to 100 kbp is obtained when meiotic pachytene chromosomes are used. McFISH on tomato pachytene chromosomes using a subtelomeric repeat, TGR1, and telomere sequences visualizes the higher structure at the termini of the pachytene chromosomes. Telomere signals were usually detected in the middle of TGR1 signals indicating that the termini of the tomato pachytene chromosomes bend inside the chromosomes [40]. Exactly the same phenomenon was observed in rice pachytene chromosomes when TrsA and telomere sequences were employed in McFISH. The bending structure could be useful to protect the termini of pachytene chromosome [41]. Thus, careful observation of the fluorescent signals enables to detect not only the positions of telomere sequences and subtelomeric repeats, but also their higher structure at the termini.

Further resolving power of FISH on DNA fibers

Fransz et al. [42] developed a FISH method on extended DNA fibers (EDFs) extracted from isolated plant nuclei. It guarantees tremendously high space-resolving power reaching down to the DNA level. Figure 9a shows the EDFs extracted from isolated rice nuclei stained with DAPI. DAPI staining often visualizes a bundle of DNA fibers. Another DNA binding dye, YOYO-1, could visualize single DNA fibers more clearly than that with DAPI [41].

McFISH on rice EDFs using TrsA and telomere sequences revealed TrsA (red fluorescent dots) and telomere (green fluorescent dots) regions separately on a single DNA fiber (Fig. 9b). The quantitative measurement of the length

Figure 8. (See color plate XVIII between pages 256 and 257.) McFISH with TrsA and telomere sequences to chromosomes (a) and an interphase nucleus (b) of *O. sativa* ssp *indica*. Blue, red and green fluorescence indicate genomic or chromosomal DNA, TrsA and telomere sequences, respectively. Overlapped disposition of TrsA and telomere is indicated by arrowheads. Bar indicates 2.5 μm [32].

Figure 9. (See color plate XVIII between pages 256 and 257.) Extended DNA fibers (EDFs) from isolated nuclei of *O. sativa* ssp. *indica* stained with DAPI (a). McFISH with TrsA and telomere sequences to an EDF (b). Red and green signal tracks indicate TrsA and telomere sequences, respectively [41].

of the TrsA and telomere signal tracks by CHIAS III, gives a precise copy number for each repetitive sequence [41]. A signal track of 1 μm corresponds to 3.27 kbp [42], although it is a little bit less extended than that of Watson and Crick's B model (1 μm = 3.05 kbp).

Further space resolution is realized by a scanning electron microscope (SEM) and scanning probe microscope (SPM). Figure 10a shows barley chromosomes observed with a high resolution SEM. Globular surface structure and paralleled fibrous structure are observed on the chromosomal arms and centromeres, respectively [43]. A combination of electron microscopy and *in situ* hybridization with gold-labeled probes enables detailed structural analyses maintaining the chromosomal structure. The atomic force microscope (AFM), a derivative of the SPM, visualizes the structure of even a single plasmid DNA without any coating treatment as is usual in electron microscopy. The resolution of the AFM reaches to the nano-meter level. Figure 10b depicts a single plasmid DNA visualized by the AFM [44]. The DNA fiber measured is 1.7 nm in height and about 10 nm in width indicating that the height was close to Watson and Crick's B model DNA (2 nm), however, the width was overestimated. This overestimation could be rectified by improvement in the tip of the fine cantilever (probe) on the AFM. Moreover, the AFM manipulates the

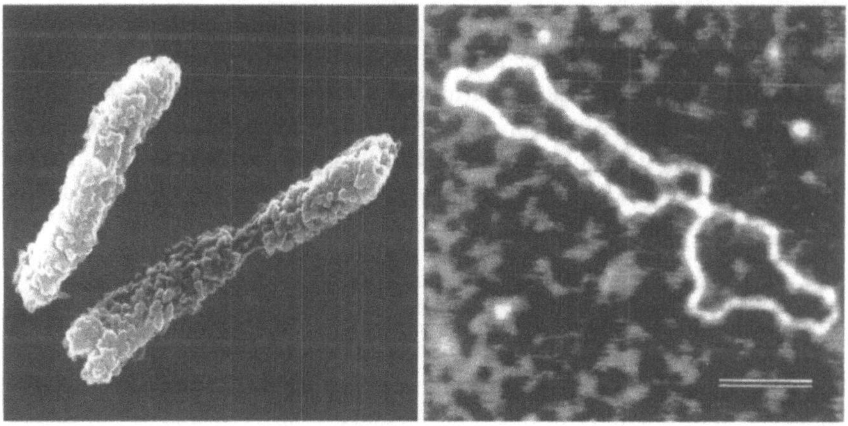

Figure 10. Barley chromosomes and a plasmid DNA depicted by using high-resolution scanning electron microscopy (a, [43]) and atomic force microscopy. Bar shows 20 nm (b, [44]).

DNAs and chromosomes using a pointed probe. With both these functions, the SPM would have great possibilities in the future of chromosome research.

Future prospects of chromosome research based on the visual sense

Localization of functional units of a chromosome within an intact nucleus

Three dimensional (3D) analyses of plant genomes and chromosomes has become common in chromosome research (e.g. [45–47]). Now, quality 3D images can directly be obtained by optically splitting the left- and right-eye components of the objective lens and channeling them directly to the respective eyepiece (High definition direct-view 3D microscope, Edge Scientific Instrument Co.). Laser confocal microscopy and a deconvolution system are both commonly used in reconstruction of 3D microscopic images. Advantages and disadvantages in the observation of biological objects by these two systems are summarized in a recent paper [48].

3D localization of functional units of a chromosome, such as telomeres and centromeres, within an intact nucleus directly reveals the 3D structure of a nucleus. 3D McFISH with the centromere and telomere probes using the deconvolution system reveals the higher organization of barley genomes. The centromeres locate at one pole of the barley nucleus with a ring-like shape and the telomeres lie scattered in the opposite hemisphere.

Detection of functional dynamics in an intact nucleus

By using the 3D observation systems, acetylation and deacetylation of histone H4 and phosphorylation and dephosphorylation of histone H3 are dynamically detected in intact barley nuclei by immunostaining methods [49]. It is revealed that the nucleolar organizing regions (NORs) are subjected to dynamic acetylation and deacetylation during mitosis. It is also found that the dynamic acetylation and deacetylation of the NORs depend on the positions of lysine residues at the N-terminal of histone H4. It is now possible to visualize the dynamic physiological changes in an intact cell or nucleus by choosing and/or combining suitable techniques.

Manipulation of genomes and chromosomes by visual and optical means

Visual perception combined with laser optics enables us to open new fields in chromosome research [50]. Rapid development of various micro-laser irradiation systems enables dissection of plant chromosomes [51, 52] and even dissection of the C-band positive regions of *Crepis* chromosomes [53].

Technology in recovering a tiny chromosome fragment dissected under a microscope in conjunction with modified PCR methods enables recovery of the DNA sequences from any chromosomal region of interest. Combination of a series of methods including image analysis of a NOR chromosome, laser-dissection of the NOR chromosomes recovery of the chromosomes, PCR amplification and FISH using amplified rDNA, confirms the accuracy and efficiency of the laser-dissection method [54]. Physical manipulation using the SPM enables the more delicate manipulation of chromosomes and DNA fibers. Visual sense again plays a vital role in this type of experiment.

Conclusion

As has been described, smallness is no longer disadvantageous in plant chromosome research. Counter-intuitively, small chromosomes are easier identified than large chromosomes as demonstrated in rice, Brassicas, and wild sugarcane chromosomes. If the pretreatment that is routinely applied without sufficient care is omitted, uneven condensation patterns can be observed on the small plant chromosomes. Secondly, visual information should be digitally analyzed by using imaging methods to avoid human native and individual biases in interpretation of the results. All the necessary software for CP analysis in the image analysis system, CHIAS III, is obtainable via the Internet. The digitized data are the basis of further chromosome research. Thirdly, physical means provide us with effective new tools in observing chromosomes. An appropriate combination of 3D microscopy and other technologies enables us to visualize the functional dynamics within an intact cell or nucleus. They also provide us with effective methods to manipulate chromosomes and even DNA fibers. Therefore, it can be concluded that the visual sense will become more and more important and be indispensable for future chromosome research.

Acknowledgments

The authors thank Prof. Ettore Olmo, University of Ancona, Ancona, Italy and all the members of the organizing committee of the 13th International Chromosome Conference for their kind invitation and giving us an opportunity to present our work mainly carried out at Joetsu Laboratory of the Chromosome Link.

References

1 Fukui K (1996) Plant chromosomes at mitosis. *In*: K Fukui, S Nakayama (eds): *Plant Chromosomes, Laboratory Methods*. CRC Press, Boca Raton, FL, 1–17
2 Bennett MD, Smith JB (1976) Nuclear DNA amounts in angiosperms. *Phil Trans R Soc Lond B* 274: 227–274
3 Uozu S, Ikehashi H, Ohmido N et al (1997) Repetitive sequences: cause for variation in genome size and chromosome morphology in the genus *Oryza*. *Plant Mol Biol* 35: 791–799

4 Kakeda K, Fukui K, Yamagata H (1991) Heterochromatic differentiation in barley chromosomes revealed by C- and N-banding techniques. *Theor Appl Genet* 81: 144–150
5 Miyamoto J, Kurita S, Fukui K (1991) Image analysis of C-banding pattern in two herbs, *Paris tetraphylla* A. Gray and *Paris verticillata* M. Bieb, (Liliaceae). *Jpn J Genet* 66: 335–345
6 Fukui K, Kamisugi Y (1995) Mapping of C-banded *Crepis* chromosomes by imaging methods. *Chromosome Res* 3: 79–86
7 Yanagisawa T, Tano S, Fukui K et al (1991) Marker chromosomes commonly observed in the genus *Glycine*. *Theor Appl Genet* 81: 606–612
8 Olin-Fatih M (1996) The morphology, cytology, and C-banded karyotypes of *Brassica campestris, B. oleracea* and B. napus plants regenerated from protoplasts. *Theor Appl Genet* 93: 414–420
9 Olin-Fatih M, Heneen WK (1992) C-banded karyotypes of *Brassica campestris, Brassica oleracea* and *Brassica napus*. *Genome* 35: 583–589
10 Fukui K, Iijima K (1991) Somatic chromosome map of rice by imaging methods. *Theor Appl Genet* 81: 589–596
11 Fukui K, Mukai Y (1988) Condensation pattern as a new image parameter for the identification of small chromosomes in plants. *Jpn J Genet* 63: 359–366
12 Iijima K, Fukui K (1991) Clarification of the conditions for the image analysis of plant chromosomes. *Bull Natl Inst Agrobiol Resource* 6: 1–58
13 Kuwada Y (1910) A cytological study of *Oryza sativa* L. *Bot Magazine (Tokyo)* 26: 267–281
14 Fukui K Nakayama S, Ohmido N et al (1998) Quantitative karyotyping of three diploid *Brassica* species by imaging methods and localization of 45S rDNA loci on the identified chromosomes. *Theor Appl Genet* 96: 325–330
15 Karpechenko GD (1922) The number of chromosomes and the genetic correlation of cultivated Cruciferae. *Bull Appl Bot Genet Plant Breeding* 13: 3–14
16 Ha S, Moore PH, Heinz D et al (1999) Quantitative chromosome map of the polyploid *Saccharum spontaneum* by multicolor fluorescence *in situ* hybridization and imaging methods. *Plant Mol Biol* 39: 1165–1173
17 Panje RR, Babu CN (1960) Studies in *Saccharum spontaneum* distribution and geographical association of chromosome numbers. *Cytologia* 25: 152–172
18 Kamisugi Y, Furuya N, Iijima K et al (1993) Computer-aided automatic identification of rice chromosomes by image parameters. *Chromosome Res* 1: 189–196
19 Fukui K (1986) Standardization of karyotyping plant chromosomes by a newly developed chromosome image analyzing system (CHIAS). *Theor Appl Genet* 72: 27–32
20 Fukui K (1996) Quantitative chromosome maps as a basis of biosciences. *In*: PE Brandham and MD Bennett (eds): *Kew Chromosome Conference*. Royal Botanic Gardens, Kew, 201–213
21 Fukui K, Kakeda K (1990) Quantitative karyotyping of barley chromosomes by image analysis methods. *Genome* 33: 450–458
22 Fukui K, Kamisugi Y (1991) Simulation of microscopic images by a desktop imaging system. *In*: M Doyama, T Suzuki, J Kihara, R Yamamoto (eds): *Computer Aided Innovation of New Materials*. Elsevier Science Publishers, Amsterdam, 101–104
23 Kamisugi Y, Fukui K (1990) Automatic karyotyping of plant chromosomes by imaging techniques. *BioTechniques*. 8: 290–295
24 Fukui K, Nakayama S (1996) Analysis of chromosome information. *In*: K Fukui, S Nakayama (eds): *Plant Chromosomes, Laboratory Methods*. CRC Press, Boca Raton, FL, 241–255
25 Kato S, Fukui K (1998) Condensation pattern (CP) analysis of plant chromosomes using a newly developed chromosome image analyzing system, CHIAS III. *Chromosome Res* 6: 473–479
26 Kato S, Hirose T, Akiyama Y et al (1997) Manual on the chromosome image analyzing system III, CHIAS III. *Research Report on Development of Agriculture at the Hokuriku Area* 36: 1–76
27 Kamisugi Y, Nakayama S, O'Neill CN et al (1998) Visualization of the *Brassica* self-incompatibility *S*-locus on identified oilseed rape chromosomes. *Plant Mol Biol* 38: 1043–1052
28 Ito M, Ohmido N, Akiyama Y et al. Quantitative chromosome map of *Arabidopsis thaliana* L. by imaging methods; *submitted*
29 Fukui K, Iijima K (1992) Manual on rice chromosomes (2nd edition). *Miscellaneous Publications of National Institute of Agrobiological Resources* 4: 1–25
30 Iijima K, Kakeda K, Fukui K (1991) Identification and characterization of somatic rice chromosomes by imaging methods. *Theor Appl Genet* 81: 597–605
31 Ohtsubo H, Umeda M, Ohtsubo E (1991) Organization of DNA sequences highly repeated in tandem in rice genomes. *Jpn J Genet* 66: 241–254

32 Ohmido N, Fukui K (1997) Visual verification of close disposition between a rice A genome-specific DNA sequence (TrsA) and the telomere sequence. *Plant Mol Biol* 35: 963–968

33 Ohmido N, Kijima K, Akiyama Y et al (2000) Quantification of total genomic DNA and selected repetitive sequences reveal concurrent changes in different DNA families in *indica* and *japonica* rice. *Mol Gen Genet; submitted*

34 Jiang J, Gill B S (1994) Nonisotopic *in situ* hybridization and plant genome mapping: the first 10 years. *Genome* 37: 717–725

35 Leitch IJ, Heslop-Harrison JS (1991) Physical mapping of four sites of 5S rDNA sequences and one site of the alfa-amylase-2 gene in barley (*Hordeum vulgare*). *Genome* 36: 517–523

36 Fukui K, Ohmido N, Khush GS (1994) Variability in rDNA loci in genus *Oryza* detected through fluorescence *in situ* hybridization. *Theor Appl Genet* 87: 893–899

37 Ohmido N, Akiyama Y, Fukui K (1998) Systematic mapping of unique nucleotide sequences on identified rice chromosomes. *Plant Mol Biol* 38: 1081–1087

38 Lawrence JB, Singer RH, McNeil JA (1990) Interphase and metaphase resolution of different distance within the human dystrophin gene. *Science* 249: 928–932

39 Van den Engh G, Sachs R, Trask BJ (1992) Estimation genomic distance from DNA sequence location in cell nuclei by a random walk model. *Science* 257: 1410–1412

40 Zhong X, Fransz PF, Wennekes-van Eden J et al (1998) FISH studies reveal the molecular and chromosomal organization of individual telomere domains in tomato. *Plant J* 13: 507–517

41 Ohmido N, Kijima K, Ashikawa I et al. Visualization of the terminal structure of rice chromosomes using multicolor FISH on chromosomes and extended DNA fibers; *submitted*

42 Fransz PF, Alonso-Blanco C, Liharska TB et al (1996) High resolution physical mapping in *Arabidopsis thaliana* and tomato by fluorescence *in situ* hybridization to extended DNA fibers. *Plant J* 9: 421–430

43 Iwano M, Fukui K, Takaichi S et al (1997) Globular and fibrous structure in barley chromosomes reveals by high-resolution scanning electron microscopy. *Chromosome Res* 5: 341–349

44 Ushiki T, Hitomi J, Ogura S et al (1996) Atomic force microscopy in histology and cytology. *Arch Histol Cytol*, 59: 421–431

45 Cremer T, Dietzel S, Elis R et al (1995) Chromosome territories, nuclear matrix filaments and interchromatin channels: a topological view on nuclear architecture and function. *In*: PE Brandham, MD Bennett (eds): *Kew Chromosome Conference IV*. Royal Botanic Gardens, Kew, 63–81

46 Montijn MB, ten Hoopen R, Fransz PF et al (1998) Characterization of the nucleolar organising regions during the cell cycle in two varieties of *Petunia hybrida* as visualised by fluorescence *in situ* hybridisation and silver staining. *Chromosoma* 107: 80–86

47 Noguchi J, Fukui K (1995) Chromatin arrangements in intact interphase nuclei examined by laser confocal microscopy. *J Plant Res* 108: 209–216

48 Wako T, Fukuda M, Furushima-Shimogawara R et al (1998) Comparative analysis of the topographical distribution of acetylated histone H4 using confocal microscopy and a deconvolution system. *Analyt Chim Acta*, 365: 9–17

49 Houben A, Wako T, Shimogawara-Furushima R et al. (1999) The cell cycle dependent phosphorylation of histone H3 is correlated with the condensation of plant mitotic chromosomes. *Plant J* 18: 675–680

50 Sawahel W, Fukui K (1995) Gene cloning in plants: Innovative approaches. *BioTechniques* 19: 106–111

51 Fukui K, Minezawa M, Kamisugi Y et al (1992) Microdissection of plant chromosomes by argon-ion laser beam. *Theor Appl Genet* 84: 787–791

52 Fukui K, Nomiya A, Nishiguchi M et al (1995) Application of laser optics to plant cell and molecular biology. *Zool Stud* 34 (Suppl. 1) 35–37

53 Kamisugi Y, Sakai F, Minezawa M et al (1993) Recovery of dissected C-band regions in *Crepis* chromosomes. *Theor Appl Genet* 85: 825–828

54 Nakamura M, Fukui K (1997) A chromosome-oriented approach to genome analysis in a woody plant – *Sequoiadendron giganteum* (Lindl.) Buchholz. *In*: Z Borzan, SE Schlarbaum (eds): *Cytogenetic Studies of Forest Trees and Shrub Species*. University of Zagreb, Croatia, 89–102

55 Khush GS, Kinoshita T (1991) Rice karyotype, marker genes, and linkage groups. *In*: GS Khush, T Kinoshita (eds): *Rice Biotechnology*. C.A.B. International, Oxon, 83–108

Conclusion

Herbert C. Mcgregor

Conclusion

Chromosomes Today, Vol. 13
ed. by E. Olmo and C.A. Redi

The future of chromosomes

Herbert C. Macgregor

Department of Biology, University of Leicester, Leicester LE1 7RH, England, UK

It seemed appropriate, at the end of the 13th International Chromosome Conference and approaching the end of a millennium, to reflect on the recent trends in chromosome science and make a few guesses as to what's in store for future generations of our kind.

The past 20 years of chromosome science have been dominated by gene mapping, widely encouraged by the Human Genome Mapping Project, DNA sequencing, fluorescence *in situ* hybridization (FISH) and, more recently, the truly remarkable developments of comparative genome *in situ* hybridization (GISH) and chromosome painting.

Of the 32,000 papers published with the word "chromosome" in either the title or as a key word since the start of 1995, 14,700 have involved *in situ* hybridization in one form or another. This particular approach to chromosome science has a long way to run. Whether it has the potential to generate more truly novel discoveries remains to be seen. At present, it seems that FISH has passed the peak of its usefulness as an experimental tool (Fig. 1) and is now

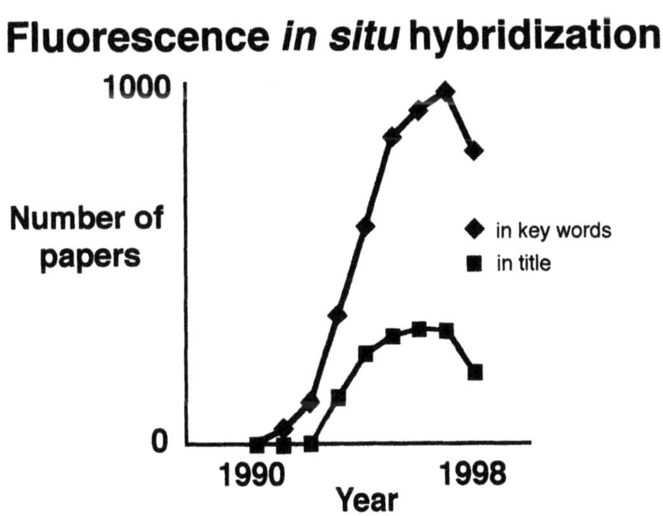

Figure 1.

more often applied in evolutionary cytogenetics than for experiments in molecular cytogenetics.

Genome mapping still occupies hundreds of people and costs millions of dollars and it will doubtless do so for some time yet. Will the bold prophecies of men like Jim Watson, "The money spent on the Human Genome Mapping Project is going to make other aspects of biological science more efficient by bringing them out from under the umbrella of complete ignorance" and Bob Moyzis, "This project (The Human Genome Project) will provide the intellectual framework for the next century of biological understanding" [1] prove to be true? Of course, it might be said that genome mapping is not an area of chromosome science, but whatever we think about this, there is no doubt that the two overlap and today's chromosome journals could easily fill their pages with the positions of genes on chromosomes if the editors were to allow it. This is especially true now that genome (and chromosome) mapping projects for a wide range of commercially important domestic animals are well underway.

So what areas of chromosome science are showing special promise for the future? My own perspective, built up over 40 years in the field, is that the last 50 years have been all about what chromosomes are. In 1958 we were still arguing about whether DNA extended from end to end of a chromosome as a single long molecule and, if it did, whether there was one molecule or several. The next 50 years will, I think, be all about what chromosomes do. Some of the major questions are already clear. How do chromosomes respond to signals? What stops, starts and regulates meiosis? What exactly are nuclear domains? These sorts of things will probably become much clearer as we learn more about chromatin, DNA-protein interactions, DNA sequences and their influence on molecular configurations. This is a fast-expanding area of research employing the highest of molecular and supramolecular technologies. It started to gain momentum about 15 years ago and its growth since then has been phenomenal (Fig. 2). An entire spectrum of investigation now ranges from the association of specific stretches of DNA with nuclear matrix proteins (contribution by Y. Ikemura at the 13th ICC refers) [2] and the formation of inactive chromatin domains, to the relationships between nucleotide arrangements, DNA curvature and nucleosome formation, through to sequence-specific DNA protein interactions that affect transcription. An excellent overview of the field can be found in Roel van Driel and Arie Otte's book *Nuclear Organization, Chromatin Structure and Gene Expression* [3]: in the words of a recent reviewer "Anyone who thinks that the completion of the Human Genome Project marks the end of genetics as we know it – read this book and you will see that we are really only just beginning" [4].

Plant cytogenetics, as always, forges ahead with keen independence, using all the tricks of the modern trade: tackling fundamental problems with fundamental systems and exploiting to the full the wealth of natural variation that is offered so generously by the plant kingdom. I find it especially pleasing to note

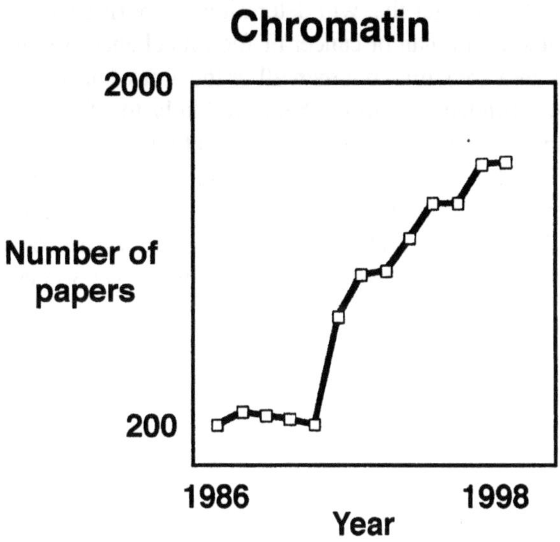

Figure 2.

that in plant cytogenetics there are currently more hypotheses and more lively disagreements than in any other field of chromosome science. Perhaps it's because they are not hung up on big genome mapping projects and not preoccupied with stopping people dying! A good overview of certain important parts of the field can be found in Gill and Friebe's article on *Plant Cytogenetics at the Dawn of the 21st Century* [5].

There are other things in the cell nucleus besides chromosomes. Most of them are decidedly dull to look at: things like nucleoli, coiled bodies, protein bodies and snurposomes as well as a host of other intranuclear bodies that usually appear as small, round, structurally featureless objects. Yet all of these things share the nuclear environment with chromosomes and many of them are associated with chromosomes at specific loci. They are likely to prove extremely interesting in years to come. They are, I guess, sites where things happen, important things, that need many complex components to be present in exactly the same place, within molecular distances of one another, at exactly the same time in order to accomplish critically important vital functions. The more we learn about what it takes to make a ribosome, for example, the more we realise that it can only be accomplished in the high density, close proximity environment that manifests itself to us as a discrete nuclear body whose existence depends upon a whole range of chromosomally derived factors. A taste of what's to come can be found in Joe Gall and his colleagues' recent beautiful work on amphibian germinal vesicle nuclei [6–8] and an excellent account of just about everything to do with nucleoli in a special issue of *Chromosoma* published in 1997 (vol. 105, numbers 7 and 8, pages 383–561).

Probably the most basic matter that will determine future directions is what will be fundable? If it's not human or cancer or food its chances of becoming a major development area are slim – but then all of the areas I have mentioned so far are already well funded and all of them are likely to offer major new insights for human and cancer cell biology. Current activity, classified according to the organisms used, seems to be very much human-orientated and perhaps more than ever targetted towards problems related to human disease (Tab. 1). I don't think that really matters, as long as there is still room to be curious and a little money to help clever young (and not-so-young) people test out good ideas.

Table 1. Favourite organisms for chromosome research, giving the numbers of papers published for each type during the period January 1995–August 1998

Human	10,185	Sheep	165
Cancer cells	7,962	Cats	144
Mice	4,362	*Drosophila*	133
Hamsters	661	other insects	72
Triticum	405	dogs	38
Arabidopsis	328	Other Amphibians	28
Cattle	251	Muntjacs	22
Xenopus	247	Horses	18
Chickens	195	Other birds	15

The difference between the total number of chromosome-related papers (approximately 32,000) and the total number represented here (approximately 25,000) is most likely a consequence of discrepancies in the ISI search techniques used by the author.

We should ask, from the standpoint of the person investigating chromosomal and/or nuclear phenomena, whether we may expect to encounter and investigate NEW principles and fundamentals, or will we spend the next 100 years filling in the picture that has been sketched during the 20th Century? I don't know. But I would like to think that there were many more uncharted areas of chromosomal and nuclear science out there to challenge our hands and minds. We shall see!

I believe that another very important question arises from all this speculation. Since virtually anything new that is discovered about chromosomes in the next millennium is bound to be of a molecular nature, is this reasonable justification for placing the emphasis in future training in chromosome science very much on the molecular level?

Technology in chromosome science

40 years ago we were all squashing chromosomes. Today we are painting them. We've come a long way!

Right now we are in the middle of a golden era in which we are enjoying and exploiting the amazing synergy of molecular biology, optical science and computing technology. There really has never been anything quite like it, except perhaps for the mid 1950s when phase contrast and electron microscopy were almost simultaneously developed as new tools for cell biologists. Today's technology is easy, colourful and quick. The formulation of the question, the design of experiment and the broad evaluation of the results are as difficult as ever. Doing the experiment, producing the pictures and collecting the data are easier than ever.

The frightening thing, however, is the extraordinary faith that we have in the systems we use and the impressive range of things that we can do with them.

"Chromosome painting requires a mixture of technology and manual dexterity"

Figure 3. Drawing kindly provided by Dr. Stanley K. Sessions of the Department of Biology, Hartwick College, Oneonta, New York, USA

Take chromosomes, FISH, a CCD camera, a digital imaging system and Adobe Photoshop, for example. There's not a lot that we cannot do. Our primary image is recorded as a grey-scale digitized image that we can then "false colour" according to our personal liking, usually with some regard to the truth: yellow for FITC, blue for DAPI, red for ethidium and so on. When it comes to detail, well, if there's one chromosome too many – we could take one out. If you don't like the contrast, you could enhance it. If signals are not where they should be, you could delete them. If you don't like the arrangement, you could change it. I think the message here has to be: be careful. Be scrupulously honest with yourself. Always take the trouble to look down the microscope long and hard at the real preparation. For only with that reference point can we know how far we dare go with our computer in constructing what we consider to be a pretty picture. Remember, above all, that Nature is notoriously nonconformist and the clues to her secrets are encoded in her diversity. To mask out that diversity is to blot out the clues.

The other important message which I consider to be of paramount importance for the future of chromosomes is that, more than ever, we must take the trouble to know what we're doing. Find out about our machines. Investigate their algorithms. Know their limitations. Test their accuracy and dependability within the particular framework of our experiment. If possible, find out how the software that you are using was developed. Two examples of publications describing the development of systems for FISH and GISH analysis can be found in [9, 10]. These were recommended to me by Applied Imaging International Ltd. as examples of independent publications upon which the development of their present range of equipment was based.

Training in chromosome science

Those of us who belong to or are affiliated with universities have a dual responsibility. We are expected to make science and to teach science – thus ensuring that there will be a next generation of chromosomologists and safeguarding the future of chromosomes. Unfortunately, in my country, the encouragements and rewards for good university teaching have largely disappeared under a policy that places university departmental funding at the mercy of research performance. Teaching and the effort needed to do it properly is time away from the bench and therefore something to be avoided, if at all possible.

Hence the opportunities for becoming a chromosomologist in the UK are declining and our national population of good teaching scientists in this field is literally dying off and not being replaced. Only in the area of plant cytogenetics are there still vigorous and well integrated programmes and opportunities for undergraduate and postgraduate training but, even in this field, are there enough young professionals being brought up as good teachers, as well as good researchers, to step into line when the old boys and girls die off?

The other major threat to good university teaching is the ever thickening web of rules and regulations that cramp our style in the teaching laboratory. Work with human tissues is now widely disallowed. You cannot even prick your own finger in the presence of a group of students without contravening Health and Safety laws. Every wet bench activity demands such an assortment of safety equipment that not only is it prohibitively expensive but it's no longer fun. Work on live animals, even of the most harmless and non-intrusive nature, is strictly controlled and students have a right to decline to handle any freshly killed animal material. At which point I am usually inclined to ask – how can one ever be a true biologist if you've never had blood on your hands? Happily, none of this applies to plants!

Given that we must operate within and respect the laws that safeguard us from things like AIDS and hepatitis and protect animals from abuse, we do now have a problem when it comes to providing practical instruction in working with animal chromosomes. At the end of my book *An Introduction to Animal Cytogenetics* [13] I included an appendix of 4 teaching laboratory protocols, each of which had been used successfully to train young cytogeneticists for over 20 years. Only one of these, a study of the giant polytene chromosomes of *Drosophila* and *Chironomus*, is now allowed under UK law, and even that has proved to be offensive to students who don't like cutting the heads off dipteran larvae.

So what do we do? The answer, I think, is plain. Most importantly, we concentrate on finding ways of creating a sense of curiosity and adventure in our students, making them want, above all else, to work with chromosomes. Alongside this we encourage initiative and an eagerness and ability to communicate. After that, I believe the best of them will take care of themselves. If we are able to complement our teaching with interesting, successful and rewarding hands-on experience of working with chromosomes then so much the better. But, above all, teach with joy in your heart and unmitigated enthusiasm for these ancient and wonderful objects that hold the innermost secrets of all living systems.

Controversies

The 13th International Chromosome Conference was a good one, noteworthy for the high quality of the science and a truly international flavour. I sensed, however, that one thing was in short supply: controversy. What are the major controversies in chromosome science right now? Where are the hypotheses? Ancona was outstanding for its aura of consensus. In the 1950s almost everything was debatable: Richard B. Goldschmidt's book, *Theoretical Genetics* [12], is a wonderful account of the excitement of those early years. In the

1970 s it was the C value paradox. In the 1980s people were arguing about karyotypic evolution, co-adapted genomes, and molecular drive – great debates, championed by great gladiators, that led on to great experiments.

Debates are about matters that cannot be resolved by plain hard work and technical ingenuity. There is unlikely to be a big debate, for example, about chromosome imprinting. How does a chromosome "know" it has a paternal origin? What is the molecular basis of chromosome inactivation? We will certainly answer these questions once and for all by choosing the right organisms, applying the right technologies and doing the right experiments. It will take time but we will get there. Sex determination, on the other hand is a much more difficult and complex matter. How *do* chromosomes determine sex, for here we have the prime example of genes and chromosomes controlling a major developmental switch mechanism? I anticipate that this will be one of the focal points for the next ICC in Würzburg, Germany. Significantly, most of the matters debated by Goldschmidt 45 years ago in his chapter on the Genetics of Sex Determination, are still unresolved.

At the 13th ICC in Ancona, I personally liked Nancy Kleckner's talk best of all [13]. It was about meiosis. It was fundamental. It was controversial. It presented the idea of "chromosome morphogenesis". As one who has spent many years wondering how a 15 micrometer metaphase chromosome can be transformed into a 1 mm lampbrush chromosome with 500 chromomeres and several thousand hyperactive transcription units, chromosome morphogenesis is the ultimate challenge for the future.

References

1 Cooper NG (ed.) (1994) *The Human Genome Project, Deciphering the Blueprint of Heredity.* University Science Books, Mill Valley California

2 Ikemura T, Ohno M, Uemura T, Sasaki H, Tenzen T (1998) Genome and sub-nuclear organization of non-B forming DNAs in mammalian chromosomes. Abstract. *Cytogenet Cell Genet* 81: 101–102

3 van Driel R, Otte A (1997) *Nuclear Organization, Chromatin Structure and Gene Expression.* Oxford Univ. Press, Oxford

4 Turner B (1998) Book Review, Nuclear Organization, Chromatin Structure and Gene Expression. *Chromosome Res* 6: 423–424

5 Gill BS, Friebe B (1998) Plant Cytogenetics at the dawn of the 21st Century. *Curr Opin Plant Biol* 1: 109–115

6 Gall JG, Callan HG (1989) The sphere organelle contains small nuclear ribonucleoproteins. *Proc Natl Acad Sci USA* 86: 6635–6639

7 Wu C-HH, Murphy C, Gall JG (1996) The Sm binding site targets U7 snRNA to coiled bodies (spheres) of amphibian oocytes. *RNA* 2: 811–823

8 Bauer DW, Gall JG (1997) Coiled bodies without coilin. *Mol Biol Cell* 8: 73–82

9 Kleinschmidt P, Mitterreiter I (1994) Improved chromosome classification using monotonic functions of Mahalanobis distance and the transportation method. *Math Method Oper Res* 40: 305–323

10 Muller S, Rocchi M, Ferguson Smith MA, Wienberg J (1997) Toward a multicolor chromosome bar code for the entire human karyotype by fluorescence *in situ* hybridization. *Hum Genet* 100: 271–278

11 Macgregor HC (1993) *An Introduction to Animal Cytogenetics*. Chapman and Hall, London and New York
12 Goldschmidt RB (1955) *Theoretical Genetics*. University of California Press, Berkeley and Los Angeles
13 Kleckner N (1998) Interactions between and along meiotic chromosomes. Abstract. *Cytogenet Cell Genet* 81, 100

Subject Index

Synonymous words

alphoid DNA, α satellite, alphoid sequence, alpha satellite, alpha satellite DNA
chromatin packaging, chromatin condensation
DNA metabolic gene, genes involved in DNA metabolism
DSB (double strand break)-independent pairing, DSB-independent homolog pairing
FISH to extended DNA fibre, FISH method on extended DNA fibre
nucleolar organizer region, NOR
polypurine/polypyrimidine tract, polypurine/polypyrimidine sequence
quality-control checkpoint (means mitosis checkpoint), spindle checkpoint, checkpoint
ransposon, transposable element
retrotransposon, retroelement, retroposon
ribosomal DNA, rDNA
satellite DNA, satellite repeat
triple helix structure, triplex, triplex DNA, triplex-forming DNA

MIX
Papier aus verantwortungsvollen Quellen
Paper from responsible sources
FSC® C105338

If you have any concerns about our products,
you can contact us on
ProductSafety@springernature.com

In case Publisher is established outside the EU,
the EU authorized representative is:
Springer Nature Customer Service Center GmbH
Europaplatz 3, 69115 Heidelberg, Germany

Printed by Libri Plureos GmbH
in Hamburg, Germany